Rise of the Dragon

Rise of the Dragon

READINGS FROM *NATURE* ON THE CHINESE FOSSIL RECORD

Edited by Henry Gee
With a Foreword by Zhe-Xi Luo

The University of Chicago Press
Chicago and London

The Chinese inscription, "Rise of the Dragon," was penned by President Jiang Zemin of the People's Republic of China. The Publisher is grateful to the National Natural Science Foundation of China for their assistance in obtaining this inscription.

HENRY GEE is a senior editor at *Nature*. He is the author of *Before the Backbone: Views on the Origin of the Vertebrates* and *In Search of Deep Time: Beyond the Fossil Record to a New History of Life* and the editor of *Shaking the Tree: Readings from Nature in the History of Life*, the last also published by the University of Chicago Press.

The University of Chicago Press, Chicago 60637
The University of Chicago Press, Ltd., London
© 2001 by Nature/Macmillan Magazines, Ltd.
Introduction © 2001 by Henry Gee
Foreword © 2001 by The University of Chicago
All rights reserved. Published 2001
Printed in the United States of America
10 09 08 07 06 05 04 03 02 01   5 4 3 2 1

ISBN (CLOTH): 0-226-28490-5
ISBN (PAPER): 0-226-28491-3

Library of Congress Cataloging-in-Publication Data

Rise of the dragon : readings from *Nature* on the Chinese fossil record /
   Edited by Henry Gee with a foreword by Zhe-Xi Luo.
      p.   cm.
   Includes bibliographical references and index.
   ISBN 0-226-28490-5 (cloth : alk. paper) – ISBN 0-226-28491-3 (pbk. :
alk. paper)
      1. Animals, Fossil—China.   I. Gee, Henry.
   QE756.C6 R57   2001
   560'.915—dc21   2001003677

♾   The paper used in this publication meets the
minimum requirements of the American National Standard for Information Sciences—
Permanence of Paper for Printed Library Materials, ANSI Z39.48-1992.

Contents

Foreword   Zhe-Xi Luo

In Chinese mythology, dragons are icons of power, fortune, and majesty. Fossils have traditionally been seen as tangible proof for the existence of dragons.[1,2] The earliest reference to fossils can be found in *Shan-Hai Jin* (*Scripture of Mountains and Seas*), a historical text of the seventh to fourth centuries B.C.[2] In reality, dragons are icons only, works of the human imagination: in the words of Henry Gee, who edited the essays in this book, "Fossils are mute. Unable to tell us their own stories, we tell their stories for them. We tell their stories to flatter our own prejudice."[3]

But times have changed. The draconic totems of ancient societies are now seen as powerful material evidence for biological evolution. The stories about fossils from modern paleontological studies are even more fascinating than those from Chinese mythology, for the history of life that gives scientific meaning to our own existence is a far more important legacy than are the dragons of the imagination.

Rise of the Dragon, the title of this collection, alludes to the increasing importance of China—homeland of dragons—in paleontological science. The volume includes the latest contributions from China to the understanding of biological evolution in the last decade of the twentieth century, during which time a large number of paleontological discoveries from China were published in influential international scientific journals. The new findings, especially those from the Mesozoic Jehol biota of Liaoning Province, and the Precambrian and Cambrian soft-bodied biotas from South China, focused the attention of the world scientific community and news media on China, her wonderful fossils, and the scientists who helped to discover them.

The rapid pace of these significant discoveries is unprecedented, and this seems to have taken the international scientific community by (pleasant) surprise. But China's recent prominence in paleontology is by no means an overnight success story. The worldwide awareness of the paleontological importance of China followed years of painstaking exploration and arduous research by many dedicated Chinese scientists, working under difficult conditions and without the attention their discoveries deserved.

For example, the Mesozoic Jehol biota from Liaoning Province in northeastern China was known long before it became famous for its exquisite fossils of feathered dinosaurs, birds, mammals, flowering plants, and pollinating insects. The fauna and its basic stratigraphy were first described as long ago as 1928, by A. W. Grabau, one of the founding fathers of Chinese paleontology.[4] Fossil vertebrates were studied by Japanese scientists Yabe, Endo, and Shikama[5] during World War II. By the 1950s, most fossils of the Jehol Biota had been described, and their biostratigraphical sequence established,[6] although the dating of the absolute age of the main stratigraphical formations has always been controversial.

By the 1980s, the Jehol biota was well known among geologists for its distinctive and abundant fossil fish (*Lycoptera* and *Beipiaosteus*), conchostracan arthropods (*Estheria*), diverse insects (such as *Ephemeropsis*), and the primitive frilled dinosaur *Psittacosaurus*.[7–10] These fossils were so widespread that they were used for biostratigraphical correlation in terrestrial sediments from northern and eastern China to central Asia.

The latest episode of discoveries in vertebrates began with the finding of the fossil bird *Sinornis*[11] in 1987, followed by the discovery of several new birds at the Boluochi site by Z.-H. Zhou in 1990, and then discoveries of the bird *Confuciusornis*[12] and the mammal *Zhangheotherium*[13] in 1993. While significant and exquisite in their own right, these were harbingers of still more remarkable discoveries.

The discovery of *Sinosauropteryx* and *Protarchaeopteryx*[14,15]—two dinosaurs with feather-like filaments and feathers—was a watershed event in the history of Chinese vertebrate paleontology. Within a few years, a spectacular diversity of significant fossils was uncovered in impressively large quantities from the lagerstätte in Sihetun and from adjacent sites. The feathered dinosaurs made headlines in the world's scientific journals and news media, for they provided graphic evidence for a close relationship between birds and theropod dinosaurs. From a historical perspective, however, it was the culmination of the research and exploration of several decades.

Research into the Precambrian and Lower Cambrian biotas of southwest China have a similar history to that of the Jehol biota. Precambrian and Lower Cambrian strata of south China had been explored for phosphorites by geologists as early as the 1940s[16–18]—long before these strata become known for their exceptional soft-bodied fossils, such as the earliest known vertebrates, and for their fossilized animal embryos.

The Lower Cambrian Chengjiang fauna of Yunnan in south China is now widely seen as the Chinese answer to the famous Middle Cambrian Burgess Shales of Canada. This important fauna came to light following the discovery of abundant soft-bodied fossils by X.-G. Hou in 1984. This sparked a rapid burst of publication on the fossils by scientists at the Nanjing Institute of Geology and Palaeontology. In 1987, *Acta Palaeontologica Sinica*, a leading journal in paleontology in China, published six back-to-back reports in a single issue on these newly discovered soft-bodied fossils. Eight more papers on the new organisms from the Chengjiang fauna appeared in the same journal between 1988 to 1990.

Systematic studies of the fossils from the Precambrian Doushantuo Formation at the Weng-An site of Guizhou province were first published by late Y. Zhang[19] of Beijing University, and then by Y.-S. Xue et al.[20,21] of Nanjing Institute of Geology and Palaeontology. M.-E. Chen,[22] of Geological Institute (Beijing), and many others also studied the biota of the Doushantuo Formation extensively exposed in the Yangtze River Gorge area.

The majority of these early descriptive works, emphasizing algal and shelly fossils,[21,22] were published in Chinese. Although well known to specialists, the broader significance of these faunas on the early evolution of life through the Proterozoic–Cambrian transition was largely unappreciated outside China. It was not until the early 1990s that taxonomic and stratigraphic information of these rich faunas became available to a broader international audience through reviews and summaries published in English (for example, refs. 23–26). Ten years after the Chengjiang soft-bodied fossil assemblage was discovered, the first putative chordate fossil was reported;[27] a further five years elapsed before its first vertebrates came to light.[28,29] The history of research on the Doushantuo formation reveals a similar story: the stunning and insightful reinterpretation of the earliest known animal embryos on the basis of the new and three-dimensional fossils[30] came 14 years after the stratigraphy of Doushantuo fossil sites in Guizhou was established.[17] The recent success of paleontology in China was an evolution by itself. It is more of a climax in a scientific drama after a long prelude, than thunder from a blue sky.

China is a vast country, endowed with rich fossil resources. But the new fossils from the last decade are exceptional even by Chinese standards. Their exquisite preservation of soft-tissues, their exceptional completeness, their abundance, and their breathtaking beauty left many paleontologists—accustomed to scarce and often fragmentary fossils—lost for words. After all, who would have expected the ontogeny of organisms from 570 million years ago to have been caught in action? Yet this is precisely what the fossils from the Doushantuo phosphorites have shown us. Who would have expected to see dinosaur skeletons ornamented with all kinds of fancy feathers—not just a couple of accidental finds, but as many as six feathered dinosaurs represented by more than a dozen exceptional fossils? Yet feathered dinosaurs are now found almost routinely in Liaoning province. Who would have expected that there are now more than a thousand skeletons of the bird *Confuciusornis*—a total equivalent to the number of Mesozoic birds accumulated in the entire world for the past 150 years multiplied fifty-fold?

China's emergence on the international paleontological scene would not have been possible without her fantastic fossils. But fossils by themselves are not enough. These fossils chanced to arrive when science was ready to appreciate their great value. The newly found fossils provided fresh evidence and timely tests for such long-standing issues of evolutionary biology as the evolution of birds from dinosaurs, the diversification of main mammalian groups, and the early anatomical evolution of flowering plants. Fossilized embryos from the Doushantuo phosphorites and the early chordates and vertebrates of Chengjiang hit the headlines at a time of increasing interest in the evolution of ontogeny and the application of results from a wide range of disciplines to research into early metazoan phylogeny.

The recent success of Chinese paleontology can also be related to social, political, and economic reforms beginning in 1979. One change resulting from the reforms has been growing competition between rival institutions, spurring increased quality and publication in international journals. This competition has been healthy, although occasionally intense, and fueled by the fact that competing institutions have access to the same fossil resources. In the most dramatic case, the shale that bore the type specimen of the feathered dinosaur *Sinosauropteryx* was split into a slab and counter-slab, with each half of the same fossil obtained by a different research institution. An advantage of this competition is the rapid dissemination of important research results, although rivalries can sometimes make already heated scientific controversy even more emotive. Nevertheless, the lively discussion on the relationship of feathered dinosaurs and the age of their fauna has been the wellspring for scientific productivity, stimulating more research.

Another change has been the emergence of an "open-door" national policy that has greatly facilitated international research collaboration and effective dissemination of scientific information to a worldwide audience. This volume of original research papers and their companion commentaries represents 25 authors from six Chinese institutions, plus 20 co-authors from eighteen institutions in six other countries. It is symbolic of the fruitful scientific collaboration of Chinese paleontologists and their international colleagues, and is a powerful argument in favor of the view that truly great science transcends political boundaries. With free scientific exchanges and effective international collaboration in paleontology, some of these wonderful discoveries in China and their studies have reached a broader international audience at a much faster pace.

The essays in this collection provide some good snapshots of a much larger research panorama on new fossils from China. They are the tips of gigantic icebergs: a voluminous body of literature is now available, not just in brief reports, as in this book, but also in massive and detailed descriptive taxonomic studies (see, e.g., refs. 10,31–33). Great progress was already made in the advancement of paleontology in China. Undoubtedly, even greater future progress will be made as the essays in this book will stimulate more exploration and scientific work.

Rise of the Dragon, as the title of a book dedicated to China's fossils, resonates with the icon of the dragon as a part of a cultural tradition three thousand years old. Now that these mythical "dragon bones" have acquired a new, scientific meaning, we can look forward to a period of increasing, worldwide interest in Chinese fossils, which have always been cherished in the homeland of the dragon.

References

1. Mayor, A. *The First Fossil Hunters* (Princeton, NJ: Princeton University Press, 2000).
2. Dong, Z.-M., and Cheng, Y.-N. *The Land of the Dragon: The Varied Dinosaurs from China and Mongolia* (Taichung, Taiwan: Yen-Nien Cheng, 1995).
3. Gee, H. *In Search of Deep Time: Beyond the Fossil Record to a New History of Life* (New York: Simon & Schuster, 1999).
4. A. W. Grabau. *Stratigraphy of China*, pt. 2: Mesozoic (Peking: Geological Survey of China, Ministry of Agriculture and Commerce, 1928).
5. Hasegawa, Y., ed. *Selected Papers by Professor Tokio Shikama* (Yokohama University, Department of Earth Science, 1978).
6. Gu, Z.-W. *The Jurassic and Cretaceous of China* (Beijing: Science Publishing House, 1962).
7. Wang, S.-E. *The Jurassic of China* (Beijing: Geological Publishing House, 1985). [In Chinese, with English summary.]
8. Hao, Y.-C., Su, D.-Y., Yu, J.-X., et al. *The Cretaceous of China* (Beijing: Geological Publishing House, 1986). [In Chinese, with English summary.]
9. Ren, D., Lu, L.-W., Guo, Z.-G., and Ji, S.-A. *Faunae and Stratigraphy of Jurassic-Cretaceous in Beijing and the Adjacent Areas* (Beijing: Seismology Publishing House, 1995). [In Chinese, with English summary.]
10. Chen, P.-J., and Fan, J., eds. *Jehol Biota* (Hefei, China: Press of the University of Science and Technology of China, 2000). [In Chinese, with English abstracts.]
11. Sereno, P. C., and Rao, C.-G. Early evolution of avian flight and perching: New evidence from Lower Cretaceous of China. *Science* 255, 845–848 (1992).

12. Hou, L.-H., Zhou, Z.-H., Gu, Y.-C., and Zhang, H. *Confuciusornis sanctus:* A new Late Jurassic sauriurine bird from China. *Chinese Science Bulletin* **40**, 1545–1551 (1995).

13. Hu, Y.-M, Wang, Y.-Q, Luo, Z.-H., and Li, C-K. A new symmetrodont mammal from China and its implications for mammalian evolution. *Nature* **390**, 137–142 (1997).

14. Ji, Q., and Ji, S.-A. On the discovery of the earliest bird fossil in China and the origin of birds. *Chinese Geology* **1996(10)**, 30-33 (1996).

15. Ji, Q., and Ji, S.-A. Protarchaeopteryx, a new genus of Archaeopteridae in China. *Chinese Geology* **1997(3)**, 38–41 (1997).

16. Luo, H.-L., Jiang, J.-W., Wu, X.-C., et al. *Sinian-Cambrian Boundary in Eastern Yunnan* (Kunmin, China: People's Publishing House of Yunnan, 1982).

17. Wang, Y., et al., eds. *The Upper Precambrian and Sinian-Cambrian Boundary in Guizhou* (Guiyang, China: People's Publishing House of Guizhou, 1984).

18. Chang, W.-T. Early Cambrian Chengjiang fauna and its trilobites. *Acta Palaeontologica Sinica* **26**, 223–235 (1987).

19. Zhang, Y. Multicellular thallophytes and fragments of cellular tissue from Late Proterozoic phosphate rocks, South China. *Lethaia* **22**, 113–132 (1989).

20. Xue, Y.-S., Tang, T.-F., and Yu, C.-L. Discovery of oldest skeletal fossils from Upper Sinian Doushantou Formation in Weng'An, Guizhou, and its significance. *Acta Palaeontologica Sinica* **31**, 530–539 (1992).

21. Xue, Y.-S., Tang, T.-F., Yu, C.-L., and Zhou, C.-M. Large spheroidal Chlorophyta from Doushantou Formation phosphoric sequence (Late Sinian), central Guizhou, south China. *Acta Palaeontologica Sinica* **34**, 688–706 (1996).

22. Chen, M.-E., and Liu, K.-W. The geological significance of newly discovered microfossils from the Upper Sinian (Doushantou age) phosphorites. *Scientia Geologica Sinica* **1986(1)**, 46–53 (1986).

23. Simonetta, A. M., and Conway Morris, S., eds. *The Early Evolution of Metazoa and the Significance of Problematic Taxa* (Cambridge: Cambridge University Press, 1991).

24. Lipps, J. H., and Signor, P. W., eds. *Origin and Early Evolution of the Metazoa* (New York: Plenum Press, 1992).

25. Hou, X.-G., Ramsköld, L., and Bergström, J. Composition and preservation of the Chengjiang fauna: A Lower Cambrian soft-bodied biota. *Zoologica Scripta* **20**, 395–441 (1991).

26. Zhang, Y., and Yuan, X. New data on multicellular thallophytes and fragments of cellular tissues from Late Proterozoic phosphate rocks, South China. *Lethaia* **25**, 1–18 (1992).

27. Chen, J.-Y., Dzik, J., Edgecombe, G. D., Ramsköld, L., and Zhou, G. Q. A possible Early Cambrian chordate. *Nature* **377**, 720–722 (1995).

28. Shu, D.-G., Luo, H.-L., Conway Morris, S., Zhang, X.-L., Hu, S.-X., Han, J., Zhu, M., Li, Y., and Chen, L.-Z. Lower Cambrian vertebrates from south China. *Nature* **402**, 42–46 (1999).

29. Chen, J.-Y., Li, C.-W., and Huang, D.-Y. An early Cambrian craniate-like chordate. *Nature,* **402**, 518–522 (1999).

30. Xiao, S.-H., Zhang, Y., and Knoll, A. H. Three-dimensional preservation of algae and animal embryos in a Neoproterozoic phosphorite. *Nature* **391**, 553–558 (1998).

31. Chen, J.-Y., Cheng, Y.-N., and Iten, H. V., eds. The Cambrian explosion and the fossil record. *Bulletin of the National Museum of Natural Science* **10**, 1–319 (Taichung, Taiwan, 1997). [In English, with Chinese abstracts.]

32. Hou, L.-H., *Mesozoic Birds of China* (Nantou, Taiwan: Taiwan Provincial Feng- Huang Gu Bird Park [Phoenix Valley Aviary], 1997).

33. Chiappe, L. M., Ji, Q., and Ji, S.-A. Anatomy and systematics of the Confuciusornithi-dae (Theropoda: Aves) from the Late Mesoaoic of northeastern China. *Bulletin of American Museum of Natural History* **242**, 1–89 (1999).

Here Be Dragons   Henry Gee

In rather less than a decade, paleontology in China has risen from relatively modest beginnings[1] to being a dominant force on the international scientific scene with a series of spectacular discoveries that have immediately enriched our understanding of key episodes in the history of life.[2] *Nature* has been pleased to have published many of these findings. This book is an anthology of this research, bringing together 16 original reports, some of which are discussed with commentary originally published in *Nature*'s "News and Views" section.

It is a book of contrasts. The fossils described in these pages span almost 700 million years of geological time, but the earliest paper in this collection dates from as recently as 1997. And the story is far from complete.

Celebrating the fossil riches of China could be seen as celebration enough: of China, her scientific expertise and cultural heritage; of the fossils themselves; and (false modesty being a much overrated virtue) the part played by *Nature* in bringing this work before the scientific and general public. But when we have stopped congratulating ourselves, we must ask for more enduring reasons for creating a collection like this.

Happily, I have one. It lies in the property of fossils of confronting our expectations of morphology, thereby extending them. Every fossil described in these pages is a challenge to convention. In the early 1980s, the school of so-called transformed cladists held that the topology, or branching order, of a phylogeny created using data from extant creatures alone could not be overturned by data from extinct organisms.[3] An extensive analysis showed this idea to be practically flawed—data from fossils could change the branching order of cladograms.[4] It is easy, now, to see why—given the paucity of data from the fossil record generally, especially in regard to vertebrate diversity—fossil species have the potential to outnumber extant ones. But there is another reason: fossil forms may embody combinations of characters that are seen as separate features characteristic of two or more groups nowadays considered to be distinct. "Feathered" dinosaurs like *Caudipteryx* are cases in point, because they challenge the idea that birds and dinosaurs followed separate evolutionary courses; the fish *Psarolepis*, combining the head of a lobe-finned fish with the fin spines of a placoderm, force a reassessment of long-held views of early vertebrate evolution.

This chimerical quality of fossils (a quality that has led to misleading *clichés* such as "missing links") is more than simple evidence for the occurrence of evolution. It can be used constructively, to trace the order in which features of a group were acquired as the group evolved. This order, in turn, allows us to constrain hypotheses and scenarios about the evolution of morphological innovation.

The evolution of flight is a good example. It had long been suspected from detailed consideration of morphology that birds and theropod dinosaurs were closely related. Less clear, however, was how terrestrial cursors, such as theropods, acquired flight and became birds. The problem is all the more acute because birds, as extant organisms, form a neat, well-defined group, recognizable using a host of distinctive features found in no other organisms. Many of these features are interpreted as adaptations for flight, making it hard to disentangle two distinct problems—the origin of birds, and the origin of flight. Because of this confusion, it is a challenge even to begin to understand how the unique assemblage of characters we use to recognize birds came together. Traditionally, researchers have resorted to post-hoc evolutionary scenarios ("palaeontological stories")[5] to "explain" the origin of birds.[6]

When looking at the distinctive features of birds, we might list such things as feathers, powered flight using wings, a pronounced sternal keel, hollowed bones, a beak, endothermic metabolism, and so on. It would be an impressively long list, and this is why explaining the origin of birds (as distinct from flight) appears to pose a problem. However, the problem is hard only if we try to imagine all the features that constitute birds appearing more or less simultaneously. The problem is less daunting once we appreciate what must have happened: if features did not arrive all together, then they must have appeared one at a time. Fossils that display some but not all of these characters, or unexpected combinations of characters, can be arranged phylogenetically to deduce the order in which they were acquired.

The fossil record of China has increased our understanding of the origin of birds by providing theropods of several groups that have feathers or structures that can plausibly be regarded as related or even precursory to feathers.[7]

This evidence shows that feathers were present in a diverse range of theropods, presumably for reasons unrelated to flight. In sum, feathers were acquired in the lineage leading to birds long before flight, so we can suppose that the earliest birds took to the air with a reasonable complement of feathers—these structures were not later elaborations to flaps of integument, say, in animals already partially or wholly airborne. In this way, fossils—interpreted through sound phylogenetic reconstruction—can help us choose between several possible or plausible evolutionary scenarios. Much of the increase in understanding offered by the fossils described in this volume can be attributed to another feature that many of them have in common. That is, abundant evidence for soft-tissue preservation. This is no more true than for the fossils described in the first of four thematic sections into which the book naturally falls, looking at the remarkable cellular preservation of microfossils in the Proterozoic Doushantuo phosphorites of Guizhou province. This is followed by reports on the Lower Cambrian Chengjiang fauna of Yunnan. Staying in Yunnan, our third theme looks at the Silurian and Devonian fishes that shed light on the earliest days of bony fishes. Moving both northward in space and forward in time, the fourth theme and the main part of the book concerns the feathered dinosaurs, birds and mammals of the Early Cretaceous Yixian formation (and related strata) of Liaoning Province, with some notes about the biogeographical issues raised by the fauna.

The minutely preserved embryos of the 670-million-year-old Doushantuo phosphorites offer the earliest direct evidence of animal life. Perhaps just as important is the psychological impact these fossils have on paleontologists inured to the long silence of the Precambrian record. This was the silence that troubled Darwin and which has been scarcely lifted since, inasmuch as it concerns the fossil record of animals. As Stefan Bengtson (chap. 2 in this volume) writes in his commentary, "a spell seems to have been broken." We know for certain that Precambrian fossils can be found, if we know where to look. The previous lack of animal fossils is, therefore, not just capricious preservation. Indeed, the Doushantuo fossils are of the very things we would usually expect not to find as fossils—microscopic, soft-bodied embryos. Other finds have since been published.[8] The remains of algae and sponges have been recovered from the phosphorites, and the animal embryos are tantalizing. They suggest that the protostomes and deuterostomes, the major metazoan clades, had become distinct by 670 million years ago. The big question is what these embryos would have come to look like once they had grown up. It could have been that the earliest metazoans remained small and planktonic until the evolution of genetic and developmental mechanisms to create large bodies.[9] Perhaps these embryos could have come to look like some of the unusual creatures of the so-called Ediacaran radiation, immediately before the Cambrian.

On the other hand, they could have emerged as one or another of the unusual creatures of the Lower Cambrian Chengjiang fauna of Yunnan: China's version of the Burgess Shales of Canada brought to public attention by Stephen Jay Gould.[10] Here we provide our first direct confrontation with the strangeness of fossil organisms. Strange, however, only from our parochial perspective. Presumably, any creature, no matter how odd, must have "worked" as an integrated whole, just like any modern and mundane horse or hound. Yet to us, whose images of life are conditioned by the limited range of creatures to which we have been exposed, these fossils are hard to relate to any one extant group.

The Burgess Shales are rich in fossils of arthropods—jointed-limbed creatures—and have less to report on the immediate relatives of vertebrates, the group of animals that includes ourselves. The one exception is *Pikaia*, a sliver of a creature thought to resemble the modern amphioxus, *Branchiostoma*, the closest living invertebrate relative of vertebrates. *Pikaia*, however, may have less in common with the amphioxus than was at first thought. In any case, direct fossil evidence for the early history of the vertebrates has come in greater abundance from Chengjiang. Immediately, the problem of interpreting unusual morphologies has become the focus of debate. *Xidazoon*, for example, described by Shu et al. (chap. 3 in this volume), can be compared only with other enigmatic fossils. One strange creature, *Yunnanozoon*, has been variously interpreted as similar to chordates, more similar to hemichordates, or even as a distant cousin of all deuterostomes. J.-Y. Chen and colleagues (see below, chap. 4), who have interpreted *Yunnanozoon* as closest to chordates, now present the similar *Haikouella* as an extremely primitive vertebrate-like creature. *Haikouella* is known from abundant fossils.

Very much scarcer are the two fish-like organisms *Myllokunmingia* and *Haikouichthys*, described by Shu et al. (chap. 5 in this volume), and yet these remarkable fossils are arguably far more like modern-looking vertebrates, akin to jawless fishes such as hagfishes and lampreys. These finds set the record for modern-looking vertebrates right back to the early Cambrian. In his commentary, Philippe Janvier (chap. 6), suggests that these two creatures are "the most convincing Early Cambrian vertebrates ever found." Nevertheless, they still confront paleontologists with a mixture of the familiar and the unexpected. In some ways, says Janvier, the two fish-like fossils look just like what paleontologists had imagined the earliest vertebrates to look. In other ways, they are "completely at odds" with our parochial expectations.

In the early Paleozoic Era, South China formed a distinctive zoological province. One of its distinctive creatures was a fish called *Psarolepis*. At around 400 million years old, this fish was originally thought to be a very primitive sarcopterygian (lobe-finned bony fish), a member of the group that includes lungfishes, the coelacanth, and all land vertebrates, as well as many extinct fishes. Chapter 7, by Zhu et al. reports more complete evidence, showing that this is a very odd sarcopterygian indeed. As Ahlberg discusses in the accompanying commentary (below, chap. 8), it also bears features of the teeth and jaws more like those seen in actinpterygians (ray-finned bony fishes) as well as skeletal features reminiscent of acanthodians and placoderms—two primitive and extinct groups of jawed vertebrates. This creature is indeed strange by modern lights, but it is not alone. An early and as-yet-unnamed braincase of an actinopterygian from Australia[11] is surprising in that the eyeball seems to have been held in place by a stalk, as in cartilaginous fishes and placoderms. A primitive cartilaginous fish, *Pucapampella* from Bolivia,[12] and another stalk-eyed fish from Australia,[13] have also been thrown into what seems to be a discussion that could result in the fundamental reappraisal of the basal relationships of ray- and lobe-finned bony fishes (collectively, the Osteichthyes) and jawed vertebrates as a whole.

The feathered dinosaurs, from the early Cretaceous of Liaoning Province, have made headlines. They are probably what most people think of when Chinese paleontology is mentioned, an iconic status that places them at the heart of this book. As is discussed above, their scientific importance lies in their presentation of a suite of characters that allows us to comprehend the origin of birds in a very graphic way. For example, the dinosaur *Caudipteryx*—now regarded as an oviraptorosaur,[14] more distant from the direct ancestry of birds than supposed by Ji et al. in the paper included here—clearly has feathers, like birds, but it also has jaws with teeth instead of a beak, a long, bony tail, and its forelimbs did not form wings. To say that it is a "feathered dinosaur" does not help us much in our effort to understand the origin of birds. Rather, we should consider it as a "stem-group" bird, branching off the "stem" lineage after the evolution of feathers, but before the appearance of a beak, the loss of teeth, and so on. (The stratigraphic incongruity that places *Caudipteryx* in rocks 25 million years younger than the more derived—that is, advanced—*Archaeopteryx* is more apparent than real from a phylogenetic standpoint, and I discuss this fully elsewhere).[15]

Apart from the obvious feathers of *Caudipteryx*, and the tail-feathers of *Protarchaeopteryx* (also described by Ji et al.), interpreting the not-quite-feathery, not-quite-hairy integument of the other "feathered" dinosaurs presents a more difficult task. Once again, we are defied by the richness of past worlds, accustomed as we are to the poverty of the present. Consider, for example, the range of skin-covering seen in present-day amniotes. Apart from naked skin, there are reptilian scales, mammalian hair, avian feathers, and—well, that's it. This limited choice does not rule out the past existence of other kinds of integument, such as the carpety clothing worn by animals such as *Sinosauropteryx*, *Sinornithosaurus*, and *Beipaiosaurus*. It is possible, indeed likely, however, that the integumentary structures seen in these forms are closely related to feathers. The argument comes partly from morphology,[7] but mainly from phylogeny. The feathered dinosaurs described so far are all theropods, but otherwise phylogenetically highly diverse. Given that the lineage leading to extant birds emerged from within the theropods, it is likely that the integumentary structures described from the primitive *Sinosauropteryx* and the otherwise specialized *Beipaiosaurus* (a therizinosaur), *Sinornithosaurus* and, lately, *Microraptor*[16] (both dromaeosaurs) represent a more inclusive kind of feather. Feathers are such remarkable and complex structures that it is hard to imagine how they evolved. The feathered dinosaurs show us stages in the evolution of these structures, nowadays—because of our limited perspective—firmly associated with extant birds.

The Liaoning beds have produced more than just feathered dinosaurs. They have produced a cornucopia of other creatures including *Confuciusornis* (a volant bird belong to an extinct group, the Enantiornithes), and a number of remarkable mammals. Two important mammals from the Liaoning beds have been described in *Nature*, and both reports are reprinted here, along with a commentary from Tim Rowe (chap. 17 in this volume). One, from Hu et al. (chap. 18), describes the symmetrodont *Zhangheotherium*; the other, from Ji et al. (chap. 16), reports the triconodont *Jeholodens*. Symmetrodonts form an extinct group of mammals close to the base of the Therian mammals—that is, the marsupials and placentals. In other words, *Zhangheotherium* branches from the stem lineage above the branching point leading to extant monotremes, but before the divergence of marsupials and placentals. Triconodonts are very much more primitive: as perhaps the most complete known triconodont, *Jeholodens* provides what may be the closest picture we have of what the latest common ancestor of mammals looked like.

The last few contributions to this book are archival, but nonetheless important for all that. They go into the puzzling age of the Liaoning beds. In the late 1990s, it emerged that the beds were laid down around 125 million years ago, well into the Cretaceous Period. Many elements of the fauna, however, seemed antique by comparison with contemporary faunas from other parts of the world, holdovers from the Jurassic Period, as if Liaoning Province represented a kind of Lost World of its time. This finding echoes the biogeographical uniqueness of Southern China during the Silurian and Devonian Periods, and shows us that the fossil heritage of China has only just begun to emerge. Given the effusion of remarkable forms uncovered in the past few years, the most we can say is that the next few years will be full of surprises.

In the meantime, we can share the anticipation, puzzlement and frustration of medieval cartographers who, frightened and intrigued by the vastness of the empty spaces in their maps, filled the void with a telling phrase—"Here Be Dragons."

As editor of this book I get to enjoy seeing my name on the front, but it would not be there but for the dedication of many for whom more praise is due. The authors of the papers in this volume gave *Nature* the opportunity of publishing their finds, and the several commentators undertook the enjoyable but sometimes difficult task of setting the finds in context. I should like to thank, in particular, Zhe-Xi Luo, Qiang Ji and many researchers and adminsitrators in China for their encouragement and support. My colleagues at *Nature*, including Philip Campbell, Antoine Bocquet, and Peter Wrobel, have offered unfailing encouragement; James Porteous and his colleagues on the *Nature* Web site team retrieved the materials you see here from *Nature*'s digital vaults. As with *Shaking The Tree*, our earlier collaboration, Christie Henry and her colleagues at the University of Chicago Press took these materials to create the fine volume you hold in your hands. Finally, I wish to express my gratitude to President Jiang Zemin of the People's Republic of China for his kindness in penning the inscription that appears on the title page.

References

1. Luo, Z. Foreword to this volume.
2. Stokstad, E. Exquisite Chinese fossils add new pages to book of life, *Science* **291**, 232–236 (2001); Normile, D., Research kicks into higher gear after a long, uphill struggle, *Science* **291**, 237–238 (2001); Normile, D., Internal fights, looting hinder work in the field, *Science* **291**, 239–241 (2001); Lei, X., Fruitful collaborations follow a two-way street, *Science* **291**, 241 (2001).
3. Rosen, D. E., Forey, P. L., Gardiner, B. G., and Patterson, C. Lungfishes, tetrapods, paleontology and plesiomorphy, *Bulletin of the American Museum of Natural History*, **167**, 154–276 (1981).
4. Gauthier, J., Kluge, A. G., and Rowe, T. Amniote phylogeny and the importance of fossils. *Cladistics* **4**, 104–209 (1988).
5. Forey, P. L. Neontological analyses versus palaeontological stories. Pp 119–157 in *Problems of Phylogenetic Reconstruction*, ed. K. A. Joysey and A. E. Friday. Systematics Association Special Volume **21** (London: Academic Press, 1982).
6. Feduccia, A. *The Origin and Evolution of Birds* (New Haven: Yale University Press, 1996).

7. Xu, X., Zhou, Z. H., and Prum, R. O. Branched integumental structures in *Sinornithosaurus* and the origin of feathers. *Nature* **410**, 200–204 (2001).

8. Chen, J-Y., et al. Precambrian animal diversity: Putative phosphatized embryos from the Doushantuo Formation of China. *Proceedings of the National Academy of Sciences (U.S.)* **97**, 4457–4462, 2000.

9. Davidson, E. H., Peterson, K. J., and Cameron, R. A. Origin of bilaterian body plans: Evolution of developmental regulatory mechanisms, *Science* **270**, 1319–1375 (1995).

10. Gould, S. J. *Wonderful Life: The Burgess Shale and the Nature of History* (New York: Norton, 1989).

11. Basden, A. M., Young, G. C., Coates, M. I., and Ritchie, A. The most primitive osteichthyan braincase? *Nature* **403**, 185–000, 2000.

12. Maisey, J. A primitive chondrichthyan braincase from the Middle Devonian of Bolivia. Pp. 263–288 in *Major Events in Vertebrate Evolution*, ed. P. E. Ahlberg (London: Taylor and Francis, 2001).

13. Zhu, M., Yu, X.-B., and Ahlberg, P. E. A primitive sarcopterygian fish with an eyestalk. *Nature* **410**, 81–84 (2001).

14. Sereno, P. The origin and evolution of dinosaurs. *Annual Reviews of Earth and Planetary Sciences* **25**, 435–489 (1997).

15. Gee, H. *In Search of Deep Time: Beyond the Fossil Record to a New History of Life* (New York: Free Press, 1999).

16. Xu, X., Zhou, Z., and Wang, X. The smallest known non-avian theropod dinosaur. *Nature* **408**, 705–000 (2000).

1

Three-Dimensional Preservation of Algae and Animal Embryos in a Neoproterozoic Phosphorite
Shu-Hai Xiao, Yun Zhang, and Andrew H. Knoll

Phosphorites of the late Neoproterozoic (570 ± 20 Myr BP) Doushantuo Formation, southern China, preserve an exceptional record of multicellular life from just before the Ediacaran radiation of macroscopic animals. Abundant thalli with cellular structures preserved in three-dimensional detail show that latest-Proterozoic algae already possessed many of the anatomical and reproductive features seen in the modern marine flora. Embryos preserved in early cleavage stages indicate that the divergence of lineages leading to bilaterians may have occurred well before their macroscopic traces or body fossils appear in the geological record. Discovery of these fossils shows that the early evolution of multicellular organisms is amenable to direct palaeontological inquiry.

Most of the fossils that document the first 85% of evolutionary history are microscopic. Not until the Phanerozoic eon (<544 Myr BP) do the remains of large animals, algae and, later, plants become conspicuous constituents of the sedimentary record. The most important biological event that connects these palaeobiologically distinct eras is the evolution of complex multicellularity in eukaryotes. Multicellular organisms arose at least six times: in animals, fungi and several groups of algae.[1] Macroscopic remains of uncertain systematic affinities occur in rocks as old as 1,800–2,100 Myr BP[2] and cellularly preserved microfossils of red, green and stramenopile (brown and related) algae indicate that multicellularity was achieved in these groups by about 1,000 Myr.[3] Multicellularity may have evolved comparably early in minute ancestral animals, but until now any pre-Ediacaran animal history has been contentious and thought by many to be unrecognized and perhaps unrecognizable by palaeontologists.

Phosphorites of the Doushantuo Formation in southern China contain three-dimensionally preserved fossils that record in exquisite cellular detail the anatomy and reproductive biology of diverse multicellular algae that lived in the late Neoproterozoic ocean. Doushantuo phosphorites also contain large populations of globular fossils which we interpret to be embryos of early animals. The quality of preservation and evolutionary importance of these fossils rival those of younger Lagerstätten such as the Burgess Shale[4,5] or Rhynie Chert,[6,7] and shed unprecedented palaeontological light on the early evolution of multicellular organisms.

Geological Setting and Fossil Preservation

In its type area along the Yangtze Gorges, the Doushantuo Formation comprises a 250-m succession of carbonates, shales and phosphatic shales that lie disconformably above the glaciogenic rocks of the Nantuo Tillite and conformably beneath the carbonates of the Dengying Formation[8,9] (Fig. 1). Dengying successions contain rare Ediacaran fossils[8,10] and, in their uppermost part, basal Cambrian shelly fossils.[11] Depositional age is only broadly constrained by U–Pb dates on volcanic rocks: tuffs in the underlying Liantuo Formation date from 748 ± 12 Myr,[8] whereas bentonites in the overlying Cambrian Zhongyicun Formation have an older age limit of 539 ± 34 Myr.[12] Diverse acritarchs and a distinctive carbon-isotopic signature, however, allow unambiguous correlation with better-dated successions elsewhere, indicating that Doushantuo sediments accumulated 570 ± 20 Myr ago.[13] Most diverse Ediacaran assemblages are younger than ~550 Myr,[14,15] although frond-like fossils in Newfoundland occur in association with 565 ± 3 Myr ash beds.[16] Globally, acritarch assemblages of the type found in Doushantuo rocks occur in strata that underlie Ediacaran macrofossils. Thus available data indicate that the Doushantuo Formation is older than the diverse Ediacaran assemblages of South Australia and northern Russia, and may antedate all known Ediacaran-type assemblages, except for the simple centimetre-scale discs reported from pre-Varanger rocks in northwestern Canada.[17]

About 600 km to the southwest of the Yangtze Gorges, in the Weng'an region of Guizhou Province (Fig. 1), the Doushantuo Formation has similar acritarchs but markedly different lithologies.

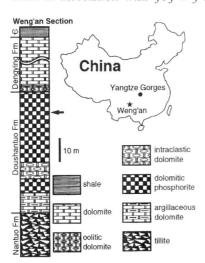

Figure 1 Location and generalized stratigraphy of the fossiliferous Weng'an section in Guizhou Province, South China. The arrow indicates the principal stratigraphic horizon containing phosphatized algae and embryos. Fm, Formation.

Here, Doushantuo sections are only 40–50 m thick and consist predominantly of phosphorites that lie unconformably above older metasediments and, locally, thin Nantuo diamictites[13] (Fig. 1). Above a thin (0.5 m) basal conglomerate and up to 12 m of massive dolostone lie 6–18 m of phosphatic mudstones and subordinate grainstones. The phosphorites exhibit parallel to slightly undulose lamination and are interbedded with fine-grained siliciclastic rocks in the lower part of the unit. Phosphatic beds coarsen upwards, become increasingly dolomitic, and locally contain stromatolites. Phosphatic and dolomitic grainstones, the latter characterized by cross-bedding and phosphorite nodules, are capped by a coarsely conglomeratic unit associated with subaerial erosion. Above this exposure surface, the cycle roughly repeats, with silicified phosphatic dolomicrites and dolarenites of the uppermost Doushantuo Formation abruptly overlain by oolitic dolostones of the basal Dengying Formation. The environment of phosphorite deposition is interpreted as a shallow subtidal platform subject to episodic storms.

Fossils occur in both the lower and upper phosphorite sequences. *In situ* collophane crusts contain oriented algal thalli along with generally uncompressed microfossils and fine-grained organic detritus. Fossils also occur in interbedded, locally derived phosphatic grainstones and gravelstones. Phosphatization was penecontemporaneous with deposition, as shown by the local reworking of phosphatized fossils to form thin bioclastic sandstones.[13]

The chemistry of phosphate permineralization is not well understood, but it has been shown that soft tissues can become phosphatized within days of death.[18] In general, phosphogenesis requires both that phosphate be supplied to surface sediments in relatively high concentrations and that physical and/or biological processes operating within sediments further increase pore-water PO_4^{3-} concentration beyond the point of saturation with respect to apatite.[19-21] Divergent models have been advanced to explain the genesis of bedded phosphorites on ancient shelves and platforms, but most share several features. High biological productivity, commonly related to the upwelling of nutrient-rich deep water, is commonly invoked, as is an 'iron-pumping' mechanism in which particulate FeOOH scavenges PO_4^{3-} from the water column and delivers it to sediments, where it is reduced, liberating PO_4^{3-} and increasing pore-water PO_4^{3-} concentration. Dysaerobic bottom waters are thought to facilitate phosphogenesis because of their elevated PO_4^{3-} concentration. On ancient shelves and platforms, delivery of suboxic waters has commonly been attributed to transgression,[19-21] and the two episodes of phosphorite deposition recorded in Doushantuo successions certainly reflect marked transgressions across the Yangtze Platform.[13] Continental weathering in warm, postglacial environments may also have contributed to regional phosphate enrichment. Low rates of sedimentation, the absence of bioturbation, and extensive cover by microbial mats and thalloid algae would all have facilitated phosphate precipitation.

Multicellular Algae

Zhu *et al.*[22] first reported fossils in Doushantuo phosphorites, but it was Zhang[23,24] who established the superb preservation and diversity of this assemblage. In continuing collaborative investigations, we have documented eight cyanobacterial and 31 acritarch taxa in the formation as a whole, as well as eight formally named and numerous other fragmentary remains of multicellular algae.[13] Several other authors have reported evidence of animal remains in Doushantuo rocks. Rare triact spicules in Doushantuo cherts may record early sponges,[8,9] but structures originally interpreted as microburrows[25] appear to be oblique sections through large, multilamellate cyanobacterial filaments.[13]

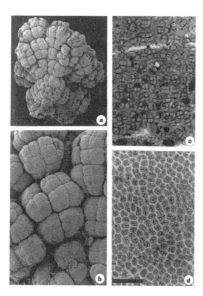

Doushantuo algal thalli range from simple colonies of undifferentiated cells to erect, branching forms characterized by tissue differentiation and specialized reproductive structures. Among the simplest multicellular entities are stacked cuboidal cell packets (Fig. 2a, b) of a type referred to as "incipient tissues."[26] Very similar structures occur today in several clades of green algae.[27,28] Discoidal parenchymatous thalli (Fig. 3e) also occur in Doushantuo phosphorites. These fossils are morphologically simple as well, but in this case deceptively so; parenchymatous growth represents an evolutionary departure from the plesiomorphic states in red, green and brown algae, where developmentally regular multicellularity is based fundamentally on filamentous growth.[27,28]

Figure 2 Algal thalli from the Doushantuo phosphorite and the modern bangiophyte red alga, *Prophyra suborbiculata*. **a–c**, Thalli from Doushantuo phosphorite. **a, b**, Scanning electron micrographs of a thallus composed of cuboidal cell packets similar to those of modern chlorosarcinacean green algae; **b**, the upper left quadrant of **a** at higher magnification. **c**, Photomicrograph of cruciate cell tetrads embedded in a foliose thallus; **d**, Carposporangia within the thallus of living *Porphyra*. Scale bar (in **d**): 200 μm for **a**; 50 μm for **b**; and 100 μm for **c** and **d**.

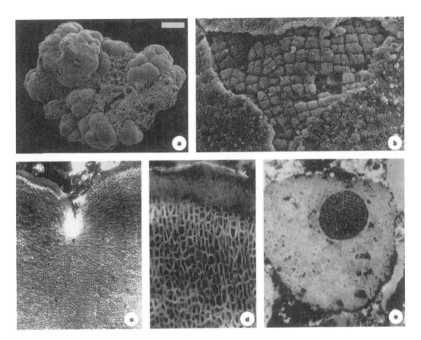

Figure 3 Parenchymatous and pseudoparenchymatous thalli from the Doushantuo phosphorite. **a**, Scanning electron micrograph showing the three-dimensional morphology of a pseudoparenchymatous thallus. **b**, Higher-magnification view of a portion of the same thallus (the arrow in **a** denotes position) showing the cell-surface pattern. **c**, Photomicrograph of a phosphatized thallus in thin section, showing pseudoparenchymatous 'cell fountain' anatomy. **d**, Higher magnification of the same thallus, showing the details of the cell arrangement; the apparently distinct cells at the top of the figure reflect taphonomic and not anatomical differentiation. **e**, Parenchymatous thallus preserved within a phosphatic intraclast. Scale bar (in **a**): 140 μm for **a**; 7.5 μm for **b**; 55 μm for **c**; 20 μm for **d**; and 100 μm for **e**.

Pseudoparenchymatous thalli, formed by the coordinated growth of closely packed filaments, are abundant in the Doushantuo assemblage (Fig. 3a–d). A distinctive characteristic of pseudoparenchymatous growth is the cell fountain, comprising vertical rows of cells that have expanded and diverged upwards to form a fountain-like array in longitudinal section. Cell fountains are common features of Doushantuo thalli, and similar features typify several clades of living red algae, although they also occur in brown algae such as *Ralfsia*.[26] Rhodophytic affinities are supported for at least some Doushantuo thalli by the differentiation of distinct medullary and cortical tissues coupled with reproductive structures similar to the carposporangia and spermatangia of living red algae (Fig. 4).

Compact thalli found abundantly in Doushantuo phosphorites[13,24] contain ellipsoidal clusters of large (8–18 μm) dark cells distributed along

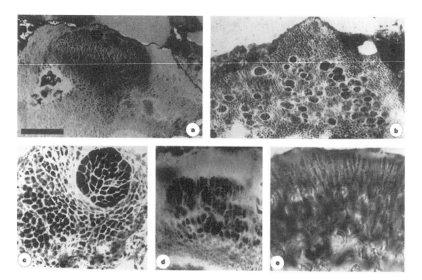

Figure 4 Reproductive structures in Doushantuo thalli and a modern red alga. **a–d**, Doushantuo thalli: **a**, Conceptacle in which carposporangia (clusters of dark, large cells) arise from gonimoblast filaments or supporting cells. **b**, Carposporangia (packets of dark cells) embedded in an anatomically preserved thallus. **c**, Higher-magnification view of carpospores within encompassing vegetative tissue. **d**, Linearly arranged, dark, elongate cells interpreted as possible spermatangia. **e**, Spermatangial sori of the modern florideophyte red alga *Gracilaria* sp., for comparison with **d**. Scale bar (in **a**): 100 μm for **a**, **b** and **d**; 50 μm for **c**; and 30 μm for **e**.

thallus peripheries and occassionally protruding from the thallus surface (Fig. 4a–c). One or more layers of light-coloured, elongate cells (1–4 μm wide and up to 10 μm long) curve around and encompass the clusters. The cell clusters are interpreted as carposporangia and empty spheroidal regions that occupy a similar anatomical position are interpreted as empty conceptacles or cystocarps. Distinctive cell rows at the base of carposporangial clusters compare closely with the gonimoblast filaments that support carposporangial development in living red algae (Fig. 4a). Other thalli contain distinctive tissues of large, dark, oblong or rod-like cells arranged in parallel rows, with apical cells distinctly elongated and separated from subjacent cell rows by transverse septa (Fig. 4d); these structures are interpreted as male reproductive tissues rather like the spermatangial sori of some extant rhodophytes (Fig. 4e).

Such features suggest affinities to the red-algal class Florideophyceae. Other Doushantuo fossils indicate that the class Bangiophyceae had also diversified markedly by the end of the Proterozoic eon: foliose thalli with regularly arranged cruciate cell tetrads (and, less commonly, octads) are indistinguishable from the carposporangial thalli of the extant bangiophyte alga *Porphyra* (Fig. 2c, d).

The oldest known fossil red algae are silicified *Bangia*-like filaments found in 1,200–900 Myr-old tidal flat carbonates from arctic Canada.[29] Doushantuo fossils show that by the end of the Proterozoic eon the red algae were structurally complex and taxonomically diverse.

A comparable evolutionary pattern seems to characterize the photosynthetic stramenopiles. Compressed seaweeds in uppermost Doushantuo shales in the Yangtze Gorges area include a large population attributable to the Fucales, one of the most derived of all brown algal orders,[30] whereas microscopic, coenocytic algae in the 1,000–900 Myr-old Lakhanda Group of eastern Siberia[31] are indistinguishable from species of the extant xanthophyte *Vaucheria*. Similarly, the possible green algae in Doushantuo phosphorites and shales[32] complement 750–700 Myr-old compression fossils from Spitsbergen that include populations similar to *Cladophora*,[33] among the shallowest branching of all ulvophyte green algae. Thus, by the time large animals enter the fossil record, the three principal groups of multicellular algae had not only diverged from other protistan stocks but had evolved a surprising degree of the morphological complexity exhibited by living algae.

Neoproterozoic Embryos

Since Haeckel[34] it has been broadly accepted that the earliest metazoans must have been microscopic organisms similar in development and morphology to the embryos or larvae of living animals. Davidson *et al.*[35] have expanded this view of animal 'prehistory' by hypothesizing that animals not only originated but underwent substantial early cladogenesis as minute, little-differentiated metazoans similar in form, function and ontogeny to the larvae of living invertebrates. Only later, after developmental toolkits and physiological tolerances had been well established, did macroscopic size and the adult body plans of extant phyla evolve within already discrete clades. Molecular clock estimates have suggested that the divergence of protostome and deuterostome animals took place as early as 1,000–1,300 Myr BP.[36] These data are currently the subject of intensive methodological scrutiny, although Neoproterozoic algae provide support for the broad hypothesis that animals originated long before they became conspicuous elements of the geological record. As noted above, algal fossils document rapid eukaryotic divergence beginning at least 1,000 Myr BP, and molecular phylogenies imply that the ancestors of animals diverged from other eukaryotes as part of this radiation.[37] Three important groups of algae evolved multicellular forms early in their history, and it is reasonable to suggest that animals did so too. The conventional metazoan fossil record can be reconciled with molecular hypotheses by making the simple assumption that clade divergence and the evolution of large size and adult body plans within clades are distinct events separated by a considerable interval of time.[38,39]

Palaeontological tests of such conjectures require Lagerstätten of the first order. Reports of phosphatized invertebrate embryos in Cambrian carbonates suggest that older phosphatic Lagerstaätten may be the best places to search for early records of animal evolution.[40,41] The exquisite preservation of Doushantuo algae specifically invites a search for microscopic animal remains, and this search has yielded positive results.

Xue *et al.*[42] have described spheroidal microfossils containing geometrically arranged cells from Weng'an phosphorites, interpreting them as volvocacean green algae similar to extant *Pandorina*. Our collections, however, demonstrate that comparisons to volvocacean or any other algae are unlikely, given the size, geometry of cell division and structure of encompassing vesicles in the fossil population. Instead, we argue that these fossils are preserved embryos.

The specimens in our sample population are globular, measuring about 500 μm in diameter ($\bar{x} = 584$ μm, $s_{\bar{x}} = 12$ μm, $n = 115$). Individuals contain one, two, four, eight or more closely packed internal bodies, the size and orientation of which suggest they are cells that underwent successive binary divisions with little or no intervening growth (Fig. 5). The diminishing size of internal bodies as their number multiplies is suggestive of early embryonic development. The geometric arrangement of the internal bodies is precise and is strikingly similar to the early cleavage stages of metazoan embryos.[43] An ornamented external covering 10 μm thick encompasses single cells, which we interpret as resting zygotes within egg cases. We further interpret fossils containing two, four, eight or more internal bodies with polygonal or faceted geometries as cleaving embryos, with the internal bodies being blastomeres. Cross-sections (Fig. 5g) suggest that they are stereoblastulas. Developing embryos are not enveloped by a thick egg case, but instead are bounded by a thin wall rather like the zygotic membrane of living invertebrates. Judging from their egg size, we infer that the Doushantuo embryos underwent direct or lecithotrophic larval development.

At the four-cell stage, blastomeres are arranged in a modified tetrahedron, with opposite pairs of cells meeting across cross-furrows oriented perpendicular to one another at either end of the embryo (Fig. 5c). Structures similar to the cross-furrows of early embryos can also be seen at the 8- and 16-cell stages (Fig. 5e–f). Therefore, we interpret the Doushantuo fossils as holoblastic and equally cleaving stereoblastulas. Gastrulas or later developmental stages have not yet been identified.

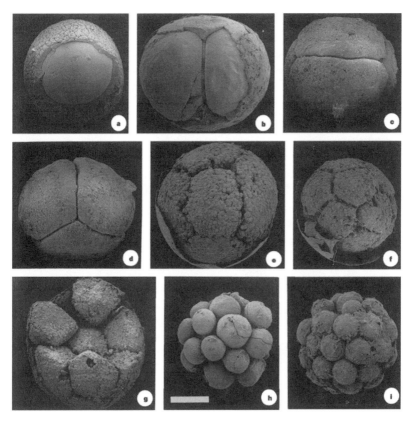

Figure 5 Fossil embryos preserving different stages of cleavage from the Doushantuo phosphorite. **a**, Fertilized (?) egg with thick membrane. **b**, Two-cell stage. **c**, **d**, Four-cell stage, **c** and **d** show different views of the same specimen, illustrating the tetrahedral geometry. **e**, Eight-cell stage. **f**, **g**, Later cleavage stages showing faceted cell geometry and, in **g**, the three-dimensional distribution of cells. **h**, **i**, Multicellular structures that record later cleavage stages or, especially possible for **h**, colonial protists. Scale bar (in **h**): 200 μm for **a**, **e**, **f**, **g**, **h** and **i**; 150 μm for **b**; and 240 μm for **c** and **d**.

The Doushantuo fossils could be broadly equivalent to blastaea or planuloids *sensu* Haeckel,[34] embryos of microscopic animals as envisioned by Davidson *et al.*,[35] or the earliest developmental stages of some as yet unrecognized macroscopic metazoan. Tetrahedral geometries comparable to those in the Doushantuo population are unusual in modern animal embryos, but not unknown.[43–46] In fact, some living crustacean arthropods,[43] have large eggs, direct development, a tetrahedral four-cell stage, and equal holoblastic cleavage in combination. Thus, the Doushantuo fossils are most probably bilaterian, and they may document cladogenesis within the Bilateria before the appearance of diverse Ediacaran assemblages. Given the age and architectural simplicity of these remains, however, phylogenetic interpretation is best approached with caution.

Despite the many uncertainties that surround the interpretation of the Doushantuo embryos, they provide the first direct geological evidence in support of the hypothesis that the main metazoan clades diversified before the emergence of a conspicuous animal fossil record. More generally, they show that palaeontological investigation can tell us a great deal about the early history of metazoan evolution.

Conclusion

More than a century ago, Agassiz[47] recognized the 'three-fold parallelism' of patterns in ontogeny, systematics and biostratigraphy. The remarkable phosphatic thalli and embryos of the Doushantuo Formation show that unanticipated palaeontological observations, together with insights from molecular phylogeny and developmental genetics, can facilitate a modern integration of phylogeny, development and palaeontology that extends deeply into evolutionary history to address the early evolution of multi-cellular life.

References

1. Buss, L. W. *The Evolution of Individuality* (Princeton Univ. Press, NJ, 1987).
2. Han, T.-M. & Runnegar, B. Megascopic eukaryotic algae from the 2.1 billion-year-old Negaunee Iron-Formation, Michigan. *Science* 257, 232–235 (1992).
3. Knoll, A. H. The early evolution of eukaryotes: a geological perspective. *Science* 256, 622–627 (1992).
4. Whittington, H. B. *The Burgess Shale* (Yale Univ. Press, New Haven, CT, 1985).
5. Briggs, D. E. G., Erwin, D. H. & Collier, F. J. *The Fossils of the Burgess Shale* (Smithsonian Institution Press, Washington DC, 1994).
6. Kidston, R. & Lang, W. H. On Old Red Sandstone plants showing structure from the Rhynie chert bed, Aberdeenshire. Parts I–Iv. *Trans. R. Soc. Edinb.* 51, 761–784; 52, 603–627; 62, 643–680; 52, 831–854 (1917–1921).
7. Remy, W., Gensel, P. J. & Hass, H. The gametophyte generation of some early Devonian land plants. *Int. J. Plant Sci.* 154, 35–58 (1993).
8. Zhao, Z., Xing, Y., Ma, G. & Chen, Y. *Biostratigraphy of the Yangtze Gorge Area, (1) Sinian* (Geological Publishing House, Beijing, 1985).
9. Zhao, Z. et al. *The Sinian System of Hubei* (China University of Geosciences Press, Wuhan, 1988).
10. Sun, W. Late Precambrian pennatulids (sea pens) from the eastern Yangtze Gorge, China: *Paracharnia gen. nov. Precambrian Res.* 31, 361–375 (1986).
11. Qian, Y., Chen, M. & Chen, Y. Hyolithids and other small shelly fossils from the Lower Cambrian Huangshandong Formation in the eastern part of the Yangtze Gorge. *Acta Palaeontol. Sinica* 18(3), 207–232 (1979).
12. Compston, W., Williams, I. S., Kirchvink, J. L., Zhang, Z. & Ma, G. Zircon U-Pb ages for the Early Cambrian time-scale. *J. Geol. Soc. Lond.* 149, 171–184 (1992).
13. Zhang, Y., Ying, L., Xiao, S. & Knoll, A. H. Permineralized fossils from the Terminal Proterozoic Doushantuo Formation, South China. *Paleontol. Soc. Mem.* (in the press).
14. Grotzinger, J. P., Bowring, S. A., Saylor, B. Z. & Kaufman, A. J. Biostratigraphic and geochronologic constraints on early animal evolution. *Science* 270, 598–604 (1995).
15. Kaufman, A. J., Knoll, A. H. & Narbonne, G. M. Isotopes, ice ages, and terminal Proterozoic earth history. *Proc. Natl Acad. Sci.* USA 94, 6600–6605 (1997).

16. Benus, A. P. Sedimentologic context of a deep-water Ediacaran fauna (Mistaken Point Formation, Avalon zone, eastern Newfoundland). *Bull. N. Y. State Mus.* **463**, 8–9 (1988).

17. Hofmann, H. J., Narbonne, G. M. & Aitken, J. D. Ediacaran remains from intertillite beds in northwestern Canada. *Geology* **18**, 1199–1202 (1990).

18. Briggs, D. E. G., Kear, A. J., Martill, D. M. & Wilby, P. R. Phosphatization of soft-tissue in experiments and fossils. *J. Geol. Soc. Lond.* **150**, 1035–1038 (1993).

19. Krajewski, K. P. *et al.* Biological processes and apatite formation in sedimentary environments. *Ecolog. Geol. Helvet.* **87**, 701–745 (1994).

20. Glenn, C. R. *et al.* Phosphorus and phosphorites: Sedimentology and environments of formation. *Eclog. Geol. Helvet.* **87**, 747–788 (1994).

21. Föllmi, K. B. The phosphorus cycle, phosphogenesis and marine phosphate-rich deposits. *Earth Sci. Rev.* **40**, 55–124 (1996).

22. Zhu, S. & Wang, Y. in *The Upper Precambrian and Sinian-Cambrian Boundary in Guizhou* (eds Wang, Y. *et al.*) 93–103 (People's Publishing House of Ghizhou, Guiyang, 1984).

23. Zhang, Y. Multicellular thallophytes with differentiated tissues from Late Proterozoic phosphate rocks of South China. *Lethaia* **22**, 113–132 (1989).

24. Zhang, Y. & Yuan, X. New data on multicellular thallophytes and fragments of cellular tissues from Late Proterozoic phosphate rocks, South China. *Lethaia* **25**, 1–18 (1992).

25. Awramik, S. M. *et al.* Prokaryotic and eukaryotic microfossils from a Proterozoic/Phanerozoic transition in China. *Nature* **315**, 655–658 (1985).

26. Bold, H. C. & Wynne, M. J. *Introduction to the Algae* (Prentice-Hall, Englewood Cliffs, NJ, 1985).

27. Fritsch, F. E. *The Structure and Reproduction of the Algae* Vols 1, 2 (Cambridge Univ. Press, 1965).

28. van den Hoek, C., Mann, D. G. & Jahns, H. M. *Algae: An Introduction to Phycology* (Cambridge Univ. Press, 1995).

29. Butterfield, N. J., Knoll, A. H. & Swett, K. A bangiophyte red alga from the Proterozoic of Arctic Canada. *Science* **250**, 104–107 (1990).

30. Xiao, S., Knoll, A. H. & Yuan, X. Morphological reconstruction of *Miaohephyton bifurcatum*, a possible brown alga from the Terminal Proterozoic Doushantuo Formation, South China. *J. Paleontol.* (in the press).

31. Hermann, T. N. *Organic World Billion Year Ago* (Nauka, Leningrad, 1990).

32. Chen, M. & Xiao, Z. Discovery of the macrofossils in the Upper Sinain Doushantuo Formation at Miaohe, eastern Yangtze Gorges. *Sci. Geol. Sinica* **4**, 317–324 (1991).

33. Butterfield, N. J., Knoll, A. H. & Swett, K. Paleobiology of the Neoproterozoic Svanbergfjellet Formation, Spitsbergen. *Fossils Strata* **34**, 1–84 (1994).

34. Haeckel, E. The gastrea theory, the phylogenetic classification of the animal kingdom and the homology of the germ-lamellae. *Q. J. Microsc. Soc.* **14**, 142–165 (1874).

35. Davidson, E. H., Peterson, K. J. & Cameron, R. A. Origin of bilaterian body plans: Evolution of develpmental regulatory mechanisms. *Science* **270**, 1319–1325 (1995).

36. Wray, G. A., Levinton, J. S. & Shapiro, L. H. Molecular evidence for deep Precambrian divergences among metazoan phyla. *Science* **274**, 568–573 (1996).

37. Sogin, M. L. in *Early Life on Earth* (ed. Bengtson, S.) 181–192 (Columbia Univ. Press, NY, 1994).

38. Vermeij, G. J. Animal Origins. *Science* **274**, 525–526 (1996).

39. Fortey, R. A., Briggs, D. E. G. & Wills, M. A. The Cambrian evolutionary 'explosion': decoupling cladogenesis from morphological disparity. *Biol. J. Linn. Soc.* **57**, 13–33 (1996).

40. Zhang, X. & Pratt, B. R. Middle Cambrian Arthropod embryos with blastomeres. *Science* **266**, 627–639 (1994).

41. Bengtson, S. & Yue, Z. Fossilized metazoan embryos from the earliest Cambrian. *Science* **277**, 1645–1648 (1997).

42. Xue, Y., Tang, T., Yu, C. & Zhou, C. Large Spheroidal Chlorophyta fossils from the Doushantuo Formation phosphoric sequence (late Sinian), central Guizhou, South China. *Acta Palaeontol. Sinica* **34**, 688–706 (1995).

43. Kumé, M. & Dan, K. Invertebrate Embryology. (NOLIT, Belgrade, 1968).

44. Anderson, D. T. *Embryology and Phylogeny in Annelids and Arthropods* (International Series of Monographs in Pure and Applied Biology, Vol. 50) (Pergamon, Oxford, 1973).

45. Brusca, R. C. & Brusca, G. J. *Invertebrates* (Sinauer, Sunderland, MA, 1990).

46. Nielsen, C. Animal Evolution: *Interrelationships of the Living Phyla* (Oxford Univ. Press, 1995).

47. Agassiz, L. *Essay on Classification* (reprinted from *Contributions to the Natural History of the United States*, vol. 1, 1857 (Harvard Univ. Press, Cambridge, MA, 1962).

Acknowledgements

We thank Y. Leiming for field assistance, E. Seling for technical help and S. Bengtson, D. McHugh, R. M. Woollacott, S. J. Gould, C. Nielsen, E. Ruppert, E. Davidson, A. Cameron and J. Henry for discussions and comments. This work was partly supported by grants from NSFC (to Y.Z.) and NSF (to A.H.K.).

2

Animal Embryos in Deep Time
Stefan Bengtson

Discoveries of spectacularly preserved embryos and tissues, in rocks that are about 570 million years old, open a new era in the study of early animal evolution.

A spell seems to have been broken—animals considerably older than the Cambrian are finally being found in the fossil record, and they are preserved in a way that reveals details down to the cellular level. This stirring claim is based on studies of the roughly 570-million-year-old Doushantuo phosphorites in southern China, and is hardly weakened by the fact that it comes from two different groups of palaeontologists. On page 553 of this issue,[1] Xiao *et al.* report on exquisitely preserved algae and animal embryos from the phosphorites, while, in *Science,* Li *et al.*[2] describe sponges and animal embryos from the same deposits.

The Cambrian Explosion, the evolutionary radiation of life forms about 550 million years ago, was one of the major turning points in the history of the Earth. Over a period of a few tens of millions of years, practically all of the principal animal lineages (phyla) appeared. At least this is what the fossil record suggests. The dearth of animal fossils below the Precambrian–Cambrian boundary has been one of the main frustrations in studies of early animal history, for there has been very little agreement on how, when and where animals lived and evolved during the Proterozoic— the period of time that ended with the beginning of the Cambrian and the Phanerozoic (see Fig. 1 for a timescale). The celebrated Ediacara biota of the Vendian Period, mostly preserved as impressions and traces in sandstones and shales, provides some clues. But it has so far produced more disputes than data with regard to early animals; most Ediacaran deposits are in any case only slightly older than the Cambrian.[3]

For a long time following Darwin, a long interval of hidden animal evolution before the Cambrian used to be assumed. But owing to the influence of Preston Cloud[4,5] and others, the more recent inclination has been not to postulate more hidden evolution than absolutely necessary. An estimate in 1996 by three palaeontologists suggested that the main divergence of animal phyla occurred no earlier than 565 million years ago.[6] At the same time, an attempt[7] to estimate the same divergence times using sequence comparisons of various animal genes landed at figures that were more than twice as high. This analysis has been criticized on methodological grounds, and a date of about 670 million years[8] has been proposed instead. Although most palaeontologists will probably be quite happy with the latter date, the fossil record has hitherto been ominously silent.

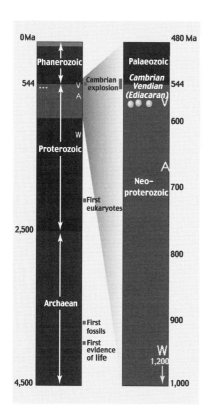

Figure 1 Timescale of Earth's history (left), and the interval between 1,000 and 480 million years ago (Ma), showing some significant events in the fossil record. The Doushantuo fossils described by Xiao et al.[1] and Li et al.[2] are indicated by three spherules. V, A and W show the alternative approximate ages (without error bars) of the main radiation of animals as proposed by Valentine et al.,[6] Ayala et al.[8] and Wray et al..[7] Specifically, these are alternative dates for the divergence between protostomes and deuterostomes.

The origin of animals has thus been as elusive as that of life itself: there are many possibilities but few facts to test them against. The Doushantuo fossils promise to change all that—not only do they give the first convincing glimpse of pre-Ediacaran animal life, but the quality of preservation is almost unheard of, even in much younger fossils. This preservation offers insights into cell-level anatomy, embryological development and life cycles—such matters have not normally been considered to be open to investigation in fossils. Above all, the usual explanation for the missing fossil record has been that the animals were soft and small and therefore would not be preserved as fossils. Now such fossils are beginning to appear. Even better, there seems to be nothing very unusual about the Doushantuo phosphorites, and they may therefore show the way to many other such sites.

The key to the exquisite preservation is calcium phosphate. This mineral is known for its faithful replication of delicate tissues,[9,10] although early phosphatization in sediments tends to be very patchy and frequently only fossils of millimetre size or below are preserved.[11,12]

The specimens described by Li *et al.*[2] are interpreted as sponges. The siliceous or calcareous spicules of sponges are common in the fossil record, and sponges are among the simplest of multicellular creatures. So they would not be unexpected members of the earliest animal assemblages. The needle-shaped spicules in the examples depicted by Li *et al.* are regularly arranged in distinct bodies built up of cell-like objects, some of which adhere to the spicules in the same way as sclerocytes (spicule-forming cells) do in living sponges. Details of the proposed interpretations are open to question, but it will be difficult to disprove that these are indeed sponges.

The Doushantuo sponges are small, about 150–750 μm in maximum dimension. Although Li *et al.* interpret them as fully formed adults, they could also represent propagules or embryos. But the authors quite reasonably interpret other fossils as embryos of different kinds of animals, and here they find themselves in good agreement with Xiao *et al.*,[1] who have isolated a suite of globular fossils from the Doushantuo which they identify as animal embryos in the early stages of cell division (cleavage).

Animal embryos are small and delicate. A few years ago, the thought of finding fossilized embryos of anything but bony pre-hatchlings of dinosaurs and the like was preposterous. However, recent discoveries of Cambrian phosphatized embryos of animals in various developmental stages[13,14] suggested that this might be a fruitful search strategy for the missing record of Proterozoic animals.[14] It indeed seems that all we needed was to open our eyes to the possibility: the fossils now identified as embryos had actually been described in the literature but were interpreted as colonial green algae.[15]

Xiao and colleagues' fossils are about half a millimetre in diameter, and are compartmentalized into two, four, eight or more bodies which are proposed to be blastomeres (cells in a cleavage embryo; see the stunning picture on the cover and those on page 556 of this issue). The constant size of the fossils, irrespective of the number of compartments, fits a pattern of developing early embryos with a constant cytoplasmic volume. This would not be expected in colonial algae or in objects formed by non-biological processes.

So what information can we get from these kinds of fossils? The two studies,[1,2] although preliminary in nature, already yield some insights. The previously known oldest sponges were late Ediacaran hexactinellids (glass sponges),[16,17] but the spicule morphology and cell configuration of the Doushantuo sponges are very different from those of hexactinellids. The early-cleavage embryos have a tetrahedral blastomere configuration that is today known in some animals such as nematodes, flatworms and arthropods. One should be careful about drawing evolutionary conclusions from physically simple patterns like these, but the observation clearly shows the potential of such material. Only finds of later developmental stages will tell in which direction, and how far, these embryos developed.

The Doushantuo phosphorites cover an area of 57 km^2, and they undoubtedly contain further secrets. Phosphate is thus pay dirt. But whereas digging up the basal roots of animals may have its particular appeal, let's not forget about the rest of animal history. Developmental and evolutionary biology are complementary but largely separate sciences, and the fossil record might help in bringing them together. Palaeoembryology may be a science of the past, but it could have a brilliant future.

References

1. Xiao, S., Zhang, Y. & Knoll, A. H. *Nature* **391**, 553–558 (1998).
2. Li, C.-w., Chen, J.-y. & Hua, T.-e. *Science* **279**, 879–882 (1998).
3. Grotzinger, J. P., Bowring, S. A., Saylor, B. Z. & Kaufman, A. J. *Science* **270**, 598–604 (1995).
4. Cloud, P. E. *Evolution* **2**, 322–350 (1948).
5. Cloud, P. E. Jr in *Evolution and Environment* (ed. Drake, E. T.) 1–72 (Yale Univ. Press, New Haven, CT, 1968).
6. Valentine, J. W., Erwin, D. H. & Jablonski, D. *Dev. Biol.* **173**, 373–381 (1996).
7. Wray, G. A., Levinton, J. S. & Shapiro, L. H. *Science* **274**, 568–573 (1996).
8. Ayala, F. J., Rzhetsky, A. & Ayala, F. J. *Proc. Natl Acad. Sci. USA* **95**, 606–611 (1998).
9. Müller, K. J. *Phil. Trans. R. Soc. Lond. B* **311**, 67–73 (1985).
10. Martill, D. M. *Nature* **346**, 171–172 (1990).
11. Walossek, D. *Fossils & Strata* **32**, 1–202 (1993).
12. Briggs, D. E. G. & Wilby, P. R. *J. Geol. Soc. Lond.* **153**, 665–668 (1996).
13. Zhang, X.-g. & Pratt, B. *Science* **266**, 637–639 (1994).
14. Bengtson, S. & Yue, Z. *Science* **277**, 1645–1648 (1997).
15. Xue Y.-s., Tang T.-f., Yu C.-l. & Zhou C.-m. *Acta Palaeontol. Sinica* **34**, 688–706 (1995).
16. Gehling, J. & Rigby, J. K. *J. Paleontol.* **70**, 185–195 (1996).
17. Brasier, M., Green, O. & Shields, G. *Geology* **25**, 303–306 (1997).

3

A Pipiscid-like Fossil from the Lower Cambrian of South China

De-Gan Shu, S. Conway Morris, Xing-Liang Zhang, Liang Chen, Y. Li, and J. Han

Exceptional fossil preservation is critical to our understanding of early metazoan evolution. A key source of information is the Burgess Shale-type faunas.[1-5] Fossils from these deposits provide important insights into metazoan phylogeny, notably that of stem-group protostomes,[2,3,6] and related topics such as trophic specialization.[7] Metazoan relationships are also being significantly reappraised in terms of molecular-based phylogenies,[8,9] but integration of these data with palaeontological systematics is not straightforward.[10,11] Moreover, molecular phylogenies are silent concerning the anatomies of stem-groups and the functional transitions that underpin the origin of different body plans.[2,6] Some hitherto enigmatic fossils possess unique character–state combinations that, although they can be shoe-horned into extinct phyla,[12] may be more profitably interpreted as defining major stem-groups.[2,3] Here we describe a possible pipiscid, a metazoan previously known only from the Upper Carboniferous,[13,14] from the Lower Cambrian of south China. Pipiscids are currently interpreted as being agnathan chordates,[13-15] but this discovery from the Chengjiang fossil-Lagerstätte indicates that the assignment of pipiscids to the Agnatha deserves to be reconsidered.

Phylum Uncertain

Xidazoon Shu, Conway Morris & Zhang gen. nov.

Xidazoon stephanus Shu, Conway Morris & Zhang sp. nov.

ETYMOLOGY.   Genus name an abbreviation of Chinese name for Northwest University at Xi'an. Species name *stephanos* (Greek) for crown.

HOLOTYPE.   Early Life Institute, Northwest University, Xi'an. ELI-0000194.

STRATIGRAPHY AND LOCALITY.   Qiongzhusi (Chiungchussu) Formation, Yu'anshan member (*Eoredlichia* Zone); Lower Cambrian. Specimen collected from Haikou, Kunming, located about 50 km west of Chengjiang.

DIAGNOSIS.   Body with two-fold division, reminiscent of *Banffia* but anterior section more inflated and possessing prominent mouth circlet.

Anterior section with faint transverse divisions towards front, otherwise smooth. Mouth defined by circlet of about 25 plates, divided into inner and outer regions, otherwise unarmed. Circlet similar to plated mouth of *Pipiscius*, although in the latter taxon the plates are more cuticularized and inner circlet folded into pharynx. Posterior section tapering towards front and back, segmented with cuticularized region of about six segments succeeded anteriorly by about three less well-defined segments. Posterior section similar to arthropodan metameres, but lacking evidence of appendages. Cuticular segments also reminiscent of posterior region in *Yunnanozoon*, but in latter taxon segments are ventrally incomplete. Short terminal spines at posterior tip. Alimentary canal with terminal openings, anterior region possibly expanded and rectum with ?dilator muscles.

DESCRIPTION.   *Xidazoon stephanus*, new genus and species, is known from two, or possibly three, specimens on a single slab (Figs 1a, 2). The most complete specimen is about 8.5 cm long and a second individual shows details of the anterior (Fig. 1d). The body comprises two main regions. The anterior section is moderately inflated, and the prominent circlet of the presumed anterior is interpreted as a feeding apparatus surrounding a voluminous mouth (Fig. 1c). The apparatus itself consists of plate-like structures, transversely folded to define inner and outer circlets. The edges of the inner circlet of plates are ridged (Fig. 1c), but they do not bear teeth or other extensions. In the second specimen (Fig. 1d) the plates appear to be separated adorally by narrower recessed areas. These may represent flexible inter-plate membranes. The anterior of the second specimen is incomplete, and that of the first is too crushed to give more than an estimate of the total number of plates. The better-preserved half-circumference displays about 13 plates, and an allocation of typical plate width around the circumference (~45 mm) gives a total of about 25. The mouth is gaping, but apart from the circlet of plates lacks evidence of jaws or other associated structures. Behind the feeding apparatus the anterior region bears faint, widely separated transverse divisions that may be segmental.

The posterior section tapers in either direction from an expanded central zone. It consists of about six well-defined segments (Figs 1b, 2), and in the anterior direction there is a series of more faint transverse annulations. The surface appears to have been lightly cuticularized. The segment boundaries vary from tightly adpressed to separated, indicating originally relatively wide and flexible intersegmental membranes. At the posterior tip there are two or three spinose projections (Figs 1b, 2).

Little is known of the internal anatomy. A gut trace is present in the mid and posterior sections, and near the terminal anus diverging strands may represent dilator muscles (Figs 1a, b, 2). Towards the anterior of the

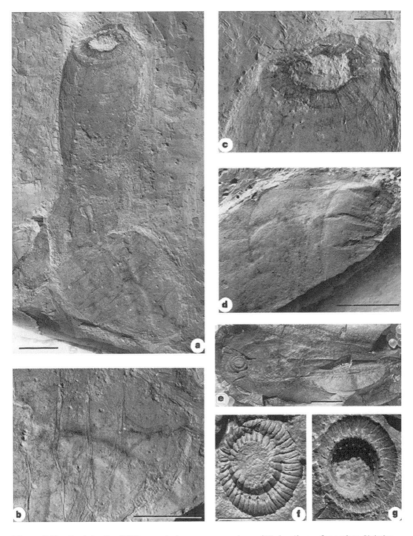

Figure 1 The Cambrian fossil *Xidazoon stephanus*, new species and Carboniferous ?agnathan *Pipiscius zangerli*. **a**, Entire specimen and (to lower left) incomplete individual of *Xidazoon* (compare to Fig. 2); **b**, detail of posterior section showing segmental divisions, gut trace, ?dilator muscles and posterior spines (right-hand side); **c**, detail of feeding apparatus of complete specimen; **d**, detail of anterior and incomplete feeding apparatus of second specimen. **e**, Entire specimen of holotype of *Pipiscius*, part (PF 8345); **f**, detail of feeding apparatus of part; **g**, detail of feeding apparatus of counterpart. Scale bars: 10 mm (**a**, **b**, **e**), 5mm (**c**, **d**) and 2 mm (**f**, **g**).

visible gut trace it appears to expand, and in the anterior section it may have been voluminous.

PRESERVATION.   The style and quality of preservation is similar to other Chengjiang taxa, such as *Yunnanozoon*.[16–19] Thus, the extent of decay appears to be limited. Features, notably the circlet of plates and the posterior segmentation, seem to be original rather than post-mortem artefacts.

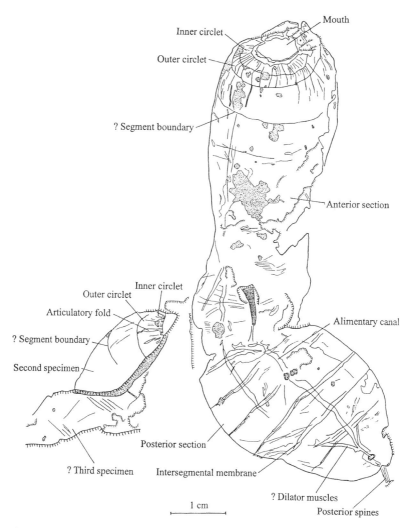

Figure 2 Camera-lucida drawing of slab containing the two (or possibly three) specimens of the Cambrian fossil *Xidazoon stephanus*, new species.

ECOLOGY.   The ecology of *Xidazoon* is problematic. It was presumably benthic, with the anterior circlet periodically contracting to ingest detritus. The inflated nature of the anterior section in the most complete specimen might be because of sediment ingestion. An alternative possibility is that the anterior organ acted as a sucker for lodgement on prey or hard substrates.

DISCUSSION.   Comparisons between *Xidazoon* and various extant metazoan groups, such as the sipunculans and the much smaller cycliophorans, are not convincing. Similarly, among the diverse Burgess Shale-type assemblages, no exact counterpart to *Xidazoon* has been recognized. There are some similarities to the otherwise enigmatic *Banffia confusa*,[5] which consists of a segmented unit, apparently posterior to an elongate and smooth section, but this taxon lacks evidence for the prominent feeding apparatus of *Xidazoon*. The better-known anomalocaridids[20] have a prominent circular feeding apparatus and a bipartite body with segmented posterior section. There are, however, many differences. The feeding apparatus occurs in a variety of forms,[3,5,20] but none is particularly similar to *Xidazoon*. Other characteristic features of the anomalocaridids, notably the anterior giant appendages and lateral lobes, have no parallel in this new fossil.

The anterior circlet of *Xidazoon* is, however, similar to the otherwise unique feeding apparatus of the putative agnathan *Pipiscius zangerli* (Fig. 1e), a rare species from the 300-Myr-old Mazon Creek fossil-Lagerstätte (Upper Carboniferous) of Illinois. The original description[13] is convoluted, but in essence the feeding apparatus is composed of two circles of sclerotized plates. The inner series ('collar lamellae' of ref. 13) total 23, a number with no apparent parallel in other metazoan organ systems. The outer circlet is also cited[13] as consisting of 23 plates. There is, however, a hitherto unrecognized duplication on the leading anterior plate, so that the total number of plates is effectively 24. This duplication defines a line of bilateral symmetry in the apparatus. The plates are separated by narrow clefts ('vanes' of ref. 13) that presumably accommodated shape changes associated with feeding. The principal similarities between the anterior apparatus of *Pipiscius* and *Xidazoon* are the double nature of the circlet with direct continuity between the inner and outer plates, the similar number of plates and evidence for articulatory zones (Fig. 1d) that seem to be comparable to the 'vanes' (Fig. 1f). The apparatus, however, are not identical. In *Pipiscius* the outer plates have a more complex structure, housing triangular insets. These latter units may have accommodated movement of the apparatus, possibly necessitated by a more pronounced sclerotization. Deep pits associated with the 'vanes', and possibly employed for muscle insertions,[13] are not evident in *Xidazoon*. Finally, the inner circle ('collar') of *Pipiscius* is directed inwards, whereas in *Xidazoon* it appears to be more rim-like.

Concerning the possible connection between *Xidazoon* and *Pipiscius*, there seems to be three alternative evolutionary scenarios. First, the annular feeding apparatus is simply an example of convergence. Among the many suctorial and other biological attachment structures similarities can be shown, for example, with the attachment organ of the ectoparasitic ciliate *Trichodina pediculus*[21] and the arm suckers of the octopus,[22] although no phylogenetic connection with *Xidazoon* can be seriously entertained. Notwithstanding the bi-annular arrangement of about 25 plates, the few similarities that otherwise exist between *Xidazoon* and *Pipiscius* make convergence a reasonable option. Second, *Xidazoon* and *Pipiscius* are related, but the assignment of the latter taxon to the agnathans[13,14] is erroneous: together they would represent a new major Palaeozoic clade of as yet unknown affinities. In this sense it would be comparable to such enigmatic groups as the typhloesids[23] and tullimonstrids.[24]

The third proposal is that *Xidazoon* is a precursor to the agnathans, including *Pipiscius*. This presupposes the homology of the circular feeding apparatus in the two taxa, and that certain features (such as fin-rays and possible myotomes) of *Pipiscius* are indicative of a chordate relationship. In this scenario *Xidazoon* would potentially provide new insights into the organization of stem-group deuterostomes. A link may also exist with the coeval *Yunnanozoon*.[16–19] This Chengjiang taxon displays putative gill slits, and the cuticular segmentation has some similarities with *Xidazoon*. One reconstruction[18] of *Yunnanozoon* also depicts a circum-oral set of plates. The bipartite nature of *Xidazoon* is more strongly developed than in *Yunnanozoon*, but the almost arthropod-like segmented posterior section could provide an intriguing phylogenetic link with the protostomes.[25] Continuing investigations of Lower Cambrian fossil-Lagerstätten may yield relatives of *Xidazoon* that will help to resolve the controversial status of these fossils in the context of metazoan phylogeny.

References

1. Conway Morris, S. *The Crucible of Creation: The Burgess Shale and the Rise of Animals* (Oxford Univ. Press, Oxford, 1998).
2. Conway Morris, S. & Peel, J. S. Articulated halkieriids from the Lower Cambrian of North Greenland and their role in early protostome evolution. *Phil. Trans. R. Soc. Lond. B* **347**, 305–358 (1995).
3. Budd, G. E. in *Arthropod Relationships* (eds Fortey, R. A. & Thomas, R. H.) *Syst. Ass. Spec. Vol.* **55**, 125–138 (1997).
4. Chen, J-Y. *et al. The Chengjiang Biota* (National Museum of Natural Science, Taiwan, c. 1996).
5. Chen, J-Y. & Zhou, G-Q. Biology of the Chengjiang fauna. *Bull. Natl Mus. Nat. Sci. Taiwan* **10**, 11–105 (1997).
6. Budd, G. E. The morphology of *Opabinia regalis* and the reconstruction of the arthropod stem-group. *Lethaia* **29**, 1–14 (1996).
7. Butterfield, N. J. Burgess Shale-type fossils from a Lower Cambrian shallow-shelf sequence in northwestern Canada. *Nature* **369**, 477–479 (1994).
8. de Rosa, R. *et al.* Hox genes in brachiopods and priapulids and protostome evolution. *Nature* **399**, 772–776 (1999).
9. Ruiz-Trillo, I. *et al.* Acoel flatworms: Earliest extant bilaterian metazoans, not members of platyhelminthes. *Science* **283**, 1919–1923 (1999).

10. Conway Morris, S. Why molecular biology needs palaeontology. *Development* (Suppl.) **1994**, 1–13 (1994).

11. Conway Morris, S. Metazoan phylogenies: falling into place or falling to pieces? A palaeontological perspective. *Curr. Op. Genet. Dev.* **8**, 662–667 (1998).

12. Gould, S. J. *Wonderful Life: The Burgess Shale and the Nature of History* (Norton, New York, 1989).

13. Bardack, D. & Richardson, E. S. New agnathous fishes from the Pennsylvanian of Illinois. *Fieldiana Geol.* **33**, 489–510 (1977).

14. Bardack, D. in *Richardson's Guide to the Fossil Fauna of Mazon Creek* (eds Shabica, C. W. & Hay, A. A.) 226–243 (Northeastern Illinois Univ. Press, Chicago, 1997).

15. Janvier, P. *Early Vertebrates* (Clarendon, Oxford, 1996).

16. Chen, J-Y. *et al.* A possible early Cambrian chordate. *Nature* **377**, 720–722 (1995).

17. Chen, J-Y. & Li, C-W. Early Cambrian chordate from Chengjiang, China. *Bull. Natl Mus. Nat. Sci. Taiwan* **10**, 257–273 (1997).

18. Dzik, J. *Yunnanozoon* and the ancestry of chordates. *Acta Palaeont. Pol.* **40**, 341–360 (1995).

19. Shu, D., Zhang, X-L. & Chen, L. Reinterpretation of *Yunnanozoon* as the earliest known hemichordate. *Nature* **380**, 428–430 (1996).

20. Collins, D. The "evolution" of *Anomalocaris* and its classification in the arthropod class Dinocarida (nov.) and order Radiodonta (nov.). *J. Paleont.* **70**, 280–293 (1996).

21. Nachtigall, W. *Biological Mechanisms of Attachment* (Springer, Berlin, 1974).

22. Packard, A. in *The Mollusca, Form and Function* Vol. 11 (eds Trueman, E. R. & Clarke, M. R.) 37–67 (Academic, San Diego, 1988).

23. Conway Morris, S. *Typhloesus wellsi* (Melton and Scott, 1973), a bizarre metazoan from the Carboniferous of Montana, USA. *Phil. Trans. R. Soc. Lond. B* **327**, 595–624 (1990).

24. Johnson, R. G. & Richardson, E. S. Pennsylvanian invertebrates of the Mazon Creek area, Illinois: The morphology and affinities of *Tullimonstrum*. *Fieldiana Geol.* **12**, 119–149 (1969).

25. Holland, L. Z. & Holland, N. D. Developmental gene expression in Amphioxus: New insights into the evolutionary origin of vertebrate brain regions, neural crest, and rostrocaudal segmentation. *Am. Zool.* **38**, 647–658 (1998).

Acknowledgements

We thank the National Foundation of Natural Sciences of China, Minister of Science and Technology of China, Royal Society, National Geographic Society, and St John's College, Cambridge for support. Access to *Pipiscius* was facilitated by P. Crane (Field Museum, Chicago), and M. P. Smith and P. Donoghue (University of Birmingham). Technical assistance by S. J. Last and D. R. Simons is acknowledged, as are critical comments by P. Janvier, D. B. Norman and S. Jensen.

4

An Early Cambrian Craniate-like Chordate
Jun-Yuan Chen, Di-Ying Huang, and Chia-Wei Li

Since the identification of the Lower Cambrian *Yunnanozoon* as a chordate in 1995 (ref. 1), large numbers of complete specimens of soft-bodied chordates from the Lower Cambrian Maotianshan Shale in central Yunnan (southern China) have been recovered. Here we describe a recently discovered craniate-like chordate, *Haikouella lanceolata*, from 305 fossil specimens in Haikou near Kunming. This 530 million-year-old (Myr) fish-like animal resembles the contemporaneous *Yunnanozoon* from the Chengjiang fauna (about 35 km southeast of Haikou) in several anatomic features. But *Haikouella* also has several additional anatomic features: a heart, ventral and dorsal aorta, an anterior branchial artery, gill filaments, a caudal projection, a neural cord with a relatively large brain, a head with possible lateral eyes, and a ventrally situated buccal cavity with short tentacles. These findings indicate that *Haikouella* probably represents a very early craniate-like chordate that lived near the beginning of the Cambrian period during the main burst of the Cambrian explosion. These findings will add to the debate on the evolutionary transition from invertebrate to vertebrate.[2]

Genus *Haikouella* gen. nov.

TYPE SPECIES. *Haikouella lanceolata* gen. et sp. nov.

ETYMOLOGY. The generic name refers to the fossil locality, Haikou.

DIAGNOSIS. A small soft-bodied chordate a few centimetres long, which is lancelet-like, elongated and pointed at both ends (Fig. 1). The body is broadly triangular in its anterior half but narrowly triangular in the rest. It differentiates into a head, a trunk and a bent caudal projection. A distinct head bears possible lateral eyes and a ventral buccal cavity at its hind end. An expanded pharyngeal cavity bears six pairs of filamentous branchial arches. A large subventral notochord extends through the entire length of the trunk into a bent caudal projection. A circulatory system comprises ventral and dorsal aortae; a globular heart at the posterior end of the ventral aorta; and an anterior branchial artery that runs in a

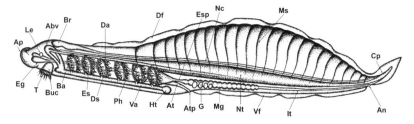

Figure 1 Anatomical interpretation of *Haikouella lanceolata* (gen. et sp. nov.) from Haikou, near Kunming. Abbreviations (also used in Figs 2–4): Abv, anterior branchial vessel; An, anus; Ap, anterior projection; At, atrio; Atp, atriopore; Ba, branchial arches; Baf, branchial arch filament; Br, brain; Buc, buccal cavity; Co, copulatory organ; Cp, caudal project; Da, dorsal aorta; Df, dorsal fin; Ds, denticular structure; Eg, endostyle glands; Es, endostyle; Esp, oesophagus; G, gonad; Hd, head; Ht, heart; It, intestine; Lb, lobated structures; Le, lateral eye; Mg, midgut; Mm, myomeres; Mo, mouth opening; Ms, myosepta; Mw, median wall; Nc, neural cord; Nt, notochord; Ph, pharyngeal cavity; T, tentacle-like structure; Va, ventral aorta; Vf, ventral fin.

circular course to connect the dorsal and ventral aorta at their anterior ends. The trunk bears a neural cord, which extends throughout the entire length of the trunk and into the head region, and forms an expanded brain anteriorly, which appears to be tripartite. The alimentary canal is differentiated into oesophagus, spiral midgut and a straight intestine. Four pairs of gonads are not metamerically arranged.

REMARK.   This new chordate shows a striking resemblance to the contemporaneous *Yunnanozoon*.[1,3] Both have a large subventral notochord, massive dorsal musculature divided by nearly straight myosepta into about two dozen myomeres, and a large pharyngeal cavity with six or seven pairs of branchial arches. Unlike *Yunnanozoon*, however, this new chordate has fewer and non-metamerically arranged gonads, a wider ventral region in the anterior part of the body, smaller and more anteriorly situated pharyngeal teeth (Ba3) and distinctive gill filaments.

Haikouella lanceolata gen. et sp. nov.

HOLOTYPE.   Complete specimen (Figs 2a and 3p), 27.5 mm long, oriented with sagittal plane subvertical to bedding in anterior but subhorizontal in the rest (EC00213a,b).

MATERIAL.   A total of 305 specimens under study include: 30 complete specimens, 32 nearly complete, 110 representing the anterior part of the body (mainly pharynx, a few with the head region), 30 with only isolated preservation of branchial arches and ventral aorta, and 103 with only a part of the trunk.

ETYMOLOGY.   The specific name refers to lancelet-shaped animal (L., lanceolatum).

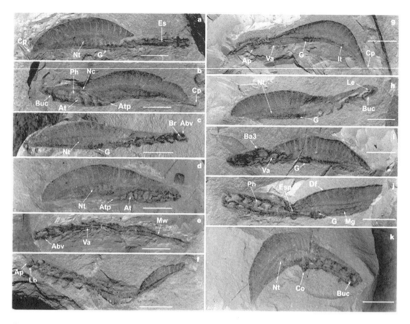

Figure 2 *Haikouella lanceolata* (gen. et sp. nov.) from Haikou, near Kunming. Abbreviations are as Fig. 1. Scale bars, 5 mm. **a**, Holotype (EC00213a), a sublateral specimen with subdorsally compressed anterior part, showing Ap, Lb, Br, Eg, Es, G, Nt, Cp and Ba. **b**, A sublateral specimen (EC00041a), showing mud-filled At and Ph, Atp, Nc, Cp and Buc. **c**, A sublateral specimen (EC0002a) but with the anterior part (before Ba6) compacted lateroventrally, showing Abv, Br, G and Nt. **d**, A sublateral specimen (EC00258a), showing mud-filled At, Atp, Nt and Ba. **e**, A dorsally compacted specimen (EC00231), showing Ap, Va, Ba and Mw running internally through the trunk. **f**, A twisted specimen (EC00047a), showing Ap and Lb. **g**, A sublateral specimen (EC00048a) showing Ap, Va, G, It and Cp. **h**, A sublateral specimen (EC00001a), showing Buc, Nt and an eye on lateral side of head. **i**, A sublateral specimen with a subventral anterior part (EC00042a), showing Ba, Va, G and It. **j**, EC00118a compacted subventrally before Ba6 with the rest compacted sublaterally, with a twist between the two parts, showing Ph, Esp, Df, Vf, G and Mg. **k**, A ventrally curved specimen (EC00012a), showing Co, Buc and Nt. See also Fig. 5.

DIAGNOSIS. Same as in the generic diagnosis.

LOCALITY AND STRATIGRAPHY. Ercai Village, Haikou, Kunming; Lower Cambrian Maotianshan Shale.

OCCURRENCE. Most *Haikouella* specimens are derived in a mud bed about 2.5 cm thick. Like most of the soft-bodied fossil-bearing layers in the Chengjiang fauna,[4,5] it represents a microturbidite deposit, which comprises a lower, graded unit and an upper, homogeneous mud unit. An abundant occurrence of algal remains in the lower unit indicates the presence of a eutrophic environment in the burial site, whereas the *Haikouella* specimens at an aggregate occurrence in the upper mud unit were evidently transported by turbidite mudflow from an adjacent shallower area.

DESCRIPTION. Most adult specimens of *Haikouella lanceolata* are 25–30 mm long, although few reached a length of 40 mm. *Haikouella* differs from

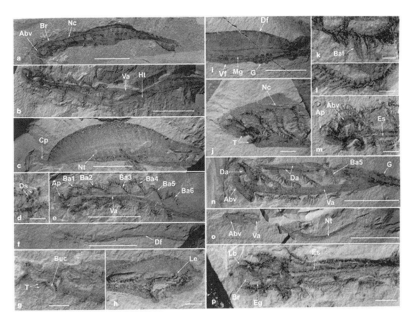

Figure 3 Views of *Haikouella lanceolata* (gen. et sp. nov.) from Haikou, near Kunming. Abbreviations are as in Fig. 1. Scale bars, 5 mm (**a–c**, **e**, **f**, **i**, **n**, **o**), 1 mm (**g**, **h**, **j**, **m**, **p**) and 0.5 mm (**d**, **k**, **l**). **a**, A twisted specimen (EC00214a), showing Abv, Br and Nc. **b**, EC00043a, showing Va and Ht. **c**, lateral specimen (EC00007a), showing Nt and a small Cp. **d**, Enlargement of Fig. 2b (EC00041a), showing Ds. **e**, Anterior part of EC00035, showing six pairs of filamentous branchial arches. **f**, A dorsal specimen (EC00031a), showing Df. **g**, Anterior part of EC00027, showing T and Buc. **h**, Enlargement of Fig. 2h (EC00001a), showing one of lateral eye. **i**, EC00118b, counterpart of Fig. 2j showing G, Mg, Df and Vf. **j**, EC00300, showing T, Ba and Nc. **k**, EC00024a, showing Baf. **l**, Enlargement of a jointed Ba (EC00093b) that consists of narrow, dark disc-like bars and light wider spaces in between. **m**, EC00072a, showing Es and Ba with paired filaments. **n**, EC00157 showing Va, Da, Abv and Ba. **o**, Ventral view of a dorsally sinuous specimen (EC00050), showing Nt and Abv. **p**, Enlargement of the anterior part of EC00213a (Fig. 2a) showing Lb, Br, Eg and Es. See also Fig. 5.

Yunnanozoon by having a broadly ventral region in its anterior part (with a dorsal angle of about 70°). The broadly triangular anterior part was buried either with its lateral side or its broad ventral region lying parallel to the plane of bedding, and thus compacted ventrolaterally or subdorsally. Behind the pharyngeal region, the trunk narrows to a dorsal angle of about 20° to 30°, usually compacted laterally. The transitional region of the two different parts of the trunk at the hind part of the pharynx is usually preserved as a twist (Figs 2a, c, g–j; 3a, b, e) as a result of *post mortem* compaction.

The dorsal compaction is extremely uncommon both in the present species (in less than 2%) and in *Yunnanozoon*. Figure 3a (see also 5j) shows a ventral view of a dorsally compacted animal that is in a laterally sinuous posture, with traces of axial structures interpreted as notochord and intestine. The dorsally compacted animal (Fig. 3f, 5l), which is relatively straight, bore a dorsal fin compacted into a ridge-like structure running mid-dorsally along the trunk. The same ridge-like structure is also seen in a dorsally compacted specimen of a possible *Yunnanozoon* in Chengjiang and interpreted as a subdorsal notochord. These dorsally compacted specimens, especially those in laterally sinuous preservation, may represent individuals that were killed by a sudden burial event. As in *Yunnanozoon*,[7] most of the *Haikouella* fossils are preserved mostly as bluish-grey flattened films.

The head region is well preserved in about 10% of the specimens, either in dorsal (Figs 2a, e–g; 3p) or in sublateral compaction (Figs 2b, h; 3h, j, m). The head, as seen in ventral view, includes an anterior bulb-shaped region (referred as an anterior projection) that is set off from the rest of the head with paired lateral constrictions (Figs 2a, e–g; 3p). The ventral side of the anterior projection is divided into a pair of ventrolateral lobated structures (Figs 2a, e, f; 3p) that are similar to the bilobed anterior end of the conodont animal.[6] The ventral side of the head bears a dark-stained indentation (Figs 2b, h, j, k; 3g, h, j, m), referred to here as a buccal cavity, that appears to be fringed with short tentacles (Fig. 3g, j) each about 0.6–0.7 mm long. The buccal cavity opens directly into the pharynx. A well defined circular structure (0.2 mm in diameter) is present on the lateral side of a head (Figs 2h and 3h), which we interpret as an indication of a lateral eye. In several specimens, the rock is split on a plane that passes through the body, revealing a find black axial structure (about 0.1 mm in diameter), interpreted as a neural cord (Figs 2b; 3a, j), which was slender in the trunk. A large brain relative to body size is well represented by an anterior thickening of the neural cord, and appears to tripartite (Fig. 4a). This elongated brain is about 4 mm long, and is differentiated longitudinally into three parts. The most posterior part is wedge-like, about 2 mm long, and lies at anterior part of the trunk. The anterior part of the brain is relatively thick, and is divided into two parts, both of which are lobate and 0.5 mm thick (Figs 2a, c; 3a, p; 4a).

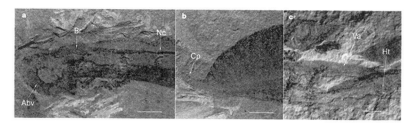

Figure 4 Enlarged view of *Haikouella laneolata* (gen. et sp. nov.) from Haikou, near Kunming. Abbreviations are as in Fig. 1. Scale bars, 1 mm. **a**, Enlargement of the anterior part of EC00214a (Fig. 3a), showing Br, Nc and Abv. **b**, Enlargement of the posterior end of EC00007a (Fig. 3c), showing Cp. **c**, Enlargement of the posterior part of the pharyngeal cavity in EC00043a (Fig. 3b), showing Ht.

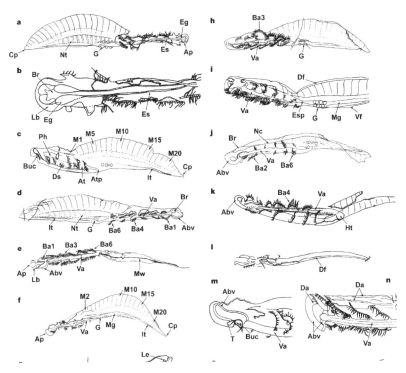

Figure 5 Camera lucida drawings of *Haikouella lanceolata* (gen. et sp. nov.) from Haikou, near Kunming. Diagrams are of the following figure parts: **a**, Fig. 2a; **b**, Fig. 3p; **c**, Fig. 2b; **d**, Fig. 2c; **e**, Fig. 2e; **f**, Fig. 2g; **g**, Fig. 2h; **h**, Fig. 2i; **i**, Fig. 2j; **j**, Fig. 3a; **k**, Fig. 3b; **l**, Fig. 3f; **m**, Fig. 3g; **n**, Fig. 3n; **o**, Fig. 3.

As in *Yunnanozoon*,[7] the notochord of *Haikouella* is large and located relatively ventrally in the greater part of the trunk and runs into the caudal projection. The notochord is widest in the middle and tapers progressively towards its anterior and posterior ends. The caudal projection (Figs 2a, b, g, h, k; 3c; 4b; 5) is a small but prominent structure, extending posteriorly, and is sometimes bent dorsally (Fig. 4b). The notochord is preserved as a flattened band which in most specimens is concealed under the muscle blocks that lie next to it (Figs 2a, c, d, h, k; 3c). As a result of the differentiated compaction of the musculature, an axial line may indicate the dorsal margin of the underlying notochord.

Fins are represented in only a few specimens. They comprise a narrow dorsal fin running along the entire length of the trunk and a narrow ventral fin running along the posterior half of the trunk, stopping at an anterior structure that may be an atriopore (Figs 2j; 3i).

An anterior mud-filled cavity is interpreted as a perforate pharynx (Fig. 2b, d, j) used for food collection and respiration. The pharynx occupies nearly the entire anterior half of the trunk before opening posteriorly into the short oesophagus (Fig. 2j). Six pairs of branchial arches (Figs 2a, c–f; 3a, b, e, n) slope forward slightly from either side of a pair of axial structures that may represent dorsal aortae (Fig. 3n). The ventral end of each arch attaches to either side of an axial structure regarded as a ventral aorta (Figs 2c, e, g, i–k; 3a, b, e, n, o). The branchial arches of *Haikouella* include a jointed structure (Fig. 3l) that resembles the striped gill-bar mucocartilage in ammocoete larvae[8] and consists of about 25 disc-like dark bars with wider light spaces in between (Fig. 3l). The branchial arches are consistently well preserved. The ventral end of each bar is attached to the ventral aorta, but the dorsal end is readily detached and relocated (Figs 2f–h; 3a). On a given branchial arch, each of the dark discs bears two posteriorly projecting gill filaments about 1 mm long that are flattened dorsoventrally (Fig. 3k, m). In the floor of the pharynx is a possible endostyle, represented by a pair of ridges (Figs 2a; 3m, p), each with an expanded, teardrop-shaped anterior end (Fig. 3p). The pharyngeal cavity narrows to form a tube-like structure interpreted as an oesophagus (Fig. 2j) at its posterior end between B6 and the first gonad (G1). The oesophagus leads directly to a spirally shaped midgut (Figs 2j and 3i) and a slender, straight intestine (Fig. 2b, c, g–i).

As in *Yunnanozoon*, there are simple cone-shaped denticular structures (Fig. 3d) within the pharynx, but those of *Haikouella* are much smaller, only 0.1 mm across and situated near B3 in a more anterior position within pharynx; in *Yunnanozoon* these structures are much larger, about 1 mm across, and situated at the posterior end of the pharyngeal cavity. We regard them as pharyngeal teeth. Midgut and intestine both lie immediately below the notochord, extending along its ventral margin to the ventral anus near to the posterior end of the trunk. The gill slits between the gill arches, instead of being exposed externally, were covered by lateral folds of the body (Figs 2k and 3c) which enclose a ventral atrial space (Fig. 2b, d). This atrium was an elongated cavity, usually mud-filled, extending ventrally with a posterior opening, the atriopore at the level of M6 (Fig. 2b, d). In *Yunnanozoon*, we also interpreted the gill slits as opening into an atrial cavity, although they have also been interpreted as opening directly to the exterior.[9] Below the notochord are four paired gonads that are arranged non-metamerically, at the posterior end of the atrium, beneath M5 to M6 (Figs 2a, c, g–j; 3a, i, n). In contrast, *Yunnanozoon* has 13 pairs of gonads, arranged metamerically with the first pair located below M6 (ref. 2).

The blood circulatory system consists of a set of closed vessels. At the posterior end of the pharyngeal region is a globular posterior enlargement of the ventral aorta, representing the heart (Figs 3b and 4c). From the ventral aorta, the blood presumably ascended through the branchial arches via the branchial arteries to the paired dorsal aortae (Fig. 3n). Each of the branchial arches has an expanded ventral end (Figs 2g and 3n), which may be equivalent to the branchial heart of lancelets.[10] At the anterior ends of the ventral and dorsal aortae, an anterior branchial artery runs forward and appears to connect them (Figs 2c, e; 3a, b, g, m–o) along the dorsal margin of the buccal cavity into the dorsal margin.

The trunk musculature, which is divided by relatively straight myosepta into about 25 myomeres (Fig. 2b, d, g, k), is situated mainly above the notochord, but also extends ventrally on either side of the notochord (Figs 2a, c, d, h, k; 3c). In contrast, the traces of segmented musculature on the flanks of notochord in *Yunnanozoon* have been reported as a gut valve structure.[11] Apart from the thin but broadly spaced M_1 and M_2, the remaining muscle segments are evenly spaced. As in *Yunnanozoon*, the septa are weakly sigmoidal, curving backwards dorsally and forwards ventrally. The musculature may be divided into two lateral sets by a median wall. A dorsally compacted animal is shown in Fig. 2a with a darker, axial line running internally through the trunk at a mid-dorsal position, which we interpret as an indication of the median wall.

There is a large papilla in front of the atriopore of a specimen (Fig. 2k). The rare occurrence of this structure indicates that it might be a possible copulatory organ, as in lamprey,[12] that is temporarily developed during the mating season.

Although *Yunnanozoon* has been interpreted as a hemichordate,[11] we believe that this hypothesis is incorrect.[7] For instance, the proposed gut contents may be a ventral view of the gonads, and the observed spiral structure is a trace of musculature next to the notochord,[7] not a part of the gut. There is little doubt that the rod in both *Haikouella* and *Yunnanozoon* is a notochord, as it lies dorsally to the pharyngeal cavity and intestine canal. It seems certain that *Haikouella* and *Yunnanozoon* are closely related and that both are indeed chordates. *Yunnanozoon* was previously classified as a cephalochordate,[1] but this was challenged by its interpretation either as a basal chordate[13] or as an independent class of the chordates.[9] Anatomic findings such as the brain and possible lateral eyes in its close relative *Haikouella* pose a question as to whether *Yunnanozoon* may also be an early craniate.

Although living chordates display an amazing diversity of body forms, extant lancelets are broadly accepted to be the best available proxies for the latest common ancestor of the caphalochordates and craniates.[14] Palaeontological information on *Yunnanozoon* and *Haikouella*, as well and molecular genetic data,[15] have lent strong support to this concept.

It is possible that all the craniate body forms evolved from a common ancestor resembling these fish-like species. Although it is commonly considered that the origin of the craniates was signalled by the relatively simultaneous appearance of a conspicuous brain and an endoskeleton including a cranium, our finding indicates that the craniates originated through a set of separate events over a long interval of time. The appearance of a conspicuous brain may have been the earliest of these events, occurring long before full endoskeletization. Although the lancelet has sometimes been regarded as a brainless chordate, recent micro-anatomical and molecular-genetic studies indicate that lancelets (and, by extension, the proximate common ancestor of the cephalochordates and craniates) have at least a diencephalic forebrain[16] and a relatively large hindbrain.[17] *Haikouella* had a relatively large brain and possibly a pair of lateral eyes, which indicate that this animal might be considered as an early craniate.

The origin of the craniates has been broadly accepted as being linked to defence in such forms as the armoured heterostracans and may have evolved in the late Cambrian and flourished in the Silurian–Devonian.[10,12] However, the first hard parts of the craniates appear to have been linked to the feeding function. The eel-like, soft-bodied conodonts, extended back into the late Cambrian by fossil remains of the euconodont elements, may represent one of the early craniates that were able to build tooth-like calcium phosphate in the mouth, allowing them to adopt to an active hunting life.[18] The pharyngeal teeth in the pharyngeal cavity of both *Yunnanozoon* and *Haikouella* may represent the earliest known biomineralization in chordates.

References

1. Chen, J. Y., Dzik, J., Edgecombe, G. D., Ramsköold, L. & Zhou, G.-Q. A possible Early Cambrian chordate. *Nature* 377, 720–722 (1995).

2. Holland, N. D. in *International Symposium of the Origins of Animal Body Plans* and *their Fossil Records* (eds Chen, J.-Y. et al.) 11 (Early Life Research Center, Jinning, 1999).

3. Hou, X.-G., Ramsköld, L. & Bergström, J. Composition and preservation of the Chengjiang Fauna–Lower Cambrian soft-bodied biota. *Zool. Scripta* 20, 395–411 (1991).

4. Chen, J.-Y., Zhou, G.-Q., Zhu, M.-Y. & Yeh, G.-Y. *The Chengjiang Biota—Unique Window of the Cambrian Explosion* (National Museum of Natural Sciences Press, Taichung, 1996).

5. Chen, J. -Y. & Zhou, G. -Q, in *the Cambrian Explosion and the Fossil Record* (eds Chen, J. -Y., Cheng. Y. -N. & Van Iten, H.); *Bull Natl. Mus. Net. Sci.*, 10, 11–105, (1997).

6. Aldridge, R. J. & Purnell, M. A. The conodont controversies. *Trends* 11, 463–466 (1996).

7. Chen, J. -Y. & Li, C. -W. in *The Cambrian Explosion and the Fossil Record* (eds Chen, J. -Y., Cheng, Y. -N. & Van Iten, H.); *Bull. Natl. Mus. Nat. Sci.* 10, 257–273, 1997).

8. Dohrn, A. Studien zur Urgeschte des Wirbelthierkoerpers. V. Zur Entstehung und Differenzierung der Viscceralbogen bei Petromyzon planeri. *Mitt. Zool. Station Neapel* 5, 152–161 (1884).

9. Dzik, J. *Yunnanozoon* and the ancestry of chordates. *Acta Palaeont. Pol.* 40, 341–360 (1995).

10. Bliek, A. At the origin of chordates. *Geobios* 25, 101–113 (1992).

11. Shu, D. -G., Zhang, X. -L. & Chen, L. Reinterpretation of *Yunnanozoon* as the earliest known hemichordate. *Nature* 380, 428–429 (1996).

12. Young, J. Z. *The Life of Vertebrates* 3rd edn (Clarendon, Oxford, 1981).

13. Chen, J. -Y. in *International Cambrian Explosion Symposium* (eds Chen. J. -Y., Edgecombe, G. & Ramaköld, L.) 7–9 (Early Life Research Center, Nanjing, 1995).

14. Stokes, M. D. & Holland, N. D. The lancelet. *Am. Sci.* 86, 552–560 (1998).

15. Wada, H. & Satoh, N. Details of the evolutionary history from in vertebrates to vertebrates, as deduced from the swquence of 18SrDNA. *Proc. Nail Acad. Sci. USA* 91, 383–490 (1994).

16. Lacalli, T. C., Holland, N. D. & West. J. E. Landmarks in the anterior central nervous system of smphioxus larvae. *Phil. Trans. R. Soc. Lond, B* 344, 165–185 (1994).

17. Holland, P.W. H., Holland, L. Z., & Holland, N. D. The molecular control of spatial patterning in amphioxus. *J. mar. Biol. Assoc. UK* 74, 49–60 (1994).

18. Purnell, M. A., Aldridge, R. J., Donogue, P. C. J. & Gabbott, S. E. Conodonts and the first vertebrates. *Endeavour* 19, 20–27 (1995).

Acknowledgements

We thank N. Holland, E. Davidson, and D. Walossek for discussions and comments, and F. Gao for technical assistance. This work was supported by National Department of Sicence and Technology, National Foundation of Natural Sciences, Jiangsu Provincial Committee of Science and Technology. The research of L. C. W. is supported by National Science Council.

Lower Cambrian Vertebrates from South China

De-Gan Shu, Hui-Lin Luo, S. Conway Morris, Xing-Liang Zhang, S.-X. Hu, Liang Chen, J. Han, Min Zhu, Y. Li, and L.-Z. Chen

The first fossil chordates are found in deposits from the Cambrian period (545–490 million years ago), but their earliest record is exceptionally sporadic and is often controversial. Accordingly, it has been difficult to construct a coherent phylogenetic synthesis for the basal chordates. Until now, the available soft-bodied remains have consisted almost entirely of cephalochordate-like animals from Burgess Shale-type faunas. Definite examples of agnathan fish do not occur until the Lower Ordovician (~475 Myr BP), with a more questionable record extending into the Cambrian. The discovery of two distinct types of agnathan from the Lower Cambrian Chengjiang fossil-Lagerstätte is, therefore, a very significant extension of their range. One form is lamprey-like, whereas the other is closer to the more primitive hagfish. These finds imply that the first agnathans may have evolved in the earliest Cambrian, with the chordates arising from more primitive deuterostomes in Ediacaran times (latest Neoproterozoic, ~555 Myr BP), if not earlier.

The 'Cambrian explosion' refers to the seemingly abrupt appearance of diverse metazoan groups, representing a number of extant phyla as well as some more problematic clades,[1,2] during the Cambrian period. Faunas of the type found at the Burgess Shale[1] provide exceptional information on Cambrian marine communities. Recent documentation of their faunal riches has led to a series of provocative phylogenetic analyses[3,4] which may be broadly congruent with the division of the protostomes into the Ecdysozoa and Lophotrochozoa.[5] However, the early evolution of the third triploblast superclade, the deuterostomes, is much less well understood.

In particular, the fossil record of Cambrian chordates is extremely meagre. The soft-bodied remains[6–8] consist almost exclusively of cephalochordate-like animals, although they seem to be significantly different from the Recent amphioxus. A unique and as yet unnamed specimen from the Burgess Shale has been compared to an ammocoete lamprey.[9,10]

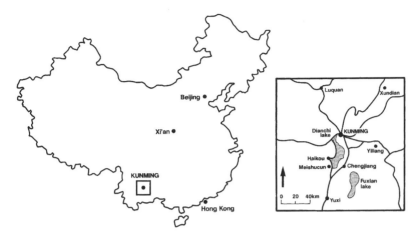

Figure 1 Map of Chengjiang area, Yunnan province and its position relative to the rest of China. The first discoveries of soft-bodied fossils were made in about 1910 near Yiliang,[1] but the principal focus of activity has been near Chengjiang, especially the localities a few kilometres north-east of Fuxian Lake, most notably Maotianshan. The vertebrate fossils described here come from close to Haikou, near Dianchi Lake.

Records of fish are confined to questionable scales[11] from the Upper Cambrian. The euconodonts, which are also chordates,[12] also appear in the uppermost Cambrian, but their precise placement relative to other agnathans remains controversial.[12,13] Transitional forms between euconodonts and paraconodonts[14] imply that this group of chordates evolved during the Middle Cambrian, but nothing is yet known of the soft-part anatomy of the early representatives of this group.

Here we describe two distinct types of agnathan from the Lower Cambrian Chengjiang fossil-Lagerstätte of the Qiongzhusi Formation at Haikou,[15] Kunming City, Yunnan (Fig. 1). The more primitive example, *Myllokunmingia* gen. nov., has well developed gill pouches with probable hemibranchs, whereas *Haikouichthys* gen. nov. has structures that we interpret as part of a branchial basket and a dorsal fin with prominent fin-radials. Shared features include complex myomeres, as well as probable paired ventral fin-folds and a pericardic cavity. These discoveries significantly predate previously published reports, but they also imply that even more primitive vertebrates had evolved before the mid-Lower Cambrian.

Phylum Chordata
Subphylum Vertebrata
Class Agnatha
Myllokunmingia Shu, Zhang & Han gen. nov.
Myllokunmingia fengjiaoa Shu, Zhang & Han sp. nov.

ETYMOLOGY.   The generic name refers to myllos (Greek, fish) and the capital of Yunnan. Species name from Chinese fengjiao, beautiful.

HOLOTYPE.   Early Life Institute, Northwest University, Xi'an: ELI-0000201.

STRATIGRAPHY AND LOCALITY.   Qiongzhusi (Chiungchussu) Formation, Yu'anshan Member (*Eoredlichia* Zone); Lower Cambrian. Specimen collected from Haikou, Kunming City, Yunnan.

DIAGNOSIS.   Fusiform, divided into head region and trunk. Dorsal fin towards anterior, ventro-lateral fin-fold arising from trunk, probably paired.

Figure 2 The Lower Cambrian agnathan vertebrate *Myllokunmingia fengjiaoa* Shu, Zhang & Han gen. et sp. nov. from Haikou, Yunnan. Specimen ELI-0000201. **a**, Entire specimen, anterior end to the right, posterior tip incomplete owing to postmortem folding; scale bar equivalent to 5 mm. **b**, Detail of the head region of the fossil, to emphasize structures interpreted as gill pouches; scale bar equivalent to 5 mm. **c**, Anterior of counterpart, to show series of ventral structures interpreted as possible extrabranchial atria; scale bar equivalent to 3 mm.

No fin-radials. Head bears series of five, possibly six, gill pouches, each with hemibranchs. Gill pouches possibly linked to extrabranchial atria. Trunk with ~25 myomeres, double V structure with ventral V directed posteriorly and dorsal V anteriorly. Internal anatomy includes alimentary canal with pharynx and intestine, notochord, and possible pericardic cavity.

DESCRIPTION.   *Myllokunmingia* gen. nov. is fusiform, with a total length of about 28 mm and maximum height (excluding the dorsal fin) of 6 mm; this point is located about 11 mm from the anterior (Figs 2a, 3). Dermal skeleton or scales appear to have been absent. The body has two regions: a head, taken here to include both snout and complex structures interpreted as gills, and a segmented trunk. A sail-like dorsal fin arises fairly close to the anterior tip and increases to a maximum observed height of 1.5 mm. Its more posterior development is uncertain. On the ventro-lateral side, and more to the posterior, there is a prominent fin-fold which in the fossil is inclined at an angle to the rest of the trunk. This deflection may indicate that originally the fin-fold was paired, with the corresponding structure concealed at a deeper level within the sediment. Neither the dorsal nor the exposed ventral fin-fold display fin-radials.

The details of the head are complex (Fig. 2b, c). Most conspicuous is a series of relatively large, posteriorly inclined structures with transverse lineations. We interpret these as gill pouches with hemibranchs. Some of the latter have a beaded appearance, possibly indicating original folding or crenulation. Four pouches are clearly visible, the posteriormost being reduced in size. Anteriorly a fifth pouch is identified with some confidence, and more tentatively a sixth. Ventral to these pouches, and more clearly preserved on the counterpart (Fig. 2c), are convex (in the part) units, each with two or three longitudinal ridges. The position of these structures relative to the gill pouches is consistent with their representing extrabranchial atria (Fig. 3). The posteriormost pouch, however, appears to have lacked a well developed atrium. No branchial basket or other skeletal units have been identified in association with the gills. Other regions of the head show scattered patches of darker tissue. Apart from some segmental structures, close to the boundary with the trunk, these patches show no coherent pattern such as might represent parts of an endoskeleton. The mouth was presumably located at the anterior end and, although there is perhaps the suggestion of an annular ring (Fig. 2a, c), precise details are difficult to discern. An approximately oval area posterior to the gills may represent the pericardic cavity.

The trunk bears prominent sigmoidal segments (Figs 2a, 3), interpreted as myocommata, which in life separated about 25 myomeres. In lateral view the dorsal V points anteriorly and the ventral one posteriorly. Accordingly, on the dorsum the myomeres of either side would have converged

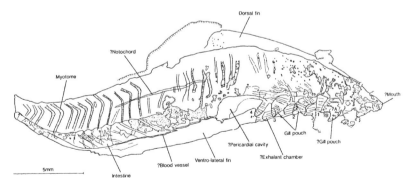

Figure 3 Camera-lucida drawing of specimen, with certain features (notably structures interpreted as extrabranchial atria) combined by reversal from the counterpart (Fig. 1c), to show interpretation. Numbers 1–4 refer to gill pouches identified with reasonable certainty; ?A and ?B refer to more tentative identification of gill pouches, of which ?A is less certain than ?B.

posteriorly, and along the venter in an anterior direction. Towards the ventral margin of the trunk a dark area with positive relief presumably represents a sediment-filled intestine. The gut can be traced, more clearly in the counterpart, to the incomplete posterior termination. It is uncertain, therefore, whether *Myllokunmingia* possessed a post-anal tail. In the mid-region of the trunk there is a fairly prominent strand (Figs 2a, 3), which probably represents the incomplete remains of the notochord. At the trunk–fin-fold boundary there are very narrow dark lines, which show at least one bifurcation. These may be blood vessels.

Haikouichthys Luo, Hu & Shu gen. nov.

Haikouichthys ercaicunensis Luo, Hu & Shu sp. nov.

ETYMOLOGY.   After a locality at Haikou, near Kunming City, and the Greek ichthyos. The specific name refers to Ercaicun, a small village close to where the specimen was found.

HOLOTYPE.   Yunnan Institute of Geological Sciences, Kunming: HZ-f-12-127.

STRATIGRAPHY AND LOCALITY.   As for *Myllokunmingia fengjiaoa.*

DIAGNOSIS.   Fusiform, more slender than *Myllokunmingia*, divided into head region and trunk. Prominent dorsal fin towards anterior, with fin-radials. Ventro-lateral fin-fold arising from trunk, probably paired. Head bears at least six, possibly up to nine, gill arches. Trunk with myomeres, double V as in *Myllokunmingia*. Internal anatomy includes cranial carti-lages, pericardic cavity, intestine and metameric ?gonads, the last being arranged along the ventral trunk.

DESCRIPTION.   *Haikouichthys* has a known length of about 25 mm (Fig. 4a, c). An important question is its orientation. The dorsal side is rec-ognized on the basis of a prominent fin with fin-radials (Fig. 4b). They are inclined towards the preserved end at about 20° to the vertical. In nearly all fish, such an inclination would unequivocally indicate that the poste-rior end is preserved. There are, however, some rare exceptions to this rule in extant agnathans. In the second dorsal fin of lamprey the fin rays are orientated forwards in the male during the mating period.[16] In the hagfish, the rays at the front of the caudal fin may be vertical or even be inclined forwards.[17] Moreover, identification of this fossil as the posterior end would be difficult to reconcile with the complex structures, identified with varying degrees of confidence, as part of the branchial basket, cranial car-tilages and a pericardic area. To our knowledge, no other living or extinct agnathan has remotely similar structures towards the posterior. The anomalous orientation of these fin-radials, therefore, is assumed to be either comparable to the similar case in the lamprey or alternatively (and less plausibly) due to post-mortem distortion.

In several respects *Haikouichthys* is similar to *Myllokunmingia*. The overall shapes are comparable, although the former is slightly more slender. The presumed head region and trunk are clearly distinguished. In the trunk the myomeres of *Haikouichthys* (Fig. 4a, c) have the same configuration as those of *Myllokunmingia*. The strongly expressed nature of the posteriorly inclined Vs may result from slightly oblique burial. The dorsal fin is more prominent than in *Myllokunmingia*, and differs in possessing closely spaced fin-radials (~7 per mm; Fig. 4b). The rather abrupt scarp between the trunk and ventro-lateral insertion of the fin-fold strongly indicates that originally the fin-folds were paired.

Several structures are visible in the anterior (Fig. 4b, c). Towards the ventral side there is a series of rod-like units, each of which consists of two components: one rod is inclined posteriorly whereas the other is more or less parallel to the ventral margin. Six sets of these units are readily discernible, two more are questionable, and a final set is only tentatively identified (Fig. 4b, c). The total number of sets, therefore, may have been as high as nine. We interpret these rods as part of a branchial basket. No associated gills or filaments are visible. Nearer to the anterior there is a complex series of mineralized masses. A preliminary attempt has been made to reconcile some of these areas with both units within the cranial cartilage of the lamprey[18] and the eye. Adjacent to this area, a very short length of the notochord may also be visible. Close to the head–trunk boundary there is an approximately triangular area with mineralized patches. This may represent either the pericardic cartilage or cavity. This, however, assumes that the structures immediately to the posterior are not part of the branchial basket. If these are branchial structures then to the posterior a rather faint pale area, consisting of two regions, is an alternative site for the pericardic cavity. Along the ventral side of the trunk there is a metameric (~13) series of roughly circular structures. These may be the remains of the gonads, although a possible alternative is slime glands comparable to those of the hagfish.[18] A faint trace within the trunk may denote the intestine, and more posteriorly a dark strand may represent its continuation.

Preservation and Ecology

The co-occurring fauna includes the trilobites *Eoredlichia* and *Yunnanocephalus*, as well as exceptionally preserved fossils such as the arthropods *Naraoia* and *Waptia*. Each fossil is preserved in lateral view (Figs 2a, 4a). *Myllokunmingia* gen. nov. is almost complete. One consequence of burial is that the tail has been bent steeply into the sediment, so that the posteriormost section is not visible. This configuration, and what appears to be sediment in the pharynx, suggest that the animal was buried alive. *Haikouichthys* gen. nov. lies parallel to the bedding. The presumed

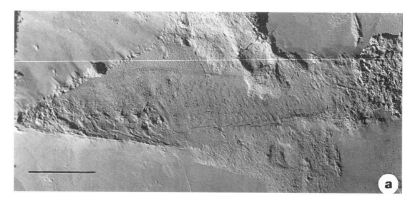

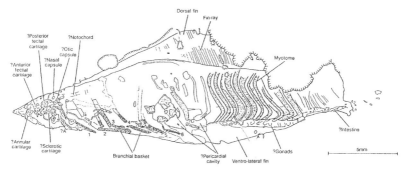

Figure 4 The Lower Cambrian agnathan vertebrate *Haikouichthys ercaicunensis* Luo, Hu & Shu gen. et sp. nov. from Haikou, Yunnan. Specimen HZ-f-12-127. **a**, Entire specimen, anterior to the left; more posterior region appears to fade out into sediment, possibly representing decay of body; attempts to excavate this area were not successful. Scale bar equivalent to 5 mm. **b**, Detail of anterior to show putative gill bars, possible elements of cranial endoskeleton, and pericardic area; scale bar equivalent to 5 mm. **c**, Camera-lucida drawing of specimen to show interpretation. Numbers 1–6 indicate units of the branchial basket that are identified with some confidence; ?A–?C refer to less secure identifications. Two possible areas representing the pericardial cavity are indicated. To the anterior of ?C a triangular area with patches of diagenetic mineralization is one possibility; a fainter region to the posterior is the alternative location.

posterior is absent (Fig. 4a, c), possibly owing to decay. There has been growth of sulphide minerals, especially in the head region. The extreme rarity (<0.025%) of chordates in the Chengjiang assemblages[6] may be because they were active swimmers and could avoid being engulfed by benthic sediment flows that are believed to have been the principal method of burial.[19]

Phylogeny

To examine the phylogenetic positions of *Myllokunmingia* and *Haikouichthys* we compiled a data set with 16 taxa (14 vertebrates and two outgroups (tunicate, cephalochordates)) and 116 characters (see Supplementary Information). Most of the characters we used were adopted from existing matrices.[20,21] Our analysis places both Chengjiang taxa firmly in the agnathans, with both more closely related to lampreys than to hagfishes (Fig. 5). The cladogram is fairly robust when a more cautious attitude is adopted for a number of character states that depend either on specific identifications (such as pericardic cavity) or an absence (for example, arcualia) that might in fact be preservational. Given these provisos, then, in detail *Haikouichthys* forms a monophyletic group with *Jamoytius* and lampreys, but the inter-relationships among them are unresolved. *Myllokunmingia*, however, is evidently more primitive and is basal to the clade comprising both the lamprey–*Jamoytius*–*Haikouichthys* and the 'ostracoderms'–gnathostomes branches. The inclusion of the two Chinese fossils in this cladistic analysis is a significant extension of our knowledge of extinct agnathans. It does not, however, substantially change the internal configuration of existing cladograms in early vertebrates.[18,20–22]

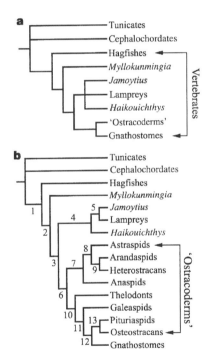

Figure 5 Phylogenetic analysis of early vertebrates, to include *Myllokunmingia* gen. nov. and *Haikouichthys* gen. nov. **a**, Strict consensus tree of the six most parsimonious trees. **b**, One of the six most parsimonious trees; the other five show minimal differences. Data set based on refs 20 and 21. Supporting characters for each node are given in Supplementary information.

The phylogenetic implications of this discovery are significant. These agnathan vertebrates predate previous records[11,23] by at least 20 and possibly 50 Myr (see refs 24 and 25 for Cambrian and early Ordovician radiometric time scales). The differences between *Haikouichthys* and *Myllokunmingia* indicate that, by Chengjiang times, there may already have been a diversity of forms. Neither animal shows evidence for unequivocal biomineralization, and this corroborates the hypothesis that the evolution of bone tissue was a relatively late development in the vertebrate lineage,[18,20] although the euconodont,[12] and possibly paraconodont,[14] record indicates that chordate biomineralization was achieved in the Cambrian. In both fossils there is evidence that the ventral fin may have been paired. The theory of lateral fin folds[26] has had considerable significance,[21] but of the fossil agnathans only the anaspids[18] and *Jamoytius*[27] have such paired fin-folds. The possible occurrence of this condition in these Lower Cambrian agnathans indicates, however, that fin-folds may be a primitive feature within the vertebrates. The occurrence of close-set dorsal fin-radials in *Haikouichthys*, on the other hand, may be a relatively advanced feature.

Wider Implications

The record of Cambrian chordates is very sporadic, and also controversial. Fossils generally accepted to be of a grade comparable to the cephalochordates include the Lower Cambrian *Cathaymyrus*[6] and Middle Cambrian *Pikaia*.[7] The former genus has been synonymized with the co-eval *Yunnanozoon*,[28] but this is very doubtful. Moreover, the status of *Yunnanozoon* as a chordate[29] is also questionable,[30] and it may be related to *Xidazoon*[31] and other stem-group deuterostomes.

The discovery of these Lower Cambrian vertebrates has implications for the likely timing of chordate evolution. The occurrence of *Myllokunmingia* and *Haikouichthys* in the Chengjiang Lagerstätte shows that even more primitive hagfish-like vertebrates had almost certainly evolved by the beginning of the Atdabanian Stage of the Lower Cambrian. Such a find may help to elucidate the transition between the cephalochordates and the first vertebrates. The derivation of the first vertebrates from the cephalochordates must have entailed a major reorganization of the body, especially by the first effective expression of neural-crest tissue.[32] Key functional steps were the elaboration of the nervous system and active pharyngeal ventilation. However, the range of anatomies seen in the Cambrian cephalochordates[6,7,9,10] indicates that our documentation of this pre-vertebrate evolutionary plexus is far from complete. It is possible that *Pikaia*, until now the cynosure of Cambrian chordates,[7,8] is peripheral to the lineage leading to the vertebrates. It has several peculiarities. The head consists of two lobes, each of which bears a prominent, slender tentacle, the mycommata have the reverse configuration of that seen in *Myllokunmingia* and *Haikouichthys*, and the notochord stops well short of the anterior.

The major steps in the early evolution of chordates may well have occurred in the late Neoproterozoic. So far, however, no suitable fossil candidates have been identified amongst the Ediacaran assemblages, although preservation in siltstones and sandstones is less conducive to survival of delicate soft-bodied taxa. Our discovery, however, gives no reason to suppose that the origin of vertebrates was hundreds of millions of years earlier,[33] and the reliability of the methods used to reach such a conclusion has been questioned elsewhere.[2,34]

Methods

PHYLOGENETIC ANALYSIS.   The analysis was performed using the software package PAUP 3.1 with a data set of 16 taxa and 116 characters (see Supplementary Information). The analysis was rooted on tunicates. All characters were weighted equally. Characters 62, 63 and 95 were ordered. Most parsimonious trees were identified using the branch and bound search algorithm, stepwise addition. The most parsimonious tree has: tree length, 196; consistency index, 0.628; retention index, 0.700.

References

1. Conway Morris, S. *The Crucible of Creation: The Burgess Shale and the Rise of Animals* (Cambridge Univ. Press, Cambridge, 1998).
2. Valentine, J. W., Jablonski, D. & Erwin, D. H. Fossils, molecules and embryos: new perspectives on the Cambrian explosion. *Development* **126**, 851–859 (1999).
3. Conway Morris, S. & Peel, J. S. Articulated halkieriids from the Lower Cambrian of North Greenland and their role in early protostome evolution. *Phil. Trans. R. Soc. Lond. B* **347**, 305–358 (1995).
4. Budd, G. E. in *Arthropod Relationships* (eds Fortey, R. A. & Thomas, R. H.) *Syst. Ass. Spec. Vol.* **55**, 125–138 (1997).
5. de Rosa, R. *et al.* Hox genes in brachiopods and priapulids and protostome evolution. *Nature* **399**, 772–776 (1999).
6. Shu, D-G., Conway Morris, S. & Zhang, X-L. A *Pikaia*-like chordate from the Lower Cambrian of China. *Nature* **384**, 156–157 (1996).
7. Conway Morris, S. in *Atlas of the Burgess Shale* (ed. Conway Morris, S.) 26 (Palaeontological Association, London, 1982).
8. Briggs, D. E. G., Erwin, D. H. & Collier, F. J. *The Fossils of the Burgess Shale* (Smithsonian, Washington, 1994).
9. Simonetta, A. M. & Insom, E. New animals from the Burgess Shale (Middle Cambrian) and their possible significance for understanding of the Bilateria. *Boll. Zool.* **60**, 97–107 (1993).
10. Insom, E., Pucci, A. & Simonetta, A. M. Cambrian Protochordata, their origin and significance. *Boll. Zool.* **62**, 243–252 (1995).
11. Young, G. C., Karatajute-Talimaa, V. N. & Smith, M. M. A possible Late Cambrian vertebrate from Australia. *Nature* **383**, 810–812 (1996).
12. Donoghue, P. C. J., Purnell, M. A. & Aldridge, R. J. Conodont anatomy, chordate phylogeny and vertebrate classification. *Lethaia* **31**, 211–219 (1998).
13. Pridmore, P. A., Barwick, R. E. & Nicoll, R. S. Soft anatomy and the affinities of conodonts. *Lethaia* **29**, 317–328 (1997).

14. Szaniawski, H. & Bengtson, S. Origin of euconodont elements. *J. Paleontol.* **67**, 640–654 (1993).

15. Luo, H. *et al.* New occurrence of the early Cambrian Chengjiang fauna from Haikou, Kunming, Yunnan province. *Acta. Geol. Sin.* **71**, 97–104 (1997).

16. Marinelli, W. & Strenger, A. *Vergleichende Anatomie und Morphologie der Wirbeltiere, 1.* Lampetra fluviatilis (*L.*) (Franz Deuticke, Vienna, 1954).

17. Marinelli, W. & Strenger, A. *Vergleichende Anatomie und Morphologie der Wirbeltiere, 2.* Myxine glutinosa (*L.*) (Franz Deuticke, Vienna, 1956).

18. Janvier, P. *Early Vertebrates* (Oxford Univ. Press, Oxford, 1996).

19. Chen, J-Y. & Zhou, G-Q. Biology of the Chengjiang fauna. *Bull. Natl Mus. Nat. Sci. Taiwan* **10**, 11–105 (1997).

20. Forey, P. L. Agnathans recent and fossil, and the origin of jawed vertebrates. *Rev. Fish Biol. Fisheries* **5**, 267–303 (1995).

21. Janvier, P. The dawn of the vertebrates: character versus common ascent in the rise of current vertebrate phylogenies. *Palaeontology* **39**, 259–287 (1996).

22. Forey, P. L. & Janvier, P. Evolution of the early vertebrates. *Am. Sci.* **82**, 554–565 (1994).

23. Young, G. C. Ordovician microvertebrate remains from the Amadeus Basin, central Australia. *J. Vert. Paleont.* **17**, 1–25 (1997).

24. Landing, E. *et al.* Duration of the Early Cambrian: U-Pb ages of volcanic ashes from Avalon and Gondwana. *Can. J. Earth Sci.* **35**, 329–338 (1998).

25. Landing, E. *et al.* U-Pb zircon date from Avalonian Cape Breton Island and geochronologic calibration of the Early Ordovician. *Can. J. Earth Sci.* **34**, 724–730 (1997).

26. Goodrich, E. S. Notes on the development, structure, and origin of the median and paired fins of fish. *Q. J. Microsc. Sci.* **50**, 333–376 (1906).

27. Ritchie, A. New evidence on *Jamoytius kerwoodi*, an important ostracoderm from the Silurian of Lanarkshire. *Palaeontology* **11**, 21–39 (1968).

28. Chen, J-Y. & Li, C. Early Cambrian chordate from Chengjiang, China. *Bull. Natl Mus. Nat. Sci. Taiwan* **10**, 257–273 (1997).

29. Chen, J-Y. *et al.* A possible early Cambrian chordate. *Nature* **377**, 720–722 (1995).

30. Shu, D-G., Zhang, X. & Chen, L. Reinterpretation of *Yunnanozoon* as the earliest known hemichordate. *Nature* **380**, 428–430 (1996).

31. Shu, D. *et al.* A pipiscid-like fossil from the Lower Cambrian of South China. *Nature* **400**, 746–749 (1999).

32. Gans, C. & Northcutt, R. G. Neural crest and the origin of vertebrates: A new head. *Science* **220**, 268–274 (1983).

33. Bromham, L. *et al.* Testing the Cambrian explosion hypothesis by using a molecular dating technique. *Proc. Natl Acad. Sci. USA* **95**, 12386–12389 (1998).

34. Ayala, F. J., Rzhetsky, A. & Ayala, F. J. Origin of the metazoan phyla: Molecular clocks confirm paleontological estimates. *Proc. Natl Acad. Sci. USA* **95**, 606–611 (1998).

For supplementary information see page 175.

Acknowledgements

Supported by the National Natural Science Foundation of China, Ministry of Sciences & Technology of China, National Geographic USA (D-G.S.; X-L.Z.; L.C.; J.H.; Y.L.), Royal Society (S.C.M.), Science & Technology Committee of Yunnan Province (H-L. L.; S-X.H.; L-Z.C.), and Chinese Academy of Sciences (M.Z.). We thank R. J. Aldridge, P. Janvier and R. P. S. Jefferies for helpful remarks, and S. J. Last, S. Capon, D. Simons and L-H. Li for technical assistance.

Catching the First Fish
Philippe Janvier

Most major animal groups appear suddenly in the fossil record 550 million years ago, but vertebrates have been absent from this 'Big Bang' of life. Two fish-like animals from Early Cambrian rocks now fill this gap.

The search for the deep past of the vertebrates has been a pet subject of palaeontologists during the past decade. Traditionally, the fossil record of fishes and their descendants is good from the present to the Early Silurian period, about 430 million years (Myr) ago, then very poor in the Ordovician period (until about 480 Myr ago), and totally lacking or controversial before that, in the earliest Ordovician and Cambrian periods (from 480 to 550 Myr ago) and earlier. On page 42 of this issue, Shu et al.[1] report the discovery of two fish-like fossils from Chengjiang in Yunnan, China, that are probably the long-awaited Early Cambrian vertebrates.

British and Australian palaeontologists have found vertebrate remains from the beginning of the Ordovician.[2] Ordovician vertebrates have been shown to be far more diverse than was previously thought, consisting of a mixture of primitive, jawless groups and comparatively advanced groups with jaws.[3] However, all the Cambrian fossils thought to be vertebrates are from the Late Cambrian, and are controversial. They are chiefly the euconodonts (often called conodonts), a group long known from

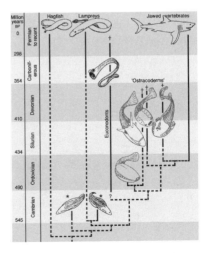

Figure 1 The current phylogenetic tree of the major living and fossil vertebrate groups.

isolated denticles (small, tooth-like fossils) of debated derivation, but now known from complete, eel-like animals that are probably vertebrates.[4] Some palaeontologists dispute this, although they accept that eucon-odonts may be related to vertebrates.[5,6]

Other putative Cambrian vertebrates are represented by small cara-pace fragments, whose microstructure recalls a group of jawless ver-tebrates with bony armour—the so-called ostracoderms—that lived from the Ordovician to the Devonian (Fig. 1). Again, some palaeontol-ogists regard these fragments as belonging to true vertebrates,[7] whereas others think they are carapace fragments from arthropods related to horseshoe crabs.[8] The exploration of the deep past of the vertebrates, before the Silurian period, is not for the faint-hearted, requiring the painstaking study of sketchy remains, with the constant danger of misinterpretation.

In fact, the bone-like fragments or denticles from the Cambrian might not lead to the 'root' of the vertebrates. The currently accepted phylogenetic trees showing the evolutionary relationships of fossil and living vertebrates suggest that mineralized tissues are a comparatively recent invention, and that the jawless ostracoderms are more closely related to the jawed verte-brates than to either lampreys or hagfish, the two living, jawless vertebrate groups, which have no bones or teeth (Fig. 1).[9] This means that the earliest history of the vertebrates can only be documented by fossils formed under certain conditions, where the imprint of soft tissues has been preserved. For the Cambrian pe-riod, there are two famous sites where an exceptional preservation of the soft tissues has occurred. One is the Burgess Shales, in Canada, and the other is Chengjiang. Fossils resembling chordates (the large animal group that includes the vertebrates) or vertebrates have been recorded from both, such as *Pikaia* from the Burgess, and *Yunnanozoon* and *Cathaymyrus* from Chengjiang.[10–12] Although probably close relatives of the ver-tebrates, none of these fossils looks entirely familiar to a vertebrate specialist.

How do you recognize a vertebrate?
All living vertebrates share a relatively small number of characters found in no other group, which are assumed to have been in-herited from a common vertebrate ancestor. All these characters are made of initially non-mineralized tissues (including cartilage) and are thus unlikely to be preserved in very early fossil vertebrates. The two Early Cambrian fossils from China display some vertebrate characters directly, whereas others can be in-ferred. Both show a dense imprint of what is probably a cartilaginous skull, one of the main vertebrate features. The gill skeleton may be indirect evidence for a neural crest, the main vertebrate character. The fin rays (radials) in one of the two forms and a large heart, lying behind the gills and possibly en-closed in a pericard, are also unique verte-brate characters. Chevron-shaped muscle blocks also occur in cephalochordates (the closest living relatives of vertebrates), but have a more complex zigzag shape in verte-brates, as can be clearly seen in at least one of the fossils.

The two new fossils described by Shu *et al.*[1] from Chengjiang are the most convincing Early Cambrian vertebrates ever found. With their zigzag-shaped muscle blocks, relatively complex and presumably cartilaginous skull, gill arches, heart, and fin supports, these fish resemble the larvae of living lampreys (Box 1). One detail, however, is at odds with what we know about vertebrates: the peculiar, forward tilt of the endoskeletal rays of the dorsal fin, which contrasts with the backward tilt of these fin rays in most fishes. Whether this is due to distortion or not awaits confirmation. Surprisingly, these animals seem to have long, ribbon-shaped, paired fins, which were thought to have appeared later in vertebrate history,[9] after the divergence of lampreys.

Shu *et al.* have analysed the phylogenetic position of these two Cambrian fossils in the framework of previous analyses of vertebrate phylogeny. Strangely, one of the two species seems to be more closely related to lampreys than to any other vertebrate group, whereas the other appears as the sister-group to all other vertebrates but hagfishes (Fig. 1). The support for this result is admittedly tenuous; morphology-based data matrices for fossil vertebrate groups include a large number of question marks, because of missing data in fossils and uncertainties as to the interpretation of some of the soft-tissue characters seen in imprints. This is true for the two Cambrian fish. Nevertheless, the result makes broad sense, because it suggests that lampreys and their fossil relatives had diverged already in the Cambrian, as predicted by previous phylogenies.

Lacking fossils, palaeontologists and anatomists have often tried to imagine the earliest vertebrates. The two new fossils from China resemble some of these imaginary reconstructions, but are completely at odds with others. Were theoretical reconstructions of the ancestral vertebrate partly accurate, or do we pay special attention to fossils that resemble the reconstructions in our minds?

References

1. Shu, D.-G. et al. *Nature* 402, 42–46 (1999).
2. Young, G. C. *J. Vert. Paleontol.* 17, 1–25 (1997).
3. Sansom, I. J., Smith, M. M. & Smith, M. P. *Nature* 379, 628–630 (1996).
4. Aldridge, R. J., Briggs, D. E. G., Smith, M. P., Clarkson, E. N. K. & Clark, N. D. L. *Phil. Trans. R. Soc. Lond. B* 340, 405–421 (1993).
5. Pridmore, P. A., Barwick, R. E. & Nicoll, R. S. *Lethaia* 29, 317–328 (1997).
6. Schultze, H.-P. *Mod. Geol.* 20, 275–285 (1996).
7. Smith, M. P., Sansom, I. J. & Repetski, J. E. *Nature* 380, 702–704 (1996).
8. Peel, J. S. *Rapp. Gmland Geol. Undersg.* 91, 111–115 (1979).
9. Janvier, P. *Palaeontology* 39, 259–287 (1996).
10. Conway Morris, S. in Atlas of the Burgess Shale (ed. Conway Morris, S.) 26 (Palaeontol. Assoc., London, 1982).
11. Chen, J.-Y. et al. *Nature* 377, 720–722 (1995).
12. Shu, D.-G., Conway Morris, S. & Zhang, X.-L. *Nature* 385, 865–868 (1996).

A Primitive Fossil Fish Sheds Light on the Origin of Bony Fishes

Min Zhu, Xiao-Bo Yu, and Philippe Janvier

Living gnathostomes (jawed vertebrates) include chondrichthyans (sharks, rays and chimaeras) and osteichthyans or bony fishes. Living osteichthyans are divided into two lineages, namely actinopterygians (bichirs, sturgeons, gars, bowfins and teleosts) and sarcopterygians (coelacanths, lungfishes and tetrapods). It remains unclear how the two osteichthyan lineages acquired their respective characters and how their common osteichthyan ancestor arose from non-osteichthyan gnathostome groups.[1,2] Here we present the first tentative reconstruction of a 400-million-year-old fossil fish from China (Fig. 1); this fossil fish combines features of sarcopterygians and actinopterygians and yet possesses large, paired fin spines previously found only in two extinct gnathostome groups (placoderms and acanthodians). This early bony fish provides a morphological link between osteichthyans and non-osteichthyan groups. It changes the polarity of many characters used at present in reconstructing osteichthyan interrelationships and offers new insights into the origin and evolution of osteichthyans.

The fossil fish *Psarolepis romeri*[3] was described on the basis of skull and lower jaw materials from the Lower Devonian strata (about 400 million years (Myr) BP) of Qujing, Yunnan, China. Materials assignable to the same genus from the Upper Silurian (about 410 Myr BP) of China[4] and Vietnam[5] made *Psarolepis* one of the earliest osteichthyans known so far. The skulls and lower jaws exhibit the overall morphology of sarcopterygians but also show characters found in primitive actinopterygians, such as a toothbearing median rostral and a lower jaw with five coronoids (rather than three as in most sarcopterygians).[3,6] *Psarolepis* was first placed within sarcopterygians, as a basal member of Dipnormorpha[3] or among the basal members of Crossopterygii.[4] The new features revealed by the shoulder girdle and cheek materials reported here (Figs. 1d, e, 2, 3) indicate that *Psarolepis* may occupy a more basal position in osteichthyan phylogeny.

Most *Psarolepis* specimens derive from four beds at the same locality in Qujing, the first bed being in the Yulongsi Formation (Pridoli), the second and third beds in the Xishancun Formation (early Lochkovian), and the fourth bed in the Xitun Formation (late Lochkovian). The specimens described here are from the second and third beds. The shoulder girdles and cheek plates from the third bed are often preserved as internal moulds (Fig. 2a) or as external moulds showing the internal casts of sensory canals and the porecanal system (Figs 2b, c, 3a). Like elements previously assigned to *Psarolepis*,[3–5] these materials exhibit the unique ornamentation with large, closely spaced pores on the cosmine surface. This unique ornamentation forms the basis of their assignment to *Psarolepis*, which is also supported by other histological and morphological features and by their association with elements that are directly comparable to the type specimen.

The shoulder girdle (Fig. 2) bears a huge pectoral spine that extends posteriorly from a conspicuous ridge between the ventral and vertical laminae of the cleithrum, resembling the condition in some placoderms and acanthodians. In addition, the symmetrical fin spine[4] indicates that *Psarolepis* may possess median spines in front of the unpaired fins (Fig. 1a, c), as in sharks and acanthodians. The median fin spine exhibits the same ornamentation as the associated parietal shield and lower jaw;[4] other taxa from the same bed (petalichthyids and galeaspids) cannot possess a similar ornamentation. Such paired and unpaired fin spines are unknown in early osteichthyans, whether sarcopterygians or actinopterygians.[1] The only possible exceptions are two questionable Silurian forms, *Lophosteus*, which has indeterminable spine-like fragments,[7] and *Andreolepis*, which has a lateral projection of the cleithrum.[8,9]

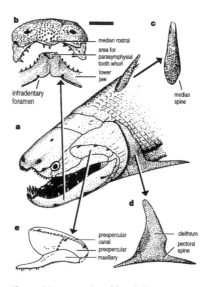

Figure 1 Reconstruction of *Psarolepis*, a 400-million-year-old sarcopterygian-like fish with an unusual combination of osteichthyan and non-osteichthyan features. **a**, Head and anterior part of the fish with tentatively positioned median fin spine. **b**, Anterior view of the skull and lower jaws (from ref. 3). Scale bar, 5 mm. **c**, Median fin spine (from ref. 4). **d**, Shoulder girdle with pectoral spine, based on specimens shown in Fig. 2. **e**, Cheek plate with maxillary and preopercular, based on specimens shown in Fig. 3. Surface ornamentation of the cheek plate is omitted to show the pattern of sensory canals. Most *Psarolepis* specimens derive from four beds at the same locality in Qujing, Yunnan, China.

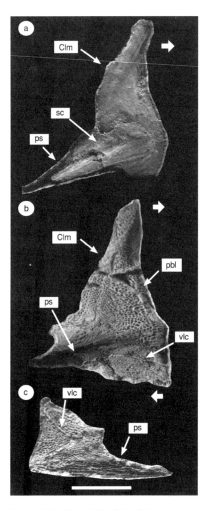

Figure 2 Shoulder girdles of *Psarolepis*. **a**, Specimen IVPP V11256.1: internal mould of a right cleithrum with pectoral spine (ps) and scapulocoracoid (sc). **b**, Specimen IVPP V11256.3: external mould of a left cleithrum with pectoral spine. **c**, Specimen IVPP V11256.4: external mould of an incomplete left cleithrum with pectoral spine. All specimens derive from the third bed in the Xishancun Formation (early Lochkovian) and are associated with undescribed cranial and lower jaw elements directly comparable to the type specimen. Short, thick arrows point to the anterior end of the fish. Scale bar, 10 mm. See Fig. 1d for a simplified reconstruction. Clm, cleithrum; pbl, postbranchial lamina; vlc, internal cast of canal on ventral lamina of cleithrum.

A critical issue is whether *Psarolepis* has the typical sarcopterygian monobasal articulation of the paired fins. Although the internal mould of the endoskeletal shoulder girdle (Fig. 2a) does not show the articulation surface of the fin skeleton clearly, its overall shape is strikingly different from that of both sarcopterygians and actinopterygians. It is a massive, plate-shaped bone pierced internally by several openings for blood vessels and nerves, in many ways like the condition in placoderms. The anterior part of the cleithrum has a denticulate postbranchial lamina (Fig. 2b), as in placoderms and actinopterygians. In contrast, the dorsal part of the cleithrum is high and pointed, as in actinopterygians, onychodonts and primitive coelacanths; it is different from the broad dorsal apex found in lungfishes, porolepiforms and osteolepiforms.

The external mould of the cheek plate from the third bed (Fig. 3a) matches the shape of a large, tilted preopercular in the cheek plate from the second bed (Fig. 3b); this latter cheek plate also shows a posteriorly expanded maxillary. The preopercular lacks a jugal canal (present in placoderms, acanthodians and chondrichthyans) but has a complete preopercular canal running along the dorsal margin of the bone (as in chondrichthyans, acanthodians and actinopterygians). Near the midpoint of the preopercular canal, a short vertical canal extends ventrally to the ventral margin of the preopercular. The dorsal portion of the preopercular carries three large foramina, similar to those found in the dermal cheeks of *Youngolepis*[10] and *Kenichthys*[11] as well as in the lower jaws of *Psarolepis*,[3,4] *Youngolepis*,[10] *Powichthys*[12] and some porolepiforms.[13] In addition to actinopterygians, sarcopterygian onychodonts[13] also have a posteriorly expanded maxillary in broad contact with the preopercular. However, the presence of a complete preopercular canal, the lack of a jugal canal and the absence of a separate squamosal distinguish *Psarolepis* from all previously known sarcopterygians.

To examine the phylogenetic position of *Psarolepis* and its impact on osteichthyan relationships, we used the data set in ref. 14 to construct an expanded matrix with 37 taxa and 149 characters (see Methods). As the codings for four characters in ref. 14 were modified by data in ref. 4, we used the phylogenetic package PAUP[15] to analyse the expanded matrix twice, first with the modified codings from ref. 4, then with the original codings from ref. 14. The strict consensus tree based on the modified codings in ref. 4 places *Psarolepis* as the sister group of all osteichthyans, whereas the strict consensus tree based on the original codings in ref. 14 places *Psarolepis* as the sister group of all previously known sarcopterygians (Fig. 4a, b). Figure 4c shows the two possible positions of *Psarolepis* and the incongruous distribution of *Psarolepis* features among the major gnathostome groups.

Although the clades common to the two competing schemes[4,14] of sarcopterygian interrelationships remain well supported, the conflicts between the two schemes remain unresolved and the exact

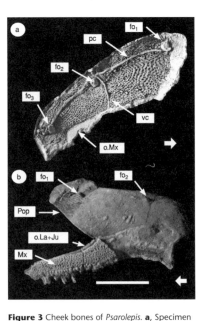

Figure 3 Cheek bones of *Psarolepis*. **a**, Specimen IVPP V11256.2: external mould of a left preopercular with the internal cast of sensory canals (pc, vc) and the pore-canal system, from the third bed in the Xishancun Formation. **b**, Specimen IVPP V11255: lateral view of a left preopercular (Pop) and maxillary (Mx), from the second bed in the Xishancun Formation. Short, thick arrows point to the anterior end of the fish. Scale bar, 10 mm. See Fig. 1e for a simplified reconstruction. *fo₁–fo₃*, foramina of unknown function; o·La + Ju, area overlapped by lacrimal and jugal; o·Mx, area overlapped by maxillary; pc, internal cast of preopercular canal; vc, internal cast of vertical canal.

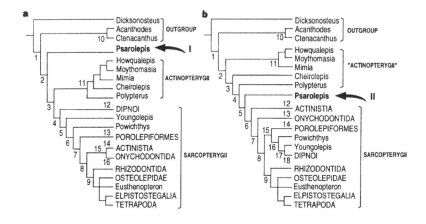

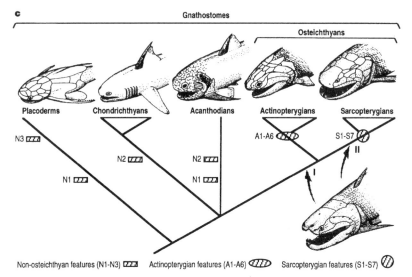

Figure 4 Phylogenetic analysis and the incongruous distribution of *Psarolepis* characters. **a**, Strict consensus tree from the expanded character matrix based on data from ref. 4 (54 most parsimonious trees at 311 steps, consistency index = 0.543, retention index = 0.804). *Psarolepis* forms the sister group of all osteichthyans. **b**, Strict consensus tree from the expanded character matrix based on data from ref. 14 (54 most parsimonious trees at 314 steps, consistency index = 0.538, retention index = 0.798). *Psarolepis* forms the sister group of all previously known sarcopterygians. The internal topologies of Sarcopterygii in **a**, **b** match the respective topologies in refs 4, 14 (see Methods for details). The trees are simplified by omitting lower-level nodes within the higher terminal taxa shown in uppercase letters. **c**, Interpreted inter-relationships of major gnathostome groups showing the incongruous distribution of *Psarolepis* characters. Our analysis uses limited non-osteichthyan characters to study the position of *Psarolepis* rather than the inter-relationships of non-osteichthyan gnathostomes. The sister-group relationship between *Acanthodes* and *Ctenacanthus* in **a**, **b** is converted to a trichotomy between chondrichthyans, acanthodians and osteichthyans to reflect our interpretation. Features: N1, bony pectoral spines; N2, median fin spine; N3, endoskeletal shoulder girdle as massive plate pierced by openings for blood vessels and nerves; A1, toothed median rostral; A2, five coronoids in lower jaw; A3, absence of squamosal bone; A4, posteriorly expanded maxillary; A5, tilted preopercular with complete preopercular canal; A6, cleithrum with pointed vertical lamina; S1, lower jaw with three infradentary foramina; S2, well-developed internasal cavities; S3, parasymphysial area carrying tooth whorls; S4, polyplocodont teeth; S5, quadrostian skull roof pattern, S6, intracranial joint; S7, cosmine. See text for details.

position of *Psarolepis* remains uncertain. The uncertainty results partly from a lack of information available for *Psarolepis* and other important stem taxa in the data set, and partly from the dificulty of selecting and polarizing characters when both osteichthyan and non-osteichthyan groups are used in the same analysis. However, whether *Psarolepis* turns out to be a stem-group osteichthyan or a stem-group sarcopterygian, its unique character combination will have a marked impact on present studies of osteichthyan evolution. For instance, porolepiform-like features found in *Psarolepis* (a lower jaw with three infradentary foramina, well developed internasal cavities and parasymphysial areas carrying tooth whorls[3]) can no longer be used to define porolepiforms[16] and/or dipnomorphs[14] (porolepiforms and lungfishes). The polyplocodont folded teeth and the quadrostian skull roof pattern of osteolepiforms[3,17] should also be regarded as primitive because of their presence in *Psarolepis*. If *Psarolepis* turns out to be a basal osteichthyan, the presence of an intracranial joint and cosmine can no longer serve as defining characters (synapomorphs) for sarcopterygians.[16–18]

Methods

PHYLOGENETIC ANALYSIS. We added *Psarolepis* and three non-osteichthyan taxa (*Acanthodes,* an acanthodian, *Ctenacanthus,* a chondrichthyan, and *Dicksonosteus,* a placoderm) to the 33 taxa in ref. 14. We also added 9 characters to the 140 characters in ref. 14 to reflect the variations found in *Psarolepis* and the three non-osteichthyan outgroups (character 141, large dermal plates; 142, paired pectoral spines; 143, median fin spines; 144, denticulate postbranchial lamina of the cleithrum; 145, wide suborbital ledge; 146, eye stalk or unfinished area for similar structure; 147, ventral and otico-occipital fissures; 148, basipterygoid articulation; 149, endochondral bone; 0 absent; 1 present). We adopted the same algorithm options as those used in refs. 4, 14 (all characters unordered and unweighted) and used the three non-osteichthyan outgroups to root the trees. See Supplementary Information for the expanded matrix and for characters supporting major nodes in Fig. 4a, b. See ref. 14 for the original 140 characters and character states; see ref. 4 for the changed codings for characters 10, 17, 78 and 108.

Sarcopterygii, Dipnoi, Porolepiformes, Actinistia, Onychodontida and Tetropodomorpha (including Rhizondontida, 'Osteolepiformes', Elpistostegalia and Tetrapoda) remain well supported in both trees. However, *Psarolepis* has changed the distribution and significance of many characters used previously to define osteichthyan groups (see Supplementary Information). In Fig. 4a, Sarcopterygii is defined by ten synapomorphies instead of fourteen as in ref. 14. Osteichthyes has no synapomorphy and is supported only by homoplasies. Actinopterygii is defined by one synapomorphy (character 6) and three reversals (characters 52, 93 and 110). In Fig. 4b, Sarcopterygii is defined by four synapomorphies, and 'Actinopterygii' appears as a paraphyletic group. One synapomorphy (character 7) defines the clade *Mimia* + (*Howqualepis* + *Moythomasia*) and five synapomorphies (characters 46, 63, 69, 98 and 134) define the clade *Polypterus* + (*Psarolepis* + Sarcopterygii). The position of *Polypterus* calls for further study, and the impact of data sampling and character coding on osteichthyan phylogeny deserves more attention.

References

1. Janvier, P. *Early Vertebrates* (Oxford Univ. Press, Oxford, 1996).
2. Rosen, D. E., Forey, P. L., Gardiner, B. G. & Patterson, C. Lungfishes, tetrapods, paleontology and plesiomorphy. *Bull. Am. Mus. Nat. Hist.* **167**, 159–276 (1981).
3. Yu, X. A new porolepiform-like fish, *Psarolepis romeri*, gen. et. sp. nov. (Sarcopterygii, Osteichthyes) from the Lower Devonian of Yunnan, China. *J. Vert. Paleontol.* **18**, 261–274 (1998).
4. Zhu, M. & Schultz, H.-P. The oldest sarcopterygian fish. *Lethaia* **30**, 293–304 (1997).
5. Tong-Duzy, T., Ta-Hoa, P., Boucot, A. J., Goujet, D. & Janvier, P. Vertébrés siluriens du Viet-nam central. *C.R. Acad. Sci. Abstr.* **32**, 1023–1030 (1997).
6. Gardiner, B. G. The relationships of the palaeoniscid fishes, a review based on new specimens of *Mimia* and *Moythomasia* from the Upper Devonian of Western Australia. *Bull. Brit. Mus. Nat. Hist.* **37**, 173–428 (1984).
7. Otto, M. Zur systematischen Stellung der Lophosteiden (Obersilur, Pisces inc. sedis). *Paläont. Z.* **65**, 345–350 (1993).
8. Schultze, H.-P. Ausgangform und Enterwicklung der rhombischen Schuppen der Osteichthyes (Pisces). *Paläont. Z.* **51**, 152–168 (1977).
9. Janvier, P. On the oldest known teleostome fish *Andreolepis hedei* Gross (Ludlow of Gotland), and the systematic position of the lophosteids. *Eesti NSV Teaduste Akadeemia Toimetised Geol.* **27**, 86–95 (1978).
10. Chang, M. M. in *Early Vertebrates and Related Problems of Evolutionary Biology* (eds Chang, M. M., Liu, Y. H. & Zhang, G. R.) 355–378 (Science Press, Beijing, 1991).
11. Chang, M. M. & Zhu, M. A new Middle Devonian osteolepidid from Qujing, Yunnan. *Mem. Ass. Australas. Palaeontols.* **15**, 183–198 (1993).
12. Jessen, H. L. Lower Devonian Poroleformes from the Canadian Arctic with special reference to *Powichthys thorsteinssoni Jessen.* Palaeontogr. Abt A **167**, 180–214 (1980).
13. Ahlberg, P. E. A re-examination of sarcopterygian interrelationships, with special reference to the Porolepiformes. *Zool. J. Linn. Soc.* **103**, 241–287 (1991).
14. Cloutier, R. & Ahlberg, P. E. in *Interrelationships of Fishes* (eds Stiassny, M. L., Parenti, L. R. & Johnson, G. D.) 445–480 (Academic, New York, 1996).
15. Swofford, D. L. PAUP: phylogenetic analysis using parsimony, version 3.1.1 (1993).
16. Jarvik, E. Basic structure and evolution of vertebrates. Vol. 2 (Academic, London, 1981).
17. Andrews, S. M. in *Interrelationships of Fishes* (eds Miles, R. S. & Patterson, C.) 137–177 (Academic, London, 1973).
18. Bjerring, H. The 'intercranial joint' vs the 'ventral otic fissure'. *Act. Zool. Stockh.* **59**, 203–214 (1978).

For supplementary information see page 179.

Acknowledgements

We thank H.-P. Schultze, M. M. Chang and P. E. Ahlberg for useful discussion; W. Harre for the photographs and P. E. Ahlberg, M. I. Coates and M. M. Smith for comments and suggestions. M. Z. acknowledges support from the Alexander von Humboldt Foundation and the Chinese Academy of Sciences. X.Y. thanks IVPP for access to specimens and Kean University for support for research and faculty development.

8

Something Fishy in the Family Tree
Per Erik Ahlberg

Where did the bony fishes come from? This simple question illustrates a theme, the quest for origins, which has been central to evolutionary biology since its birth as a discipline. Great effort has been expended on tracing the beginnings of the major groups of organisms in the fossil record. This research has had its successes, but has been curiously frustrating—time and again, the vaunted 'basal' forms it has uncovered have confounded expectations, leaving us as much puzzled as enlightened. Now, Zhu et al.[1] uncover another example of this pattern: they present a basal bony fish with such an unexpected mix of characteristics that it forces a reconsideration of large parts of the vertebrate family tree.

The 'bony fishes', Osteichthyes, a group which in fact also includes the land vertebrates or tetrapods, and thus ourselves, is by far the largest group of backboned animals. It musters some 50,000 species, as against roughly 700 sharks, rays and chimaeras, and a mere handful of jawless lampreys and hagfishes. The group's ecological importance is incalculable. It might seem odd to unite animals as different as a cod (a decent bony fish by any measure) and a human being in the same group, but all osteichthyans have a suite of common features which show beyond doubt that they share a single common ancestry. Tetrapods have evolved from more orthodox bony fishes.[2] Clearly, the origin of the osteichthyans was a momentous event in the history of life, but attempts to investigate this event always come up against a baffling barrier.

There is almost universal agreement that the Osteichthyes can be divided into two branches, the Actinopterygii or ray-fins and the Sarcopterygii or lobe-fins (Fig. 1). The actinopterygians, which include the cod and perhaps 25,000 other species, form the largest living fish group. The sarcopterygians include the phenomenally successful land vertebrates, but only four living genera of fishes—the Australian, African and South American lungfishes, and the coelacanth *Latimeria*.

However, as you trace these groups backwards in time through the fossil record, sarcopterygian fishes become more diverse until, in the Devonian Period, 408 to 363 million years ago, we find a wide range of lobe-fins alongside just a few ray-fins. But then the record stops, abruptly; no Silurian rocks have ever yielded more than isolated scales and scraps of osteichthyans, and in older deposits they are entirely absent. The Silurian forms have been given names, *Lophosteus* and *Andreolepis*, but we know almost nothing about them and cannot tell how they relate to later osteichthyans.

Phylogenetic investigations, whether based on fossils, molecular data or the morphology of recent osteichthyans, tend to hit the buffers in a similar way. They generally support the actinopterygian-sarcopterygian split, but have failed to resolve unequivocally how the main sarcopterygian groups are related to each other.[3-6] Furthermore, while almost invariably placing the Chondrichthyes (sharks, rays and chimaeras) as the living sister group of the Osteichthyes, they have not established which of the fossil non-osteichthyan groups is closest to the Osteichthyes.[7] As a result of these limitations of phylogenetic resolution and the fossil record, we don't know what the last common ancestor of the Sarcopterygii, or of the Osteichthyes as a whole, was like; we don't know which characteristics are primitive for the different subgroups; and we cannot tell from what kind of ancestry the Osteichthyes emerged.

Into this confusion drops Zhu and colleagues' discovery, the 400-million-year-old *Psarolepis* from the earliest Devonian and latest Silurian of Yunnan in south China.[1,4,8] It is the earliest osteichthyan known from reasonably complete remains of skull and shoulder girdle, and shows a strikingly incongruous suite of features.

The braincase, which is divided into separate front and back halves by a joint, looks like that of a reasonably orthodox sarcopterygian; so do the teeth, which have a characteristically sarcopterygian type of folded 'polyplocodont' dentine. Three conspicuous openings on the outer face of the lower jaw, and a matching set on the cheek, again point to sarcopterygian affinities, specifically with a group of supposed lungfish relatives known as porolepiforms.

But the tooth-bearing bones of the snout and lower jaw match those of actinopterygians, as do

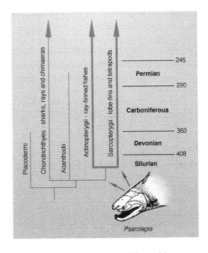

Figure 1 Fish relationships and the ambiguous position of *Psarolepis*.

the cheek bones. The shoulder girdle doesn't even look osteichthyan; a large spine sweeps outwards just in front of the pectoral fin, much as in an extinct group of fishes called placoderms. This is interesting, because it has been argued that placoderms are the closest relatives of osteichthyans.[7] However, *Psarolepis* also seems to have spines in front of its median fins, a feature more reminiscent of another group of putative osteichthyan relatives, the acanthodians.[7]

So where does all this leave us? Clearly, existing ideas about what features characterize the different osteichthyan groups are much too simplistic. Some of the supposedly unique 'actinopterygian' and 'sarcopterygian' characteristics are probably primitive for the whole Osteichthyes, and their presence side-by-side in *Psarolepis* is simply indicative of the primitive nature of this fish, while the placoderm- and acanthodian-like features may be previously unrecognized components of the primitive osteichthyan character complex. (At least one such feature, the pectoral spine, is also present in the Silurian genus *Andreolepis*.[1])

These re-evaluations will in due course have a major influence on the interpretation of osteichthyan phylogeny. For the moment, Zhu *et al.*[1] try to establish a preliminary phylogenetic position for *Psarolepis* by incorporating it, together with an acanthodian, a placoderm and a shark, into two competing published osteichthyan data sets.[3,4] The results are interesting, but also frustrating (Fig. 1); depending on which data set is used, *Psarolepis* becomes either a stem osteichthyan or a stem sarcopterygian, and the judgements about character polarities and distributions differ accordingly. It seems we are still trapped by existing phylogenetic uncertainties.

In the long term, however, the prospects look much brighter. First, the existing data sets need to be revised (I am the co-author of one of them,[3] and would certainly not regard it as a final statement). Second, a number of other early and/or primitive osteichthyans, including several more from Yunnan, are currently under study and will be described within the next few years. These should help to establish whether *Psarolepis* really is incongruous, or whether its combination of features is typical of the earliest osteichthyans.

References

1. Zhu, M., Yu, X. & Janvier, P. *Nature* **397**, 607–610 (1999).
2. Ahlberg, P. E. & Milner, A. R. *Nature* **368**, 507–514 (1994).
3. Cloutier, R. & Ahlberg, P. E. in Interrelationships of Fishes (eds Stiassny, M. L. J., Parenti, L. R. & Johnson, G. D.) 445–479 (Academic, San Diego, 1996).
4. Zhu, M. & Schultze, H.-P. Lethaia **30**, 293–304 (1997).
5. Gorr, T., Kleinschmidt, T. & Fricke, H. *Nature* **351**, 394–395 (1991).
6. Zardoya, R. & Meyer, A. Proc. Natl Acad. Sci. USA **93**, 5449–5454 (1996).
7. Janvier, P. Early Vertebrates (Oxford Univ. Press, 1996).
8. Yu, X. J. Vert. Paleontol. **18**, 261–274 (1998).

9

An Exceptionally Well-Preserved Theropod Dinosaur from the Yixian Formation of China

Pei-Ji Chen, Zhi-Ming Dong, and Shuo-Nan Zhen

Two spectacular fossilized dinosaur skeletons were recently discovered in Liaoning in northeastern China. Here we describe the two nearly complete skeletons of a small theropod that represent a species closely related to *Compsognathus*. *Sinosauropteryx* has the longest tail of any known theropod, and a three-fingered hand dominated by the first finger, which is longer and thicker than either of the bones of the forearm. Both specimens have interesting integumentary structures that could provide information about the origin of feathers. The larger individual also has stomach contents, and a pair of eggs in the abdomen.

The Jehol biota[1] was widely distributed in eastern Asia during latest Jurassic and Early Cretaceous times. These freshwater and terrestrial fossils include macroplants, palynomorphs, charophytes, flagellates, conchostracans, ostracods, shrimps, insects, bivalves, gastropods, fish, turtles, lizards, pterosaurs, crocodiles, dinosaurs, birds and mammals. In recent years, the Jehol biota has become famous as an abundant source of remains of early birds.[2,3] Dinosaurs are less common in the lacustrine beds, but the specimens described here consist of two nearly complete skeletons of a small theropod discovered by farmers in Liaoning. The skeletons are from the basal part of the Yixian Formation, from the same horizon as the fossil birds *Confuciusornis* and *Liaoningornis*.[3] Both are remarkably well preserved, and include fossilized integument, organ pigmentation and abdominal contents. One of the two was split into part and counterpart, and the sections were deposited in two different institutions. One side (in the National Geological Museum of China, Beijing) became the holotype of *Sinosauropteryx prima*, a supposed bird.[4] The counterpart and the second larger specimen are in the collections of the Nanjing Institute of Geology and Palaeontology.

The Yixian Formation is mainly composed of andesites, andesite-breccia, agglomerates and basalts, but has four fossil-bearing sedimentary intercalations that are rich in tuffaceous materials. The Jianshangou (formerly Jianshan[5,6]) intercalated bed (60 m thick) is the basal part of this volcanic sedimentary formation, and is made up of greyish–white, greyish–yellow and greyish–black sandstones, siltstones, mudstones and shales. These sediments are rich in fossils of mixed Jurassic–Cretaceous character. The primitive nature of the fossil birds of the Jianshangou fossil group has led to suggestions that the beds could be as early as Tithonian in age.[2] But although *Confuciusornis* and the other birds[3] are more advanced than *Archaeopteryx* in a number of significant features, we can only conclude that the beds that the fossils came from are probably younger than the Solnhofen Lithographic Limestones (Early Tithonian). The presence of *Psittacosaurus* in the same beds is more consistent with an Early Cretaceous age,[7] as are the palynomorphs[8] and a recent radiometric date of the formation,[9] but other radiometric dating attempts have indicated older ages.[10]

Dinosauria Owen 1842
Theropoda Marsh 1881
Coelurosauria von Huene 1914
Compsognathidae Marsh 1882
Sinosauropteryx prima JI and JI 1996

HOLOTYPE.   Part (National Geological Museum of China, GMV 2123) and counterpart (Nanjing Institute of Geology and Palaeontology NIGP 127586) slabs of a complete skeleton.

REFERRED SPECIMEN.   Nanjing Institute of Geology and Palaeontology NIGP 127587. Nearly complete skeleton, lacking only the distal half of the tail.

LOCALITY AND HORIZON.   Jianshangou-Sihetun area of Beipiao, Liaoning, People's Republic of China. Yixian Formation, Jehol Group, Upper Jurassic or Lower Cretaceous (Fig. 1).

DIAGNOSIS.   Compsognathid with longest tail known for any theropod (64 caudals). Skull 15% longer than femur, and forelimb (humerus plus radius) only 30% length of leg (femur plus tibia), in contrast with *Compsognathus* where skill is same length as femur, and forelimb length is 40% leg length. Within the Compsognathidae, forelimb length (compared to femur length) is shorter in *Sinosauropteryx* (61–65%) than it is in *Compsognathus* (90–99%). In contrast with all other theropods, ungual phalanx II-2 is longer than the radius. Haemal spines simple and spatulate, whereas those of *Compsognathus* taper distally.

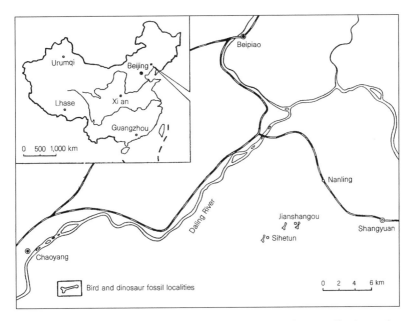

Figure 1 Map showing the localities of *Sinosauropteryx* and *Confuciusornis* in western Liaoning, northeastern China.

Description

Sinosauropteryx is comparable in size and morphology to known specimens of *Compsognathus*[11,12] from Germany and France. The smaller Chinese specimen (Fig. 2) is 0.68 m long (snout to end of tail) and has a femur length of 53.2 mm, whereas the second specimen (Fig. 3) has a femur length of 86.4 mm. The former is smaller than the type specimen of *Compsognathus longipes* (femur length about 67 mm) and the latter is smaller than the second specimen of *Compsognathus* from Canjuers (France), which has a femur length of 110 mm and an estimated length of 1.25 m. Although size and body proportions indicate that the smaller specimen was younger when it died, well-ossified limb joints and tarsals suggest that it was approaching maturity.

Sinosauropteryx and *Compsognathus* share several characteristics that indicate close relationship. These can be used to diagnose the Compsognathidae and include unserrated premaxillary but serrated maxillary teeth, a powerful manual phalanx I–I (shaft diameter is greater than that of the radius), fan-shaped neural spines on the dorsal vertebrae, limited anterior expansion of the pubic boot and a prominent obturator process of the ischium.

Other characteristics were used to diagnose *Compsognathus*, including the presence of a relatively large skull and short forelimbs. In *Compsognathus*, skull length is 30% of that of the presacral vertebral column,

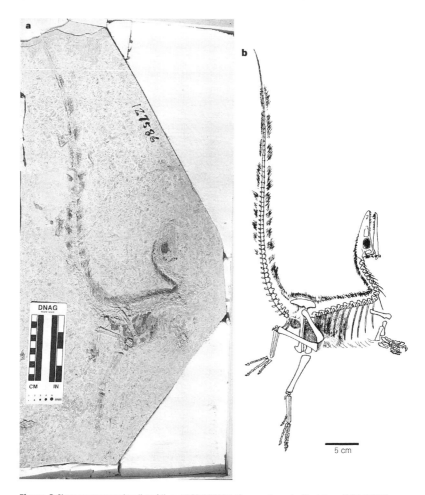

Figure 2 *Sinosauropteryx prima* Ji and Ji. **a**, NIGP 127586, the counterpart of holotype (GMV 2123). **b**, Skeletal reconstruction of NIGP 127586. The integumentary structures are along the dorsal side and tail and dark pigmentation in the abdominal region might be some soft tissues of viscera.

whereas this same ratio is 40% in the new specimen NIGP 127586 and 36% in NIGP 127587. Unfortunately, relative skull length is highly variable in theropods. Comparing skull length with femur length, which is less variable than vertebral length, most theropods have skulls 100–119% the length of the femur.[13] The *Compsognathus* skulls are 99–100% and *Sinosauropteryx* skulls are 113–117%. Compsognathids have short forelimbs,[11] 40% of the length of the hindlimb in *Compsognathus*. In *Sinosauropteryx*, the lengths of humerus plus radius divided by the sum of femur and tibia lengths produces a figure of less than 30% (Table 1). Unfortunately, such ratios are dependent on the absolute size of the animal, mostly because of negative allometry experienced by the tibia during growth or interspecific size increase. Comparing the lengths of humerus plus radius with femur length produces more useful results. The resulting figures fall within the range of most theropods (60–110%). The abelisaurid *Carnosaurus* and all tyrannosaurs have relatively shorter arms. Within the Compsognathidae, however, arm length is shorter in *Sinosauropteryx* (61–65%) than it is in *Compsognathus* (90–99%).

Both specimens of *Sinosauropteryx* have 10 cervical and 13 dorsal vertebrae. The posterior cervical vertebrae have biconcave centra. We could not determine the number of sacrals. The tail is extremely long. In the smaller specimen it is almost double the snout–vent length, and there are 59 caudal vertebrae exposed with an estimated five more than have been lost from the middle of the tail of NIGP 127586 (but present in GMV 2123). Only the first 23 vertebrae are preserved in the larger specimen, but this section is longer than the summed lengths of the cervical, dorsal and sacral vertebrae. Neither of the European specimens has a complete tail, but in both cases the tail was clearly longer than the body. When vertebral lengths are normalized (divided by the average lengths of caudal vertebrae 2–5), there are no significant differences between vertebral lengths in any of the four tails. As in *Compsognathus*, the dorsal neural spines are peculiar in that they are anteroposteriorly long but low, and often are fan-shaped.

The caudal centra increase in length over the first six segments, but posteriorly decrease progressively in length and all other dimensions. The first 10 tail vertebrae have neural spines, most of which slope posterodorsally. There are at least four pairs of caudal ribs in NIGP 127586, and more distal caudals have low bumps in this region that could also be interpreted as transverse processes. This could be another way to distinguish the Asian and European compsognathids, because the German specimen of *Compsognathus* apparently lacks caudal ribs and transverse processes.[11] Haemal spines are found on at least the first 47 caudals of NIGP 127586, and the anterior ones are simple

Figure 3 *Sinosauropteryx prima* Ji and Ji, NIGP 127587, an adult individual from the same locality as holotype.

Table 1 Comparison of size and proportions of *Sinosauropteryx* and *Compsognathus*

Species	Specimen	Skull	Humerus	Radius	Femur	Tibia	Skull/femur	Arm/leg
Compsognathus sp.	BSP ASI	70	39	24.7	71	87.6	0.99	0.40
Compsognathus sp.	MNHN	110	67	42	110		1.00	
Sinosauropteryx prima	NIGP127586	62.5	20.3	12.4	53.2	61	1.17	0.29
Sinosauropteryx prima	NIGP127587	97.2	35.5	21	86.4	97	1.13	0.31

Length measurements are given in millimetres. Data about *Compsognathus* are from ref. 11.

spatulate structures that curve gently posteroventrally. The haemal spines are oriented more posteriorly than ventrally, and are more strongly curved.

Both specimens have 13 pairs of dorsal ribs. The ribs indicate a high but narrow body. The distal ends of the first two pairs of ribs are expanded and end in cup-like depressions that suggest the presence of a cartilaginous sternum. The gastralia are well preserved with two gastralia on each side of a segment. The median gastralia cross to form the interconnected 'zig-zag' pattern characteristic of all theropods[14] and primitive birds like *Archaeopteryx* and *Confuciusornis*.

The front limb is relatively short and stout. Both NIGP 127586 and NIGP 127587 have articulated hands, something that is lacking in the two European specimens. What has been interpreted by some as the first metacarpal[11] in *Compsognathus* is the first phalanx of digit I, as was originally proposed by von Huene.[15] The first metacarpal is short (4.2 mm long in NIGP 127586, and double that length in 127587), and is probably the element identified as a carpal in the French specimen.[12] As is typical of all theropods, the collateral ligament pits of the first phalanx are much closer to the extensor surface of the bone than they are to the flexor surface. Both phalanx I–1 and the ungual that it supports are relatively large, each being as long as the radius, and thicker than the shafts and the distal ends of either the radius or the ulna. This unusual character seems to have been partially developed in at least phalanx I–1 of *Compsognathus*. Relative to the length of the radius, both these elements are longer in *Sinosauropteryx* than in any other known theropod except for *Mononykus*.[16] As indicated by the proposed phylogenetic placement[16] of *Mononykus*, there are too many anatomical differences between compsognathids and *Mononykus* to suggest a close relationship, and the similarities probably represent convergence.

The long (39 mm in NIGP 127586, 67.5 mm in NIGP 127587), low (22.2 mm high at both pubic and ischial peduncles in NIGP 127587) ilium is shallowly convex on the dorsal side in lateral aspect. The pubis, which is 82.8 mm long in NIGP 127587, is oriented anteroventrally, but is closer to vertical than it is in most non-avian theropods. The distal end expands into a pubic boot as in most tetanuran theropods. In the larger specimen, this expansion is 17.7 mm. As in *Archaeopteryx*,[17] *Compsognathus*[11] and dromaeosaurs,[18] the boot expands posteriorly from the shaft of the pubis, and the anterior expansion is moderate. The lack of the significant anterior expansion of the pubic boot may be correlated with the inclination of the shaft. The ischium is only two-thirds the length of the pubis in NIGP 127587. It tapers distally into a narrow shaft

(3.2 mm in diameter), and like *Compsognathus*, there is a slight expansion at the end (6 mm in NIGP 127587). The prominent obturator process is also found in *Compsognathus*.

The shaft of the femur is gently curved. Both tibia and fibula are elongate. The astragalus and calcaneum are present in both specimens, although not clearly seen in either. There are five metatarsals, but as in other theropods and early birds, the first is reduced to a distal articular condyle, and the fifth is reduced to a proximal splint (Fig. 2b). Metatarsals II, III and IV are closely appressed and elongate, but are not co-ossified. The second and fourth metatarsals do not contact each other. Pedal phalanges are conservative in number (2–3–4–5–0) and morphology.

Inclusions within the Body Cavity

Like the German *Compsognathus*, the larger Chinese specimen has stomach contents preserved within the rib cage. This consists of a semi-articulated skeleton of a lizard, complete with skull (Fig. 4a,b). Numerous lizard skeletons have been recovered from these beds, but have yet to be described. Low in the abdomen of NIGP 127587, anterior to and slightly above the pubic boot, lies a pair of small eggs (37 × 26 mm) (Fig. 4a,c), one in front of the other. Additional eggs may lie underneath. Gastralia lie over the exposed surfaces of the eggs, and the left femur protrudes from beneath them, so there can be no doubt that they were within the body cavity. It is possible that the eggs were eaten by the dinosaur. However, given their position in the abdomen behind and below the stomach contents, and the fact that they are in the wrong part of the body cavity for the egg shell to be intact, it is more likely that these were unlaid eggs of the compsognathid. Eggs have also been reported in the holotype of *Compsognathus*,[19] but they are more numerous and are only 10 mm in diameter. As they were also found outside the body cavity, their identification as *Compsognathus* eggs has not been widely accepted. The presence of fewer but larger eggs in *Sinosauropteryx* casts additional doubt on this identification.

Although more than two eggs may have been present in the larger specimen of *Sinosauropteryx*, it does not seem as though many could have been held within the abdomen. It may well be that these dinosaurs laid fewer eggs than most (some species are known to have produced in excess of 40).[20] However, it is more likely that their presence demonstrates paired ovulation, as has been suggested for *Oviraptor*,[21] *Troodon*[22] and other theropods. *Sinosauropteryx* therefore probably laid eggs in pairs, with a delay for ovulation between each pair.

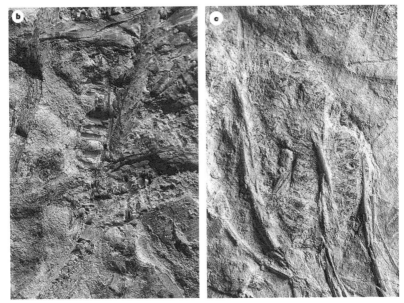

Figure 4 Body of NIGP 127587. **a**, Stomach contents are preserved within the rib cage, and include a small lizard and a pair of eggs. **b**, A close-up of the lizard skull. **c**, A close-up of a pair of the eggs.

Integumentary Structures

One of the most remarkable features of both Chinese specimens is the preservation of integumentary structures. In the larger specimen, these structures can be seen along the dorsal surface of the neck and back, and along the upper and lower margins of the tail, but in the smaller specimen the integumentary structures are clearer (Fig. 5). They cover the top of the back half of the skull, the neck, the back, the hips and the tail. They also extend along the entire ventral margin of the tail. Small patches can be seen on the side of the skull (behind the quadrate and over the articular), behind the right humerus, and in front of the right ulna. With the exception of a small patch outside the left ribs of NIGP 127587 and several areas on the left side of the tail (lateral to the vertebrae), integumentary structures cannot be seen along the sides of the body. The structures were probably present in the living animals, as indicated by the density of the covering dorsal to the body, and by the few random patches of integumentary structures that can be seen elsewhere.

In the two theropods, the distances separating the integumentary structures from the underlying bones are directly proportional to the amount of skin and muscle that would have been present. As in modern animals, the integument closely adheres to the tops of the skull and hips, and becomes progressively closer to the caudal vertebrae towards the tip of the tail. In the posterior part of the neck, over the shoulders, and at the base of the tail, the integumentary structures are more distant from the underlying skeletal elements, and in life would have been separated by greater thicknesses of muscle and other soft tissues.

The orientation and frequently sinuous lines of the integumentary structures suggest they were soft and pliable, and semi-independent of each other. They frequently cross each other, and are tangled in some areas. There is an apparent tendency for the integumentary structures to clump along the tail of the smaller specimen, but this is an artefact of the splitting plane between NIGP 127586 (Fig. 2a) and GMV 2123. As both individuals were lying in the water of a lake when they were buried, it is clear that we are not looking at the normal orientation of the integumentary structures in the fossils. Under magnification, the margins of the larger structures are darker along the edges, but lighter medially, which indicates that they might have been hollow. Overall, the integumentary structures are rather coarse for such a small animal, and the thickest strands are much thicker than the hairs of the vast majority of small mammals.[23] In NIGP 127586, integumentary structures are first seen on the dorsal surface of the skull in front of the orbit. The skull is semidisarticulated, and sediment still covers the snout region, so it is possible

Figure 5 Integumentary structures in the neck and dorsal sides of NIGP 127586.

that the integumentary structures extended more anteriorly. The most rostral integumentary structures are 5.5 mm long, and extend about 4 mm above the skull. They quickly lengthen to at least 21 mm above the distal ends of the scapulae. This axial length seems to stay constant along most of the back, but decreases sharply to 16 mm dorsal to the ilium. The longest integumentary structures seem to have been above the base of the tail, although it is impossible to measure any single structure. More distally along the tail, integumentary structures decrease more rapidly on the lower side of the tail than on the upper. By caudal 47, the ventral structures are 4.2 mm long, about half the length of the dorsal structures in that region.

The size distribution of the integumentary structures of NIGP 127587 follow the same general pattern as in the smaller specimen. Although the integument tends to look thinner on this specimen, it is simply because the integumentary structures are lying closer to the body. Individual measurements are consistently larger than those of NIGP 127586. The integumentary structures are 13 mm long above the skull, 23.5 mm above the fourth cervical, at least 35 mm over the scapulae, at least 40 mm over caudal 27 (Fig. 6), and at least 35 mm below caudal 25. Integumentary structures on the left side of the body are largely covered by ribs, gastralia, stomach contents and matrix, so it is only possible to say that each is more than 5 mm long. Those associated with the right ulna are 14 mm long.

Integumentary structures have also been reported in the theropod *Pelecanimimus*[24] from the Lower Cretaceous of Spain. These consist of subparallel fibres arranged perpendicular to the bones, with a less conspicuous secondary system parallel to them. As described, they seem to be similar to the integumentary structures of *Sinosauropteryx*.

Skin impressions have been found on most main types of dinosaurs, including sauropods, ankylosaurs, ornithopods, stegosaurs, ceratopsians, and several genera of large theropods. In all of these animals, there is no evidence of integumentary structures, and the skin

Figure 6 Integumentary structures over caudal 27 of NIGP 127587.

usually has a 'pebbly' surface texture. Integumentary structures have been claimed for both specimens of *Compsognathus*[11,25] though the interpretations have been questioned in both cases.[11] In the German specimen, there was supposedly a patch of skin over the abdominal region. The French specimen included some strange markings in the region of the forearm, that were originally identified as a swimming appendage formed either of dermal bone or of thick skin,[12] but it is clearly not well enough preserved to be positively identified. The identification of these structures as integumentary is questionable,[11] and there is nothing on the Chinese specimens to support the presence of such structures in compsognathids. Evidence of feathers in *Compsognathus* was sought[11] without success, but this lack of evidence on the German specimen of *Compsognathus* does not eliminate the possibility that they might have existed.

Discussion

The integumentary structures of *Sinosauropteryx* are extremely interesting regardless of whether they are referred to as feathers, protofeathers, or some other structure. Unfortunately, they are piled so thick that we have been unable to isolate a single one for examination. Comparison with birds from the same locality shows that the same problem exists with identifying individual feathers (other than the flight feathers) and components of feathers in avian specimens. The morphological characteristics that we describe suggest that the integumentary structures seem to resemble most closely the plumules of modern birds, having relatively short quills and long, filamentous barbs. The absence of barbules and hooklets is uncommon in modern birds, but has been noted in Cretaceous specimens.[26]

It has been proposed that the feathers of another recently discovered animal from the same locality in Liaoning are structurally intermediate between the integumentary structures of *Sinosauropteryx* and the feathers of *Archaeopteryx*.[27] The clearly preserved feathers of *Protarchaeopteryx robusta* are symmetrical, which indicates that the animal was not capable of flight. This is confirmed by the relatively short length of the forelimb. Both *Sinosauropteryx* and *Protarchaeopteryx* had been identified as birds because of the presence of feathers,[27] but much more work needs to be done to prove that the integumentary structures of *Sinosauropteryx* have any structural relationship to feathers, and phylogenetic analysis of the skeleton clearly places compsognathids far from the ancestry of birds. Despite arguments to the contrary,[28] cladistic analysis favours the notion that the bird lineage originated within theropod dinosaurs.[29,30] If this phylogenetic framework is accepted, the integumentary structures of

Sinosauropteryx could shed light on some of the many hypotheses concerning feather origins. Three main functions have been suggested for the initial development of feathers—display, aerodynamics and insulation.

The integumentary structures of *Sinosauropteryx* have no apparent aerodynamic characteristics, but might be representative of what covered the ancestral stock of birds. It is highly unlikely that something as complex as a bird feather could evolve in one step, and many animals glide and fly with much simpler structures. Even birds secondarily simplify feathers when airborne flight ceases to be their main method of locomotion, and produce structures that are intermediate between reptilian scales and feathers.[31] The multibranched integumentary structures of the Chinese compsognathids are relatively simple, but are suitable for modification into the more complex structures required for flight.

Feathers may have appeared first as display structures,[31] but the density, distribution, and relatively short lengths of the integumentary structures of *Sinosauropteryx* suggest that they were not used for display. It is conceivable that both specimens are female, and that the males had more elaborate integumentary structures for display. It is also possible that the integumentary structures were coloured to serve a display function. Therefore, the existing *Sinosauropteryx* specimens do not support the hypothesis that feathers evolved primarily for display, but do not disprove it either.

The dense, pliable integumentary structures of the Chinese compsognathids would not have been appropriate as heat shields to screen and shade the body from the Sun's rays.[32] Although they may have been effective in protecting the body from solar radiation in warm weather, they would also have been effective in preventing an ectothermic theropod from rapidly warming up by basking in the sunshine. If small theropods were endothermic, they would have needed insulation to maintain high body temperatures.[33-35] The presence of dense integumentary structures may suggest that *Sinosauropteryx* was endothermic, and that heat retention was the primary function for the evolution of integumentary structures.[36-38] Recently published histological studies suggest at least some early birds were not truly endothermic,[39] although they may have been physiologically intermediate between poikilothermic ectotherms and homeothermic endotherms.[40]

The Chinese compsognathids have integumentary structures consisting of vertical fibres running from the base of the head along the back and around the tail extending forwards almost to the legs. There are no structures showing the fundamental morphological features of modern bird feathers, but they could be previously unidentified protofeathers which are not as complex as either down feathers or even the hair-like feathers of secondarily flightless birds. Their simplicity would not have made them ineffective for insulation when wet any more than it negates the insulatory capabilities of mammalian hair. We cannot determine whether or not the integumentary structures were arranged in pterylae, but they were long enough to cover apteria, if they existed, and could therefore still have been effective in thermoregulation. Continuous distribution is not essential to be effective in this function,[28] especially if the apteria are part of a mechanism for dispersing excess heat. Finally, the aerodynamic capabilities of bird feathers are not comprised by the previous evolution of less complex protofeathers that had some other function, such as insulation.

In addition to the integumentary structures, there is dark pigmentation over the eyes of both specimens. A second region of dark pigmentation in the abdominal region of the smaller specimen might represent some soft tissues of viscera.

Multidisciplinary and multinational research is just beginning on these unique small theropods. Techniques developed to study fossil feathers[38,41] will be useful research tools as work progresses. In the meantime, the integumentary structures of *Sinosauropteryx* suggest that feathers evolved from simpler, branched structures that evolved in non-avian theropod dinosaurs, possibly for insulation.

References

1. Chen, P. J. Distribution and migration of Jehol fauna with reference to nonmarine Jurassic–Cretaceous boundary in China. *Acta Palaeontol. Sin.* **27**, 659–683 (1988).
2. Hou, L.-H., Zhang, J.-Y., Martin, L. D. & Feduccia, A. A beaked bird from the Jurassic of China. *Nature* **377**, 616–618 (1995).
3. Hou, L.-H., Martin, L. D., Zhang, J.-Y. & Feduccia, A. Early adaptive radiation of birds: evidence from fossils from northeastern China. *Science* **274**, 1164–1167 (1996).
4. Ji, Q. & Ji, S. A. On discovery of the earliest bird fossil in China and the origin of birds. *Chinese Geol.* **233**, 30–33 (1996).
5. Chen, P. J. *et al.* Studies on the Late Mesozoic continental formations of western Liaoning. *Bull. Nanjing Inst. Geol. Palaeontol.* **1**, 22–25 (1980).
6. Chen, P. J. Nonmarine Jurassic strata of China. *Bull. Mus. N. Arizona* **60**, 395–412 (1996).
7. Dong, Z. M. Early Cretaceous dinosaur faunas in China: an introduction. *Can. J. Earth Sci.* **30**, 2096–2100 (1993).
8. Li, W. B. & Liu, Z. S. The Cretaceous palynofloras and their bearing on stratigraphic correlation in China. *Cretaceous Res.* **15**, 333–365 (1994).
9. Smith, P. E. *et al.* Dates and rates in ancient lakes:[40] Ar-[39] Ar evidence for an Early Cretaceous age for the Jehol Group, northeast China. *Can. J. Earth Sci.* **32**, 1426–1431 (1995).
10. Wang, D. F. & Diao, N. C. Geochronology of Jura-Cretaceous volcanics in west Liaoning, China. *Scientific papers on geology for international exchange* **5**, 1–12 (Geological Publishing House, Beijing, 1984).

11. Ostrom, J. H. The osteology of *Compsognathus longipes* Wagner. *Zitteliana* **4**, 73–118 (1978).
12. Bidar, A., Demay, L. & Thomel, G. *Compsognathus corallestris* nouvelle espèce de dinosaurien théropode du Portlandien de Canjuers (sud-est de la France). *Ann. Mus. d'Hist. Nat. Nice* **1**, 3–34 (1972).
13. Currie, P. J. & Zhao, X. J. A new large theropod (Dinosauria, Theropoda) from the Jurassic of Xinjiang, People's Republic of China. *Can. J. Earth Sci.* **30**, 2037–2081 (1993).
14. Claesseus, L. Dinosaur gastralia and their function in respiration. *J. Vert. Palaeontol.* **16**, 28A (1996).
15. von Huene, F. The carnivorous Saurischia in the Jura and Cretaceous formations principally in Europe. *Revista Museo Plata* **29**, 35–167 (1926).
16. Perle, A., Chiappe, L. M., Barsbold, R., Clark, J. M. & Norell, M. Skeletal morphology of *Mononykus olecranus* (Theropoda: Avialae) from the Late Cretaceous of Mongolia. *Am. Mus. Novit.* **3105**, 1–29 (1994).
17. Wellnhofer, P. Das siebte Exemplar von *Archaeopteryx* aus den Solnhofener Schichten. *Archaeopteryx* **11**, 1–48 (1993).
18. Barsbold, R. Carnivorous dinosaurs from the Cretaceous of Mongolia. *Sovmestnaya Sovetsko-Mongol'skaya Paleontol. Ekspiditsiya, Trudy* **19**, 5–119 (1983).
19. Griffiths, P. The question of *Compsognathus* eggs. *Rev. Paleobiol.* Spec. issue **7**, 85–94 (1993).
20. Carpenter, K., Hirsch, K. F. & Horner, J. R. *Dinosaur Eggs and Babies* (Cambridge Univ. Press, 1994).
21. Dong, Z. M. & Currie, P. J. On the discovery of an oviraptorid skeleton on a nest of eggs at Bayan Mandahu, Inner Mongolia, People's Republic of China. *Can. J. Earth Sci.* **33**, 631–636 (1996).
22. Varricchio, D. J., Jackson, F., Borkowski, J. J. & Horner, J. R. Nest and egg clutches of the dinosaur *Troodon formosus* and the evolution of avian reproductive traits. *Nature* **385**, 247–250 (1997).
23. Meng, J. & Wyss, A. R. Multituberculate and other mammal hair recovered from Palaeogene excreta. *Nature* **385**, 712–714 (1997).
24. Pérez-Moreno, B. P. *et al.* A unique multitoothed ornithomimosaur dinosaur from the Lower Cretaceous of Spain. *Nature* **370**, 363–367 (1994).
25. von Huene, F. Der Vermuthliche Hautpanzer des *Compsognathus longipes* Wagner. *Neues Jb. F. Min.* **1**, 157–160 (1901).
26. Grimaldi, D. & Case, G. R. A feather in amber from the Upper Cretaceous of New Jersey. *Am. Mus. Novit.* **3126**, 1–6 (1995).
27. Ji, Q. & Ji, S. A. *Protarchaeopteryx*, a new genus of Archaeopterygidae in China. *Chinese Geol.* **238**, 38–41 (1997).
28. Feduccia, A. *The Origin and Evolution of Birds* (Yale Univ. Press, New Haven, 1996).
29. Gauthier, J. in *The Origin of Birds and the Evolution of Flight* (ed. Padian, K.) 1–55 (California Acad. Sci., San Francisco, 1986).
30. Fastovsky, D. E. & Weishampel, D. B. *The Evolution and Extinction of the Dinosaurs* (Cambridge Univ. Press, 1996).
31. McGowan, C. Feather structure in flightless birds and its bearing on the question of the origin of feathers. *J. Zool. (Lond.)* **218**, 537–547 (1989).
32. Paul, G. S. *Predatory Dinosaurs of the World* (Simon and Schuster, New York, 1988).
33. Ewart, J. C. The nestling feathers of the mallard, with observations on the composition, origin, and history of feathers. *Proc. Zool. Soc. Lond.* 609–642 (1921).
34. Van Tyne, J. & Berger, A. J. *Fundamentals of Ornithology* (Wiley, New York, 1976).
35. Young, J. Z. *The Life of Vertebrates* (Oxford Univ. Press, 1950).
36. Chinsamy, A., Chiappe, L. M. & Dodson, P. Growth rings in Mesozoic birds. *Nature* **368**, 196–197 (1994).

37. Chiappe, L. M. The first 85 million years of avian evolution. *Nature* **378**, 349–355 (1995).
38. Brush, A. H. in *Avian Biology* vol. 9 (eds Farner, D. S., King, J. R. & Parkes, K. C.) 121–162 (Academic, London, 1993).
39. Regal, P. J. The evolutionary origin of feathers *Quart. Rev. Biol.* **50**, 35–66 (1975).
40. Ostrom, J. H. Reply to 'Dinosaurs as reptiles.' *Evolution* **28**, 491–493 (1974).
41. Davis, P. G. & Briggs, D. E. G. Fossilization of feathers. *Geology* **23**, 783–786 (1995).

Acknowledgements

This study was supported by NSFC. We thank L.-s. Chen and P. J. Currie (Royal Tyrrell Museum of Palaeontology) for helping to prepare the fossil materials and manuscript; M.-m. Zhang, X.-n. Mu, G. Sun, J. H. Ostram, A. Brush, L. Martin, P. Wellnhofer, N. J. Mateer, E. B. Koppelhus, D. B. Brinkman, D. A. Eberth, J. A. Ruben, L. Chiappe, S. Czerkas, R. O'Brien, D. Rimlinger, M. Vickaryous and D. Unwin for assistance and comments; and L. Mazzatenta and M. Skrepnick for help producing the photographs and drawings.

10

Feathers, Filaments and Theropod Dinosaurs
David M. Unwin

One of the hottest debates in palaeontology is whether birds evolved from dinosaurs. A study of two exceptionally well-preserved specimens of a theropod dinosaur from China—complete with skin, internal organs and eggs—provides new clues to the origin of feathers.

Ever since John Ostrom resuscitated the idea in the 1970s, palaeontologists have been piling up the evidence in favour of theropod dinosaurs as the ancestors of birds. Recently, the gap between theropods and birds has been narrowed even further. Important avian characters, such as the furcula (wishbone), have been discovered in theropods thought to be close to birds.[1] And typical theropod characteristics, such as an enlarged claw on digit two of the foot, have been found in early birds.[2] The final, clinching fact would be the discovery of evidence of feathers—a defining feature of birds—in theropods.

Cue *Sinosauropteryx prima*, the so-called 'feathered' dinosaur from China (Fig. 1). Reports of this sensational discovery first appeared[3] in late 1996. Although few scientists have yet seen the fossil material, some are already incorporating *Sinosauropteryx* into models for the origin of feathers

Figure 1 Reconstruction of *Sinosauropteryx* by Michael W. Skrepnick. Three exceptionally well-preserved specimens of this dinosaur have been recovered from the Early Cretaceous Yixian Formation of China. Two of these are described by Chen *et al.*[6], who show that *Sinosauropteryx* bore discrete structures that could have been ' proto-feathers'.

and bird flight.[4] Still others argue[5] that the 'feathers' are merely an artefact of preservation. On page 147 of this issue,[6] Chen, Dong and Zhen
present the first detailed description of *Sinosauropteryx*, and show that
the integument (skin) bore discrete filamentous structures. But are they
feathers?

Chen *et al*. describe two almost complete, near-adult individuals,
with evidence of soft tissues including the integument, the eyes and possibly some internal organs.[7] They are, without doubt, the best-preserved dinosaur remains yet found. These, along with a third specimen,
were recovered by Chinese farmers in Liaoning Province, China (P. J.
Currie, personal communication). They were found in beds of the
Yixian Formation, part of a thick sequence of lake sediments intercalated with volcanic deposits. During the past six years, these sediments
have yielded many superbly preserved fossils. The result is an almost
complete Early Cretaceous (145–97.5 million years ago) continental
biota, composed of plants, insects, fish, lizards, turtles, pterosaurs,
dinosaurs, mammals and numerous birds[8]—the latter often preserved
with their plumage intact.[9]

In life, *Sinosauropteryx* was about the size of a large chicken and
distinguished by its deep, narrow body, remarkably long tail and rather
short, stout forelimbs (Fig. 1). Like its close relative *Compsognathus*,
from the Late Jurassic of Europe (163–145 million years ago),
Sinosauropteryx has a very specialized hand. Its massive first digit
bears a large claw that might have served as a killing tool. Last meals
provide further evidence of an active, predatory lifestyle—a lizard in
the gut region of one individual[6] and a tiny mammal in the third specimen (P. J. Currie, personal communication). The body of the larger
individual described by Chen *et al*. also contains another surprise—two
small, oval structures which, judging by their shape, size and position,
are almost certainly eggs. This is the first reasonably convincing record
of this type of association for any dinosaur and, if correctly identified,
provides incontrovertible evidence that this individual was female.
Moreover, the relatively small size of the eggs and the possibility of
paired oviducts[6] suggest that, unlike modern birds (which have a single
oviduct and, in general, smaller clutches of relatively large eggs),
theropods had a more reptile-like reproductive system with two oviducts
that produced larger numbers of relatively small eggs.

The most striking features of the fossils are the well-preserved remains of the integument, which forms a dark halo above the skull, neck,
back, hips and both sides of the tail. Small patches of integument also
occur on the skull, and are associated with the forelimbs, rib-cage and

legs. The halo is composed of many coarse, sinuous filaments, which are possibly hollow and up to 40 mm long.[6] The filaments seem to be discrete structures and, although often matted and tangled, are not encased in skin or remnants of the decaying dermis, as some have suggested.[5] Branching of the filaments—a construction that is also typical of feathers—has been reported,[6,10] but the topography of this branching is not yet clear.

This brings us to the critical question: are these structures some kind of 'proto-feather'? Chen *et al.*[6] and others[4,10] clearly favour this idea. They draw comparisons with the plumules of modern birds, which are relatively simple structures without barbules or hooklets. But any argument for homology between the feathers of birds and the integumentary structures of *Sinosauropteryx* needs to be supported by more than general similarities in structure and position. High-resolution microscopy and biogeochemical tests might provide some answers, but they will not solve all of the problems. Moreover, if *Sinosauropteryx* bears proto-feathers, we might expect similar (or perhaps even more feather-like) structures to have been present on at least some of the theropods that are more closely related to birds than is *Sinosauropteryx* (Fig. 2). Exceptionally well-preserved remains of the integument are now known for two of these dinosaurs—the ornithomimosaur *Pelecanimimus*,[11] and a small, unnamed maniraptoran theropod from Brazil.[12] In both cases, however, there is no evidence of the filamentous structures found in *Sinosauropteryx*.

So, it seems that we still do not have absolute proof that some dinosaurs were feathered. Or do we? The Liaoning deposits have yielded three examples of another putative dino-bird intermediate, *Protarchaeopteryx*. According to a preliminary report by Ji and Ji,[13] these individuals have well-preserved evidence of true feathers. While we wait for further details of these Chinese fossils, we might consider the irony of the present situation—nothing for hundreds of years and then, suddenly, a whole flock of evidence.

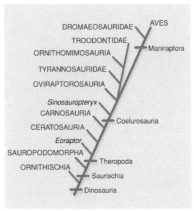

Figure 2 The relationships of *Sinosauropteryx* to other dinosaurs. *Sinosauropteryx* is not as closely related to birds as many other theropods, yet it is the first non-avian theropod that seems to show evidence of feather-like structures. (Modified from ref. 10.)

References

1. Norell, M. A., Mackovicky, P. & Clark, J. M. *Nature* **389**, 447 (1997).
2. Forster, C. A., Chiappe, L. M., Krause, D. W. & Sampson, S. D. *J. Vert. Paleontol.* **16**, (Suppl.) 34A (1996).
3. Hecht, J. *New Scientist* No. 2052, 7 (1996).
4. Padian, K. *J. Vert. Paleontol.* **17**, (Suppl.) 68A (1997).
5. Geist, N. R., Jones, T. D. & Ruben, J. A. *J. Vert. Paleontol.* **17**, (Suppl.) 48A (1997).
6. Chen, P., Dong, Z. & Zhen, S. *Nature* **391**, 147–152 (1998).
7. Ruben, J. A., Jones, T. D., Geist, N. R. & Hillenius, W. J. *Science* **278**, 1267–1270 (1997).
8. Hecht, J. *New Scientist* No. 2094, 32–35 (1997).
9. Hou, L., Martin, L. D., Zhou, Z. & Feduccia, A. *Science* **274**, 1164–1167 (1996).
10. Currie, P. J. in *Encyclopedia of Dinosaurs* (eds Currie, P. J. & Padian, K.) 241 (Academic, San Diego, 1997).
11. Briggs, D. E. G., Wilby, P. R., Pérez-Moreno, B. P., Sanz, J. L. & Fregenal-Martinez, M. *J. Geol. Soc. (Lond.)* **154**, 587–588 (1997).
12. Kellner, A. W. A. *Nature* **379**, 32 (1996).
13. Ji, Q. & Ji, S. A. *Chin. Geol.* **238**, 38–41 (1997).

A Therizinosauroid Dinosaur with Integumentary Structures from China

Xing Xu, Zhi-Lu Tang, and Xiao-Lin Wang

Therizinosauroidea ('segnosaurs') are little-known group of Asian dinosaurs with an unusual combination of features that, until recently, obscured their evolutionary relationships. Suggested affinities include Ornithischia,[1] Sauropodomorpha,[2,3] Theropoda[4–11] and Saurischia *sedis mutabilis*.[12] Here we describe a new therizinosauroid from the Yixian Formation (Early Cretaceous, Liaoning, China).[13] This new taxon provides fresh evidence that therizinosauroids are nested within the coelurosaurian theropods.[8–11] Our analysis suggests that several specialized therizinosauroid characters, such as the Sauropodomorpha-like tetradactyl pes,[1,2] evolved independently within this group. Most interestingly, this new dinosaur has integumentary filaments as in *Sinosauropteryx*.[14,15] This indicates that such feather-like structures may have a broad distribution among non-avian theropods, and supports the hypothesis that the filamentous integumentary structures may be homologous to the feathers of birds.[14,15]

> Dinosauria Owen 1842
> Theropoda Marsh 1881
> Coelurosauria *sensu* Gauthier 1986
> Therizinosauroidea Russell and Dong 1993
> *Beipiaosaurus inexpectus* gen. et sp. nov.

ETYMOLOGY.   Beipiao: the city near the locality where the specimen was found; saurus: lizard; inexpectus: referring to the surprising features in this animal.

HOLOTYPE.   IVPP V11559 (Institute of Vertebrate Paleontology & Paleoanthropology, Beijing, China; see Fig. 1).

LOCALITY AND HORIZON.   Sihetun locality near Beipiao, Liaoning, China. The lower part of the Yixian Formation, probably from the Lower Cretaceous based on latest radiometric evidence.[13]

DIAGNOSIS.   *Beipiaosaurus inexpectus* differs from other therizinosauroids in having shorter and more bulbous tooth crowns, a larger skull, a

Figure 1 *Beipiaosaurus inexpectus* (V11559, holotype). Photograph (**a**) and outline (**b**) of the skeleton (broken lines indicate features preserved in impressions). The holotype was collected in 1996 by a farmer, Li Yinxian, from the famous Sihetun locality. It was later (1997) determined to be from the lower part of the Yixian Formation. According to communication with the collector, and consistent with the close proximity, preservation and proportions of the elements, all elements (including the integumentary structures) are from a single individual. V11559 includes the partial right dentary with dentition, right postorbital, right parietal, right nasal?, right prootic, a few cervicals and dorsals, an incomplete caudal, incomplete ribs, partial scapula, coracoids and furcula, partial humerus, radius and ulna, nearly complete hands, partial ilium, pubis and ischium, complete right femur, right tibia and right fibula, incomplete left

tridactyl pes with a splint-like proximal first metatarsal, a shallow anterior iliac process, a long manus (10% longer than a femur), a long tibia
(275 mm > 265 mm of the femur), an elongated lateral articular surface
on the palmar side of manual phalanx I-1, and proximally compressed
metatarsals III and IV.

Beipiaosaurus is the largest known theropod from the Yixian Formation, with an estimated length of 2.2 m. It has a relatively large skull compared to other therizinosauroids (preserved dentary is 65% of femur
length). The anterior end of its dentary is down-turned. The dentary has a
lateral shelf, similar to other therizinosauroids and ornithischians.[1] Beipiaosaurus has a large number of teeth (more than 37, inferred from the preserved alveoli in the broken dentary). They resemble those of Protarchaeopteryx,[16] but have larger serrations (3 serrations per mm) as in other
therizinosauroids and troodontids.[9] Replacement teeth developed in oval
resorption pits next to the roots of erupted teeth (Fig. 2a), as in
Archaeopteryx.[17] Dorsally pointed, triangular interdental plates are present.

The cervical vertebrae bear low, anteroposteriorly short neural spines.
Lateral depressions are present on the lateral sides of the centra of the
fused posterior dorsals.

The coracoid is subrectangular, as in some maniraptoran theropods,
with a pronounced coracoid tubercle. Exquisite impressions show that
the furcula is a widely arched bone, oblate-shaped in cross section,
without a hypocleidium. Compared to the short and stout hindlimb, the
forelimbs are relatively long. The elongate hand is longer than the foot,
as in dromaeosaurids and primitive Avialae.[18] As in other therizinosauroids, the humerus has a pointed internal tuberosity on its
proximal end, and anteriorly positioned radial and ulnar condyles on
its distal end. A depression on the proximal surface of the humerus separates the head and internal tuberosity, as in other therizinosauroids
and Mononykus.[19] Five carpals are preserved. The largest distal carpal,
the semilunate (Fig. 2c, d), is smaller than but otherwise identical to
that of Deinonychus.[20] It primarily contacts metacarpal II but also
touches metacarpal I (Fig. 2d), unlike the condition in Alxasaurus, in
which the largest carpal is the distal carpal I.[8] Distal carpal I is large
and oval (Fig. 2c). The proximal carpals are represented by a V-shaped

femur, tibia and fibula, incomplete right foot. Some elements are represented by impressions. Sacral and
most caudal vertebrae are missing. a, astragalus; c, cervical vertebra; ca, caudal vertebra; co, coracoids;
d, dentary; dcl, distal carpal I; do, dorsal vertebra; f, femur; fi, fibula; fu, furcula; I –III, metacarpals I–III,
I-1 to III-4, manual phalanges I-1 to phalanges III-4; il, ilium; is, ischium; lh, left humerus; lr, left radius; lu,
left ulna; ?n, ?nasal; p, parietal; pe, pes; po, postorbital; pr, prootic; pu, pubis; r, rib; ?ra, ?radiale; rh,
right humerus; rr, right radius; ru, right ulna; s, scapula; sl, semilunate distal carpal; t, tibia.

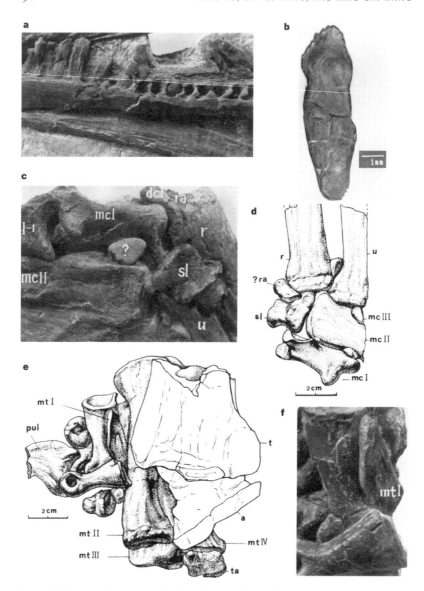

Figure 2 *Beipiaosaurus inexpectus*. **a**, Nine right dentary teeth in medial view. Note the resorption pits and replacement teeth. **b**, A dentary tooth in lateral view. **c**, Close-up of the left semilunate carpal of V11559. **d**, Drawing of part of the right manus of V11559. Note the shape and position of the semilunate, which is very similar to that of birds[17]. **e**, Drawing of the partially preserved right pes of V11559. **f**, Close-up of the first metatarsal of V11559. Note the proximally pinched theropod first metatarsal. The theropod first metatarsal is absent in other therizinosauroids, which has been argued as being strong evidence against the theropod affinities of therizinosauroids[1]. Additional abbreviations: mc I–III, metacarpals I–III; mt I–IV, metatarsals I–IV; pul, pedal ungual; r, radius; ra, radiale; ta, tarsal; u, ulna.

radiale in close contact with the radius, and a small rounded carpal between the distal ends of the radius and ulna (Fig. 2c, d). The manus is slender and elongate, proportionally similar to that of *Deinonychus*.[20] Metacarpal I has a pronounced distal flange, as in *Deinonychus*. The proximal parts of metacarpals I and II are closely appressed. Metacarpal III is slender and slightly bowed. The combined lengths of phalanges III-1 and III-2 are equal to the length of phalanx III-3, as in advanced theropods.[2] There are well developed ligament pits on the lateral sides of the distal ends of the phalanges. The manual unguals are laterally compressed and strongly curved. As in other therizinosauroids,[8] their proximal ends are deep but taper to needle-sharp points. The second manual claw is slightly longer than the first, resembling those of *Archaeopteryx* and *Protarchaeopteryx*.[21]

The ilium is shaped like a parallelogram, similar to those of dromaeosaurids and basal birds, but unlike the sauropod-like ilia of derived therizinosauroids.[1,22] The posterodorsal margin of the ilium curves ventrally in lateral view. The anterior and posterior processes are almost the same length. The posteroventral margin of the ilium is deflected laterally at a right angle to the vertical ramus, and has a shallow brevis fossa similar to those of other coelurosaurians.[23] The partial pubic peduncle of the ilium is longer than the ischiadic peduncle, similar to those of therizinosauroids, dromaeosaurids and *Archaeopteryx*.[23] Both the pubic and the ischial shafts are more rounded than flattened, unlike those of *Alxasaurus* and *Segnosaurus*. As in some theropods, the pubic apron is compressed and positioned more distally. The femur of *Beipiaosaurus* has a wing-like lesser trochanter, a cleft between the greater trochanter and the lesser trochanter, and a crest-like fourth trochanter. The tibia has a fibular crest, a feature of theropods.[2] The fibula is very slender compared to the tibia, especially the distal half. As in *Alxasaurus*[8] and the Avialae,[24] the medial surface of the fibula is flat, lacking the medial fossa of some theropods. As in other therizinosauroids, the astragalus has a tall ascending process and reduced condyles that only partly cover the distal end of the tibia. The calcaneum is suboval and disk-shaped. The metatarsus is 39% of the length of the tibia, larger than in known therizinosauroids but less than in other theropods (>45%).[8] The proximal end of metatarsal I is flattened and tapered and, as in most maniraptorans, does not contact the tarsus (Fig. 2e, f). The proximal ends of both metatarsals III and IV are compressed, especially on the medial side. Metatarsal V is slender and strap-like, being only half the length of the other metatarsals. One pedal ungual is preserved, and is shorter than any manual unguals.

Figure 3 *Beipiaosaurus inexpectus.* **a**, Partially preserved forelimb with unusual integumentary impression. **b**, Close-up of the integumentary impression.

Large patches of integumentary structures were found in close association with the ulna, radius, femur and tibia, as well as with pectoral elements. The filamentous structures are best preserved near the ulna, almost perpendicular to the bone (Fig. 3). They are similar to the integumentary structures of *Sinosauropteryx*[15] in their parallel arrangement. Unlike those of *Sinosauropteryx*, the integumentary structures of *Beipiaosaurus* contact the ulna. They are densest close to the bone. Most of the integumentary filaments are about 50 mm long, although the longest is up to 70 mm. Some filaments have shallow and faint median grooves, possibly indicating hollow cores that had collapsed, and have indications of branching distal ends as in *Sinosauropteryx.*[15] As in *Sinosauropteryx*[15] and birds from the same locality, it is difficult to isolate a single filament and thus difficult to describe the branching pattern of the integumentary filaments.

Therizinosauroidea has many perplexing features for a theropod, such as a very small head, a sauropod-like ilium and a short and broad tetradactyl pes with rudimentary metatarsal V.[1,2,12,22] Until now, no cladogram has been proposed for the relationships and morphological evolution of therizinosauroids. We ran a phylogenetic analysis with an 84-character dataset (see Supplementary Information for the character list and matrix). We left out the unnamed 'segnosaur' from the Early Jurassic Lower Lufeng Formation[25] as it is too incomplete. Using PAUP (3.1.1. Exhaustive search, Deltran optimization; Swofford, 1993), we obtained a single most parsimonious tree (tree length, 133; consistency index, 0.707; retention index, 0.645; rescaled consistency index, 0.456). Our analysis (Fig. 4) places *Beipiaosaurus* as a basal taxon within Therizinosauroidea. *Beipiaosaurus* has a relatively large skull (1.0) among therizinosaurs, a tridactyl pes (79.0) and a fibular crest on the tibia, all of which are primitive theropod features. The pelvic elements are also very similar to those of other coelurosaurians. These characteristics support the hypothesis that therizinosauroids (including *Beipiaosaurus*) are nested within the coelurosaurian theropods.[8-11] Given this phylogeny (Fig. 4), some derived characters of therizinosauroids other than *Beipiaosaurus* are most parsimoniously interpreted as having evolved convergently with some other dinosaur groups, sauropodomorphs in particular. Thus, therizinosauroids re-evolved a robust first digit in which the proximal end of metatarsal I articulates with the tarsals (79.1).

Feathers are complex structures. Their abrupt appearance in the bird fossil record has been difficult to explain, mainly because no intermediate structures are preserved in the related theropod taxa. The integumentary filaments of *Sinosauropteryx* have been considered to be 'proto-feathers' by some, but this idea has been rejected by others.[26] Such structures have not been preserved with any other theropods[26] until the discovery of *Beipiaosaurus*. The filamentous structures in *Beipiaosaurus* are similar to, but longer than, those of the compsognathid *Sinosauropteryx*. They are perpendicular to the limb bones, and are unlikely to be muscle fibres or frayed collagen.[27] Their presence in both therizinosauroids and compsognathids indicates that there may be a broader distribution of similar structures in theropod dinosaurs. This supports the idea that these simple integumentary filaments may represent an intermediate evolutionary stage to the more complex feathers of *Protarchaeopteryx, Caudipteryx*[16] and more derived Avialae. The absence of such structures in most theropod fossils is probably attributable to the lack of such ideal preservation as is found in the Yixian Formation. This again indicates that feathers preceded flight,[16] because both therizinosaurids and compsognathids apparently could not fly and did not descend from flying animals.

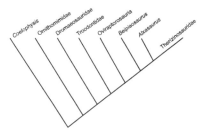

Figure 4 Phylogenetic relationships of *Beipiaosaurus inexpectus. Beipiaosaurus* and other therizinosauroids share 18 synapomorphies, including the following unique characters: a prominent dorsolateral shelf on the dentary (21.1), teeth that increase in size anteriorly (25.1), tooth crowns with sub-circular basal cross-sections that lack mediolateral compression (27.1), anteroposteriorly narrow and dorsoventrally deep pubic peduncle of ilium (46.1 and 47.1), very deep proximal end of manual unguals (70.1), short metatarsus (78.1) and reduced main body of astragalus (82.1). It is less derived than other therizinosauroids because it lacks 13 characters of Therizinosauroidea (1.1, 36.1, 38.0, 43.0, 48.1, 49.1, 51.1, 52.1, 58.1, 60.0, 66.0, 77.1, 79.1), including the following unusual characters: a very small head (1.1), the long and deep preacacetabular portion of ilium (48.1 nd 49.1) and absence of the theropod first metatarsal (79.1).

References

1. Paul, G. S. The segnosaurian dinosaurs: relics of the prosauropod-ornithischian transition. *J. Vert. Paleontol.* **4**, 507–515 (1984).

2. Gauthier, J. A. Saurischian monophyly and the origin of birds. *Mem. Calif. Acad. Sci.* **8**, 1–55 (1986).

3. Sereno, P. C. Prosauropod monophyly and basal sauropodomorph phylogeny. *J. Vert. paleontol.* (suppl.) **9**, 38A (1989).

4. Perle, A. Segnosauridae—a new family of theropods from the Late Cretaceous of Mongolia. *Trans. Joint Soviet–Mongolian Palaeontological Expedition* **8**, 45–55 (1979).

5. Perle, A. A new segnosaurid from the Upper Cretaceous of Mongolia. *Trans. Joint Soviet–Mongolian Palaeontological Expedition* **15**, 28–39 (1981).

6. Barsbold, R. & Perle, A. Segnosauria, a new infraorder of carnivorous dinosaurs. *Acta Palaeontol. Pol.* **25**(2), 187–195 (1980).

7. Barsbold, R. Carnivorous dinosaurs from the Cretaceous of Mongolia. *Trans. Joint Soviet–Mongolian Palaeontological Expedition* **19**, 1–116 (1983).

8. Russell, D. A. & Dong, Z. The affinities of a new theropod from the Alxa Desert, Inner Mongolia, China. *Can. J. Earth Sci.* **30**, 2107–2127 (1993).

9. Clark, J. M., Perle, A. & Norell, M. A. The skull of *Erlicosaurus andrewsi*, a Late Cretaceous "Segnosaur" (Theropod: Therizinosauridae) from Mongolia. *Am. Mus. Novit.* **3115**, 1–39 (1994).

10. Sues, H.-D. On *Chirostenotes*, a Late Cretaceous Oviraptorosaur (Dinosauria: Theropod) from Western North America. *J. Vert. Paleontol.* **17**, 498–716 (1997).

11. Makovicky, P. & Sues, H.-D. Anatomy and phylogenetic relationships of the theropod dinosaur *Microvenator celer* from the Lower Cretaceous of Montana. *Am. Mus. Novit.* **3240**, 1–27 (1998).

12. Barsbold, R. & Maryanska, T. in *The Dinosauria* (eds Weishampel, D. B., Dodson, P. & Osmolska, H.) 408–415 (Univ. California Press, Berkeley, 1990).

13. Swisher, C. C., Wang, Y.-q., Wang, X.-l, Xu, X. & Wang, Y. ^{40}Ar/^{39}Ar dating of the lower Yixian Fm., Liaoning Province, northeastern China. *Chinese Sci. Bull.* (suppl.) **43**, 125 (1998).

14. Ji, Q. & Ji, S. A. On discovery of the earliest bird fossil in China and the origin of birds. *Chinese Geol.* **233**, 30–33 (1996).

15. Chen, P.-j., Dong, Z.-m. & Zhen, S.-A. An exceptionally well preserved theropod dinosaur from the Yixian Formation of China. *Nature* **391**, 147–152 (1998).

16. Ji, Q., Currie, P. J., Norell, M. A. & Ji, S.-A. Two feathered dinosaurs from northeastern China. *Nature* **393**, 753–761 (1998).

17. Martin, L. D. in *Origins of Higher Groups of Tetrapods* (eds Schultz, H.-P. & Treube, L.) 485–540 (Cornell Univ. Press, Ithaca, N. Y., 1991).

18. Bellairs, A. D'A. & Jenkin, C. R. in *Biology and Comparative Physiology of Birds* Vol. 9 (ed. Marshall, A.) 241–300 (Academic, New York, 1960).

19. Perle, A., Chiappe, L. M., Barsbold, R., Clark, J. M. & Norell, M. A. Skeletal morphology of *Mononykus olecranus* (Theropod, Avialae) from the Late Cretaceous of Mongolia. *Am. Mus. Novit* **3105**, 1–29 (1994).

20. Ostrom, J. H. Osteology of *Deinonychus antirrhopus*, an unusual theropod from the Lower Cretaceous of Montana. *Bull. Peabody Mus. Nat. Hist., Yale Univ.* **30**, 1–165 (1969).

21. Ji, Q. & Ji, S. A. Protarchaeopterygid bird (*Protarchaeopteryx* gen. nov.)-fossil remains of archae-opterygids from China. *Chinese Geol.* **238**, 38–41 (1997).

22. Russell, D. A. in *Encyclopedia of Dinosaurs* (eds Currie, P. J. & Padian, K.) 729–730 (Academic, San Diego, 1997).

23. Norell, M. & Makovicky, P. J. Important features of the dromaeosaur skeleton: information from a new specimen. *Am. Mus. Novit.* **3215**, 1–28 (1997).

24. Chiappe, L., Norell, M. A. & Clark, J. Phylogenetic position of *Mononykus* (Aves: Alvarezauridae) from the Late Cretaceous of the Gobi Desert. *Mem. Queensland Mus.* 39, 557–582 (1996).

25. Zhao, X. & Xu, X. The oldest coelurosaurian. *Nature* 394, 234–235 (1998).

26. Unwin, D. M. Feathers, filaments and theropod dinosaurs. *Nature* 391, 119–120 (1998).

27. Gibbons, A. Plucking the feathered dinosaur. *Science* 278, 1229 (1997).

For supplementary information see page 186.

Acknowledgements

We thank J. Clark for advice and reviewing the manuscript; Z.-X. Luo for improving the organization and language of the manuscript as well as the use of PAUP 3.11; Z.-H. Zhou and O. Rauhut for discussions; P. Currie, M. Norell, P. Sereno, X.-C. Wu and H. Osmolska for reviews and comments; and the Liaoxi expedition members of the IVPP. Photographs were taken by J. Zhang, electronic photography by L. Oyang, and line drawings are by R.-S. Li, Y.-T. Li, H.-J. Wang and J.-Z. Ding prepared the specimen. This study was supported by research grants from the Chinese Academy of Sciences and the National Natural Science Foundation of China.

12

A Dromaeosaurid Dinosaur with a Filamentous Integument from the Yixian Formation of China

Xing Xu, Xiao-Lin Wang, and Xiao-Chun Wu

Dromaeosaurids, despite their notoriety, are poorly characterized meat-eating dinosaurs, and were previously known only from disarticulated or fragmentary specimens.[1] Many studies have denied their close relationship to birds.[2,3] Here we report the best represented and probably the earliest dromaeosaurid yet discovered, *Sinornithosaurus millenii* gen. et sp. nov., from Sihetun, the famous Mesozoic fish–dinosaur–bird locality in China.[4,5] *Sinornithosaurus* not only greatly increases our knowledge of Dromaeosauridae but also provides evidence for a filamentous integument in this group. It is remarkably similar to early birds postcranially. The shoulder girdle shows that terrestrial dromaeosaurids had attained the prerequisites for powered, flapping flight,[6] supporting the idea that bird flight originated from the ground up.[7,8] The discovery of *Sinornithosaurus* widens the distribution of integumentary filaments among non-avian theropods.[5,9,10] Phylogenetic analysis indicates that, among known theropods with integumentary filaments or feathers,[2,5] Dromaeosauridae is the most bird-like, and is more closely related to birds than is Troodontidae.

Theropoda Marsh 1881
Maniraptora Gauthier 1986
Dromaeosauridae Matthew & Brown 1922
Sinornithosaurus millenii gen. et sp. nov.

ETYMOLOGY. '*Sinornithosaurus*' derived from Sino-Ornitho-Saurus, meaning a bird-like dinosaur from China; '*millenii*' derived from Millennium, Latin for one-thousand years, referring to its discovery near the end of the twentieth century.

HOLOTYPE. IVPP (Institute of Vertebrate Paleontology and Paleoanthropology) V12811 (Figs 1, 2).

LOCALITY AND HORIZON. Sihetun, western Liaoning, China. Layer 6, lower (Chaomidianzi[2]) Yixian Formation,[4] Jehol Group; probably Early Cretaceous.[11]

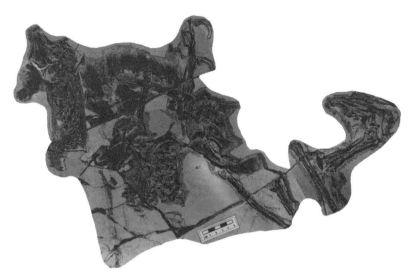

Figure 1 *S. millenii*, Holotype, IVPP V12811. The posterior part of the skull is disarticulated from the snout and lower jaws and turned 180° in the opposite direction. The postcranial skeleton is also not articulated, but the bones retain a close association. Integumentary filaments have been displaced, lacking their direct relationships to bony elements.

DIAGNOSIS.   Differing from other dromaeosaurids in the presence of ornament-like pits and ridges on the anterolateral surface of the antorbital fossa; the posterolateral process of the parietal turning sharply posteriorly; the dentary bifurcated posteriorly; unserrated premaxillary teeth; supracoracoid fenestra of coracoid; manual phalanx III-1 more than twice the length of phalanx III-2; pronounced tubercle near the midshaft of the pubis; posterodorsal process of the ischium; partially arctometatarsalian metatarsal III.

DESCRIPTION.   *Sinornithosaurus* is a small dromaeosaurid, with a skull about 13 cm long (Figs 2, 3, Table 1). It has a relatively small antorbital fenestra and a large orbit, as in other dromaeosaurids. The antorbital fossa, anterior to the antorbital fenestra, is large and well-demarcated, although its posterodorsal margin is obscured by surface damage. The anterolateral surface of the antorbital fossa bears a number of pits and ridges. The latter are well-marked and show no trace of pathological origin. Two small maxillary fenestrae, normally seen in dromaeosaurids and some other theropods, lie in the antorbital fossa. The exposure of the frontal is extensive and is more than twice the length of the parietal. The prefrontal is sutured with the lacrimal as in *Deinonychus*.[12] The frontal process of the triradiate postorbital is not upturned, unlike those in other dromaeosaurids.[13] The parietals form a narrow sagittal crest, as

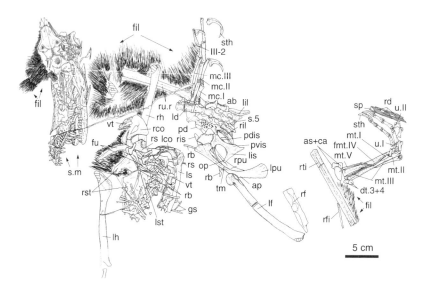

Figure 2 *S. millenii.* Drawing of the specimen shown in Fig. 1; the specimen represents an adult individual, as indicated by the five co-ossified sacral vertebrae and the partially fused astragalus and calcaneum. Abbreviations: ab, acetabulum of ilium; ap, apron of pubis; as + ca, astragalus and calcaneum; dt.3 + 4. distal tarsals 3 + 4; fil, integumentary filaments; fmt.IV, enlarged ventral flange of metatarsal IV; fu, furcula; gs, gastralia; lco, left coracoid; ld, last dorsal vertebra; lf, left femur; h, left humerus in anteromedial view; lil, left ilium; lpu, left pubis; lis, left ischium; ls, left scapula; lst, left sternum; mc.I–III, metacarpals I –III; mt.I–V, metatarsals I –V; op, obturator process; pd, pubic peduncle of ilium; pdis, posterodorsal process of ischium; pvis, posteroventral process of ischium; rb, rib; rco, right coracoid; rd, elongated prezygapophyses and chevrons of mid- and posterior caudal vertebrae; rf, right femur; rfi, right fibula; rh, right humerus; ril, right ilium; ris, right ischium; rpu, right pubis; rs, right scapula; rst, right sternum; rti, right tibia; ru.r, right ulna and radius; s.m, skull and mandible; sp, spine of caudal vertebra; sth, horny sheath; s.5, sacral vertebra 5 (the last); tm, tubercle for muscle attachment; u.I, ungual of digit I of pes; u.II, enlarged, slashing ungual of digit II of pes; vt, vertebra; III-2, phalanx 2 of digit III of manus.

in many non-avian theropods and birds. *Sinornithosaurus* is placed in Dromaeosauridae on the basis of the following shared derived features:[13–15] T-shaped lacrimal; large supratemporal fossa with a strongly sinusoidally curved anterior frontal margin; T-shaped quadratojugal (although the vertical bar is relatively shorter); widely open fenestra between the quadratojugal and quadrate; dentary with subparallel dorsal and ventral margins; ossified caudal rods increasing the lengths of prezygapophyses and chevrons (for more details, see Supplementary Information).

The shoulder girdle and forelimb of *Sinornithosaurus* closely resemble those of early birds (Fig. 4a–d). As in *Archaeopteryx*,[16,17] the articulated left scapula and left coracoid form an angle of less than 90°, and the scapula is shorter than the humerus (0.63) and ulna (0.77), and forms most of the laterally facing glenoid. The anterolateral part of the right

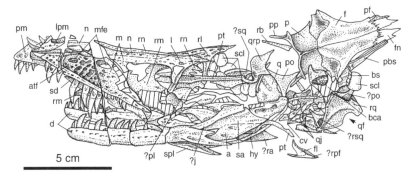

Figure 3 *S. millenii*, skull with lower jaws. Abbreviations: a, angular; atf, antorbital fossa; bca, braincase; bs, basisphenoid; cv, elements of the disarticulated first and second cervical vertebrae; d, dentary; f, frontal; fl, facet for lacrimal; fn, facet for nasal; hy, hyoid; l, lacrimal; lpm, left premaxilla in medial view; m, maxilla; mfe, maxillary fenestrae; n, nasal; p, parietal; pbs, parabasisphenoid process; pf, left prefrontal (the anterolateral part overlapped by the lacrimal was not preserved); pm, right premaxilla in medial view; po, left postorbital; pp, posterolateral process of the parietal; pt, pterygoid; q, quadrate; qf, fenestra between quadratojugal and quadrate; qj, quadratojugal; qrp, quadrate ramus of pterygoid; rb, rib; rl, right lacrimal in medial view; rm, right maxilla in medial view; rn, right nasal; rq, right quadrate in anteromedial view; sa, surangular; scl, sclerotic bones; sd, supradentary bone; spl, splenial; ?j, ?left jugal; ?pl, ?palatine; ?po, ?right postorbital; ?rpf, ?right prefrontal; ?rsq, ?right squamosal in ventral view; ?sq, ?left squamosal; ?ra, ?right angular. Most elements of the palate and braincase are obscured by cracks and dislocation.

coracoid of *Sinornithosaurus* is distorted by lateromedial folding (Fig. 4c), and was broader in life and differed little from those of other dromaeosaurids, such as *Deinonychus*[18] and an undescribed specimen of *Saurornitholestes* (TMP (Royal Tyrrell Museum of Palaeontology) 88.121.39), except for the presence of a supracoracoid fenestra. The coracoid bears a pronounced biceps tubercle and its lateral profile is almost identical to that of *Archaeopteryx*.[19] The furcula, half of which is extensively covered by other elements, gently curves in a 'boomerang'-shaped configuration (Fig. 2), differing from that of *Velociraptor*[20] but resembling that of *Archaeopteryx*.[16] Its shaft is compressed and tapers distally. A pair of ossified sterna appear similar to those of other dromaeosaurids,[21] including *Saurornitholestes* (TMP 92.36.333). The length of each sternum is slightly more than twice its width, and its constricted anterolateral side bears a series of possibly five rib attachments (Fig. 2). These costal facets on the sternum imply the presence of hinged sternocostal joints in dromaeosaurids, which is not concordant with recent arguments about the ribcage–pectoral girdle complex and the respiratory pattern of theropods.[22] As in other dromaeosaurids and birds,[1,21] the coracoid facet of the sternum faces much more anteriorly than laterally. The forelimb has the greatest relative length of those known among non-avian theropods, and it is estimated to be about 80% of the length of the hindlimb. This is also indicated by the ratios of the ulna to the scapula (1.29) and the ulna to metatarsal III (1.18). These ratios are less than 1.0 in most other non-avian theropods. Metacarpal III is bowed laterally, as in other dromaeosaurids and birds.[3]

Sinornithosaurus is also more bird-like than other non-avian theropods in the following derived features of the pelvic girdle and hindlimb (Fig. 4e, f): pubic peduncle of ilium broader than acetabulum;[3] open acetabulum tending to close off medially;[8,23] posteroventrally directed pubis bearing a short symphysis (less than half the length of the bone), and its distal end cup-like;[1] short ischium plate-like and less than half the length of the pubis, indicating, as in *Velociraptor*,[1] *Unenlagia*[8] and birds,[3] the absence of ischial symphysis; thin fibular shaft about 1/7 of the diameter of the tibia.[3]

It has been stated that the forelimbs of dromaeosaurids could not move like those of *Archaeopteryx* because of the posteroventrally facing glenoids.[8,24] This assumption is based on *Deinonychus*,[18] the scapula and coracoid of which are actually incomplete. We believe that the glenoid was formed primarily by the scapula in *Deinonychus*, judging from the glenoid part of the coracoid, and may have faced laterally as in *Sinornithosaurus*, *Velociraptor*[1] and *Saurornitholestes* (TMP 88.121.39). That the coracoid and scapula form a sharp angle, as in *Sinornithosaurus*, may also be true for other dromaeosaurids. The shoulder girdle, as exemplified in *Sinornithosaurus*, is similar to that of *Archaeopteryx*.[19] A laterally facing glenoid is consistent with an avian mode of movement (elevation and relevant rotation and abduction) of the forelimb.[8,24] The modifications of the scapulocoracoid and coracosternal articulations in Dromaeosauridae altered the orientation of the shoulder girdle, perhaps enabling a wider, more avian range of motion at the glenohumeral joint. Because *Sinornithosaurus* and other dromaeosaurids were bipedal, cursorial terrestrial dinosaurs and are closely related to birds (see below), the anatomical modification of their shoulder girdles supports a cursorial origin for avian flight.[6,7]

Table 1 Lengths of selected elements of *Sinornithosaurus millenii* (IVPP V12811)

Skull	130*
Mandible	125
Left scapula	85
Right coracoid	44
Right sternum	84*
Posterior two dorsal vertebrae	26
Five sacral vertebrae	65
Right humerus	134
Right ulna	110
Metacarpal II	63
Three phalanges of manual digit II	89
Left ilium	85
Left pubis	116
Left ischium (along posterior edge)	52
Left femur	148*
Right tibia (preserved length)	125
Metatarsal III (longest)	93
Four phalanges of foot digit III	73

*Estimation. Length measurements are given in millimetres.

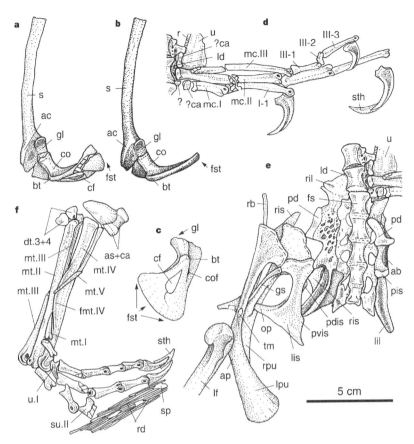

Figure 4 *S. millenii*, selected elements of the postcranial skeleton. **a, b,** Left scapula and coracoid in lateral view (original, **a**; reconstructed, **b**). **c,** Right coracoid in ventral view, with its anteromedial portion slightly folded. **d,** Right manus mainly in posteroventral view. **e,** Sacrum (in ventral view) and pelvic girdles. **f,** Tarsals, right pes (mainly in posteroventral view) and partial tail. Abbreviations: ac, acromion; bt, biceps tubercle; cf, supracoracoid fenestra; co, coracoid; cof, coracoid foramen; ?ca, ?carpal; fst, facet for sternum; fs, first sacral vertebrae; gl, glenoid of shoulder girdle; pis, ischial peduncle of ilium; r, radius; s, scapula; su.II, slashing ungual of digit II of pes; u, ulna; I-1, phalanx 1 of digit I of manus; III-1, III-2, III-3, phalanges 1–3 of digit III of manus; u, ulna; ?, possibly a carpal. For other abbreviations see Fig. 2.

The body of *Sinornithosaurus* was apparently covered by a layer of integumentary filaments (Figs 1, 2). These filaments are not in their original positions, owing to posthumous displacement, but are distributed as patches underneath or close to most bony elements, including the skull. They have been exposed by preparation only in a small part of the specimen and have sometimes been cut off near the edges of elements that needed to be exposed (see Fig. 1). The anatomical structure of these filaments is not discernible, but in appearance they differ little from the external filaments of other theropod dinosaurs[2,5,10] or even the plumulaceous feathers of *Confuciusornis* (IVPP V11307) from the same locality. The filaments generally reach 40 mm in length. Those near the postcranial elements seem longer than those around the skull. The filaments around the tibia are relatively sparse. It is uncertain whether *Sinornithosaurus* had, in life, rectrix-like structures, as seen in both *Caudipteryx* and *Protarchaeopteryx*, or remix-like structures, as seen attached to the arm of *Caudipteryx*, because evidence of the direct relationship of the preserved filaments to the relevant bones is obscured by preservation. The broad distribution of the filaments across the body of *Sinornithosaurus* further demonstrates that the clusters of fine filaments in *Sinosauropteryx*[5] do not represent internal collagenous fibres for skin support in semi-aquatic animals,[25] but integumentary derivations of terrestrial animals. *Sinornithosaurus* is the fifth kind of theropod known to possess integumentary filaments. It is possible that these filaments indicate an insulatory layer that served to maintain body heat.[2]

It has been proposed that birds are descendants of derived theropods,[6,16,26] but there has been no consensus as to which theropod group is most closely related to birds. Dromaeosauridae,[27,28] Troodontidae[3,29] and *Caudipteryx*[2] have all recently been suggested to be closely related to birds, but

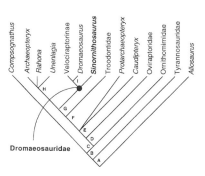

Figure 5 Cladogram showing phylogenetic relationships of Dromaeosauridae (including *Sinornithosaurus*) within the derived theropods. This is a strict consensus tree of our four most parsimonious trees based on 106 characters and 14 taxa (tree length = 183; consistency index = 0.639; retention index = 0.697). Multistate characters are unordered; tree produced using the branch-and-bound option of PAUP.[30] Synapomorphies for each node were determined by acceleration transformation (ACCTRAN). *Sinornithosaurus* is the basal member of Dromaeosauridae, which can be defined mainly by the following six unambiguous synapomorphies: characters 86 (lacrimal T-shaped); 88 (supratemporal fossa covering most of frontal process of postorbital and extending anteriorly on dorsal surface of frontal to at least level of posterior orbital margin); 90 (quadratojugal Y- or T-shaped); 91 (quadratojugal fenestra widely open); 96 (in lateral view upper and ventral margins of dentary subparallel); and 100 (ossified caudal rods extending lengths of prezygapophyses and chevrons). The close relationship of Dromaeosauridae to birds is primarily based on 10 unambiguous synapomorphies (node G): characters 39 (body of coracoid forming sharp angle with body of scapula); 46 (metacarpal III bowed laterally); 59 (pubic shaft projecting posteroventrally relative to long axis of sacral vertebrae); 65 (shape of ischial shaft mediolaterally compressed and plate-like along entire length); 104 (ulna longer than metatarsal III); 105 (glenoid of pectoral girdle facing laterally); 106 (articular facet of coracoid on sternum almost anterior); and the other three (characters 15, 92 and 93) of the 10 synapomorphies are unknown in *Sinornithosaurus*. For the list of the synapomorphies of the other clades see Supplementary Information.

previous studies have not considered all of these taxa simultaneously, making their results arguable.

We investigated the phylogenetic position of *Sinornithosaurus* using a data matrix of 106 characters and 14 taxa[3,13] (see Supplementary Information). We included all of the theropod taxa that have been proposed to be closely related to birds[2,3,8]. As with a previous study,[2] we did not include characters relative to integumentary filaments and feathers because we cannot determine whether their absence in most other derived theropods is real or merely an artefact of preservation. Our analysis indicates that *Sinornithosaurus* is a basal dromaeosaurid and strongly supports the view that Dromaeosauridae and birds (where Aves is Avialae[26]) are more closely related to each other than either is to Troodontidae (Fig. 5, node G). Our analysis also indicates that *Protarchaeopteryx* and *Caudipteryx* are more remote from birds than is Troodontidae. We re-analysed the data matrix by sequentially adding steps to test the robustness of the relationships between the taxa. The close relationship between Dromaeosauridae and birds is the most robust and did not collapse until five more steps had been added. The currently established phylogenetic relationships among derived theropods seem to support the presence of true feathers in Dromaeosauridae, but the validity of this interpretation cannot be confirmed until more direct evidence is available.

References

1. Norell, M. A. & Makovicky, P. J. A revised look at the osteology of dromaeosaurs: evidence from new specimens of *Velociraptor. J. Vert. Paleont.* **18** (**Suppl**), 66A (1998).

2. Ji, Q., Currie, P. J., Norell, M. A. & Ji, S.-A. Two feathered dinosaurs from northeastern China. *Nature* **393**, 753–761 (1998).

3. Forster, C. A., Sampson, S. D., Chiappe, L. M. & Krause, D. W. The theropod ancestry of birds: new evidence from the Late Cretaceous of Madagascar. *Science* **279**, 1915–1919 (1998).

4. Wang, X.-L. *et al.* Stratigraphic sequence and vertebrate bearing beds of the lower part of the Yixian Formation in Sihetun and neighbouring area, western Liaoning, China. *Vert. PalAsiatica* **36**, 81–101 (1998).

5. Chen, P.-J., Dong, Z.-M. & Zhen, S.-N. An exceptionally well-preserved theropod dinosaur from the Yixian Formation of China. *Nature* **391**, 147–152 (1998).

6. Gauthier, J. & Padian, K. in *The Beginnings of Birds* (eds Hecht, M. K., Ostrom, J. H., Viohl, G. & Wellnhofer, P.) 185–197 (Freunde des Jura-Museums Eichstätt, Eichstätt, 1985).

7. Ostrom, J. H. The cursorial origin of avian flight. *Mem. Calif. Acad. Sci.* **8**, 73–81 (1986).

8. Novas, F. E. & Puerta, P. F. New evidence concerning avian origins from the Late Cretaceous of Patagonia. *Nature* **387**, 390–392 (1997).

9. Currie, P. J., Norell, M. A., Ji, Q. & Ji, S-A. The anatomy of two feathered theropods from Liaoning, China. *J. Vert. Paleont.* **18** (**Suppl**), 36A (1998).

10. Xu, X., Tang, Z.-L. & Wang, X.-L. A therizinosaurid dinosaur with integumentary structures from China. *Nature* **399**, 350–354 (1999).

11. Swisher, C. C., Wang, Y.-Q., Wang, X.-L., Xu, X. & Wang, Y. 40Ar/39Ar dating of the lower Yixian Fm., Liaoning Province, northeastern China. *Chinese Sci. Bull.* **43** (**Suppl.**), 125 (1998).

12. Witmer, L. M. & Maxwell, W. D. The skull of *Deinonychus* (Dinosauria: Theropoda): new insight and implications. *J. Vert. Paleont.* **16** (**Suppl.**), 73A (1996).

13. Currie, P. J. New Information on the anatomy and relationships of *Dromaeosaurus albertensis* (Dinosauria: Theropoda). *J. Vert. Paleont.* **15**, 576–591 (1995).

14. Ostrom, J. H. Osteology of *Deinonychus antirrhopus*, an unusual theropod dinosaur from the Lower Cretaceous of Montana. *Peabody Mus. Nat. Hist. Bull.* **30**, 1–65 (1969).

15. Sues, H.-D. The skull of *Velociraptor mongoliensis*, a small Cretaceous theropod dinosaur from Mongolia. *Paläont. Zeit.* **51**, 173–184 (1977).

16. Ostrom, L. H. *Archaeopteryx* and the origin of birds. *Biol. J. Linn. Soc.* **8**, 91–182 (1976).

17. Wellnhofer, P. A. A new specimen of *Archaeopteryx* from the Solnhofen Limestone. *Nat. Hist. Mus. Los. Angeles County Sci. Ser.* **36**, 3–23 (1992).

18. Ostrom, J. H. The pectoral girdle and forelimb function of *Deinonychus* (Reptilia: Saurischia): a correction. *Postilla* **165**, 1–11 (1974).

19. Ostrom, J. H. in *The Beginnings of Birds* (eds Hecht, M. K., Ostrom, J. H. & Wellnhofer, P.) 161–176 (Freunde des Jura-Museums Eichstätt, Eichstätt, 1985).

20. Norell, M. A., Makovicky, P. & Clark, J. M. A *Velociraptor* wishbone. *Nature* **389**, 447 (1997).

21. Norell, M. A. & Makovicky, P. J. Important features of the dromaeosaur skeleton: information from a new specimen. *Amer. Mus. Nov.* **10024**, 1–28 (1997).

22. Ruben, J., Jones, T. D., Geist, N. R. & Hillenius, W. J. Lung structure and ventilation in theropod dinosaurs and early birds. *Science* **278**, 1267–1270 (1997).

23. Martin, L. in *Origin of the Higher Groups of Tetrapods* (eds Schultz, H.-D. & Trueb, L.) 485–540 (Comstock Publishing Assoc., Ithaca, 1991).

24. Jenkins, F. Jr. The evolution of the avian shoulder joint. *Amer. J. Sci.* **293**, 253–267 (1993).

25. Geist, N. R., Jones, T. D. & Ruben, J. A. Implication of soft tissue preservation in the compsognathid dinosaur, *Sinosauropteryx*. *J. Vert. Paleont.* **17** (**Suppl.**), 48A (1997).

26. Gauthier, J. A. Saurischian monophyly and the origin of birds. *Mem. Calif. Acad. Sci.* **8**, 1–46 (1986).

27. Holtz, T. R. Jr The phylogenetic position of the Tyrannosauridae: implications for theropod systematics. *J. Paleont.* **68**, 1100–1117 (1994).

28. Padian, K. & Chiappe, L. M. The origin and early evolution of birds. *Biol. Rev.* **73**, 1–42 (1998).

29. Currie, P. J. & Zhao, X.-J. A new troodontid (Dinosauria: Theropoda) braincase from the Dinosaur Park Formation (Campanian) of Alberta. *Can. J. Earth Sci.* **30**, 2231–2247 (1993).

30. Swofford, D. L. *PAUP, Phylogenetic Analysis Using Parsimony (and Other Methods), Version 4.0* (Sinauer, Sunderland, Massachusetts, 1998).

For Supplementary information see page 189.

Acknowledgements

We thank the participants of the Liaoxi Project of the IVPP; P. J. Currie and H. Osmólska for sharing unpublished information and providing references; and L.-H. Hou, Z.-H. Zhou, H. Osmólska, P. J. Currie, D. B. Brinkman, L. M. Witmer, A. P. Russell and P. Sereno for discussions. We also thank the Royal Tyrrell Museum of Palaeontology for allowing X.-C.W. access to facilities. S.-H. Xie prepared the specimen and G.-H. Cui took the photograph. This work was supported by research grants from the Chinese Academy of Sciences and Chinese Natural Science Foundation and an operating grant from the Natural Science and Engineering Research Council of Canada to A.P.R.

13

Two Feathered Dinosaurs
from Northeastern China

Qiang Ji, Philip J. Currie, Mark A. Norell, and Shu-An Ji

Current controversy over the origin and early evolution of birds centres on whether or not they are derived from coelurosaurian theropod dinosaurs. Here we describe two theropods from the Upper Jurassic/Lower Cretaceous Chaomidianzi Formation of Liaoning province, China. Although both theropods have feathers, it is likely that neither was able to fly. Phylogenetic analysis indicates that they are both more primitive than the earliest known avialan (bird), *Archaeopteryx*. These new fossils represent stages in the evolution of birds from feathered, ground-living, bipedal dinosaurs.

> Dinosauria Owen 1842
> Theropoda Marsh 1881
> Maniraptora Gauthier 1986
> Unnamed clade
> *Protarchaeopteryx robusta* Ji & Ji 1997

HOLOTYPE. National Geological Museum of China, NGMC 2125 (Figs 1, 2 and 3).

LOCALITY AND HORIZON. Sihetun area near Beipiao City, Liaoning, China. Jiulongsong Member of Chaomidianzi Formation, Jehol Group.[1] This underlies the Yixian Formation, the age of which has been determined to be Late Jurassic to Early Cretaceous.[2,3,4]

DIAGNOSIS. Large straight premaxillary teeth, and short, bulbous maxillary and dentary teeth, all of which are primitively serrated. Rectrices form a fan at the end of the tail.

DESCRIPTION. The skull of *Protarchaeopteryx* is shorter than the femur (Table 1). There are four serrated premaxillary teeth (Fig. 1c), with crown heights of up to 12 mm. Premaxillary teeth of coelophysids,[5] compsognathids[6,7] and early birds lack serrations, but premaxillary denticles are present in most other theropods. Six maxillary and seven dentary teeth are preserved (Fig. 1), all of which are less than a quarter the height of the premaxillary teeth. They most closely resemble those of *Archaeopteryx*[8] in shape (Figs 1b, c and 2b, c), but have anterior and posterior serrations (7–10 serrations per mm).

The amphicoelous posterior cervicals are the same length as the posterior dorsals, which have large pleurocoels. If the lengths of missing segments of the tail are accounted for, there were fewer than 28 caudals. Vertebrae increase in length from proximal to mid-caudals, as in most non-avian coelurosaurs.

There are two thin, flat, featureless sternal plates. The clavicles are fused into a broad, U-shaped furcula (interclavicular angle is about 60°)as in *Archaeopteryx*, *Confuciusornis* and many non-avian theropods. The forelimb is shorter than the hindlimb. The arm is shorter (compared to the femur) than it is in birds, but is longer than those of long-armed non-avian coelurosaurs such as dromaeosaurids and oviraptorids (Table 2). The

better preserved right wrist of NGMC 2125 has a single semilunate carpal capping the first two metacarpals. The hand has the normal theropod phalangeal formula of 2-3-4-x-x. The manus is longer than either the humerus or radius. Compared to femur length, the hand is more elongate than those of any theropods other than *Archaeopteryx*[9] and *Confuciusornis* (Table 2). More advanced birds such as *Cathayornis* have shorter hands.[10] Phalanges III-1 and III-2 in the hand of *Protarchaeopteryx* are almost the same size, and are about half the length of III-3. The unguals are long and sharp, and keratinous sheaths are preserved on two of them.

The preacetabular blade of the ilium is about the same length as the postacetabular blade. The pubic boot expands posteriorly. Anteriorly, the pubis is not exposed.

The tibia is longer than the femur, as it is in most advanced

Figure 1 *Protarchaeopteryx robusta.* **a**, NGMC 2125, holotype. Scale bar, 5 cm. **b**, Fourth to sixth left dentary teeth. Scale bar, 1 mm. **c**, Premaxillary teeth showing small serrations. Scale bar, 5 mm.

theropods and early birds. It is not known if the fibula extended to the tarsus.

The metatarsals are separate from each other and the distal tarsals. Metatarsal I is centred halfway up the posteromedial edge of the second metatarsal. In perching birds such as *Sinornis*,[9] metatarsal I is positioned near the end of metatarsal II and is retroverted. Its condition in *Archaeopteryx* is intermediate. Pedal unguals are smaller than manual unguals.

A clump of at least six plumulaceous feathers is preserved anterior to the chest, with some showing well-developed vanes (Fig. 3a). Evenly distributed plumulaceous feathers up to 27 mm long are associated with ten proximal caudal vertebrae. Twenty-millimetre plumulaceous feathers are preserved along the lateral side of the right femur and the proximal end of the left femur.

Parts of more than twelve rectrices are preserved[11] attached to the distal caudals. One of the symmetrical tail feathers (Fig. 3b) extends 132 mm from the closest tail vertebra, and has a long tapering rachis with a basal diameter of 1.5 mm. The well-formed pennaceous vanes of *Protarchaeopteryx* show that barbules were present. The vane is 5.3 mm wide on either side of the rachis. At midshaft, five barbs come off the rachis every 5 mm (compared with six in *Archaeopteryx*), and individual barbs are 15 mm long. As in modern rectrices, the barbs at the base of the feather are plumulaceous.

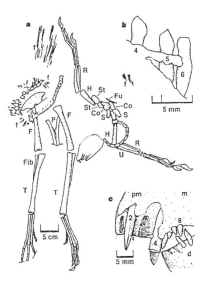

Figure 2 *Protarchaeopteryx robusta*. **a**, Outline of the specimen shown in Fig.1a. **b**, Outline of the left dentary teeth shown in Fig. 1b. **c**, Drawing of the front of the jaws, showing the large size of the premaxillary teeth compared with maxillary and dentary ones. Abbreviations: Co, coracoid; d, dentary; F, femur; f, feathers; Fib, fibula; Fu, furcula; H, humerus; m, maxilla; P, pubis; pm, premaxilla; R, radius; S, scapula; St, sternal plate; T, tibia; U, ulna. Numbers represent tooth positions from front to back.

Maniraptora Gauthier
1986
Unnamed clade

DIAGNOSIS. The derived presence of a short tail (less than 23 caudal vertebrae) and arms with remiges attached to the second digit.

Caudipteryx zoui gen.
et sp. nov.

ETYMOLOGY. '*Caudipteryx*' means 'tail feather,' '*zoui*:' refers to

Figure 3 *Protarchaeopteryx robusta*, NGMC 2125. **a**, Contour and plumulaceous feathers. Scale bar, 10 mm. **b**, Rectrices. Scale bar, 5 mm.

Zou Jiahua, vice-premier of China and an avid supporter of the scientific work in Liaoning.

HOLOTYPE.   NGMC 97-4-A (Figs 4 and 5b).

PARATYPE.   NGMC 97-9-A (Fig. 5d).

LOCALITY AND HORIZON.   Sihetun area, Liaoning. Jiulongsong Member of the Chaomidianzi Formation.

DIAGNOSIS.   Elongate, hooked premaxillary teeth with broad roots; maxilla and dentary edentulous. Tail short (one-quarter of the length of the body). Arm is long for a non-avian theropod; short manual claws. leg-to-arm ratio, 2.5.

DESCRIPTION.   The skulls of both specimens of *Caudipteryx* are shorter than the corresponding femora because of a reduction in the length of the antorbital region. The relatively large premaxilla (Figs 6 and 7) borders most of the large external naris. The maxilla and nasal are short, but the frontals and jugals are long. The lacrimal of NGMC 97-4-A is an inverted L-shaped, pneumatic bone. Scleral plates are preserved in the 20-mm-diameter orbits of both specimens. The tall quadratojugal seems to have contacted the squamosal and abutted the lateral surface of the quadrate. The single-headed quadrate is vertical in orientation. The ectopterygoid

Table 1 Lengths of elements in *Protarchaeopteryx* and *Caudipteryx*

Element	NGMC 2125	NGMC 97-4-A	NGMC 97-9-A
Body Length	690	890	725
Skull	70	76	79
Sternal plates	25	36	—
Humerus	88	69	70
Arm (humerus to end of phalange II-2)	297	214	220
Ilium	95	101	—
Ischium	—	77	—
Leg (femur to end of phalange III-4)	450	550	540
Femur	122	147	149
Tibia	160	188	182
Tibia	160	188	182
Metatarsal III	85	115	117

Length measurements are given in millimetres. NGMC 2125. *Protarchaeopteryx*; NGMC 97.4-A and NGMC 97-9-A, *Caudipteryx*.

Table 2 Relative proportions of elements in relevant avian and non-avian theropods

Element	Drom	Ov	Tro	Cx	Px	Ax	Con
Arm/F	1.8–2.6	1.5–1.8	1.8	1.5	2.4	3.7	3.9
S/H	0.8	1.0–1.2	—	1.1	—	0.6	0.8
R/H	0.7–0.8	0.8–0.9	0.6–0.7	0.9	0.8	0.9	0.8
Manus/H	0.9–1.2	1.2–1.4	1.3	1.2	1.6	1.2	1.3
Manus/F	1.0	0.7–1.0	0.8	0.6	1.2	1.5	1.6
Mcl/McII	0.4–0.5	0.4–0.6	0.3	0.4	0.4	0.3	0.4
T/F	1.1–1.4	1.2	1.1–1.2	1.2	1.3	1.4	1.1
Leg/F	3.6	3.3	3.8	3.7	3.7	3.8	3.3
Leg/arm	1.4	1.7	2.1	2.5	1.5	1.1	0.8

All data were collected from original specimens by PJ.C. Ax, *Archaeopteryx*; Con, *Confuciusornis*; Cx, *Caudipteryx*; Drom, dromaeosaurids; F, femur; H, humerus; L, length; Mc, metacarpal; Ov, oviraptorids; Px, *Protarchaeopteryx*; R, radius; S, scapula; T, tibia; Tro, troodontids.

has a normal theropod hooklike jugal process. There is a broad, beak-like margin at the symphysis of the dentaries. Posteriorly, the dentary bifurcates around a large external mandibular fenestra as in oviraptorids. A well-developed, sliding intramandibular joint is present between dentary and surangular.

There are four teeth in each premaxilla. They have elongate, needlelike crowns, and the roots are five times wider than the crowns (Fig. 7b). The lingual wall of the root of the third right tooth has been resorbed for the crown of a replacement tooth. The teeth seem to have been procumbent, with an inflection at the gumline. *Caudipteryx* had no maxillary or dentary teeth.

There are ten amphicoelous cervical vertebrae and five sacrals as in most non-avian theropods and *Archaeopteryx*.[8,12] The tail of NGMC 97-4-A is

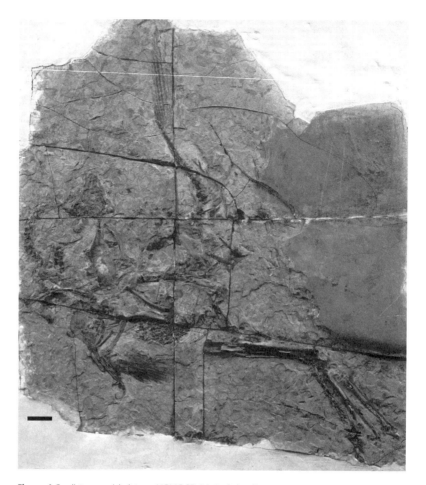

Figure 4 *Caudipteryx zoui*, holotype, NGMC 97-4-A. Scale bar, 5 cm.

articulated and well-preserved, and includes 22 vertebrae, as in
Archaeopteryx. It is shorter than the 30-segment tails of oviraptorids.
Most other non-avian theropods have much longer tails. Caudals do not
become longer posteriorly, as they do in most non-avian theropods and
Archaeopteryx. Almost two-thirds of the tail of NGMC 97-4-A is
preserved as a straight rod, but the vertebrae are not fused. The first six
haemal spines are elongate, rodlike structures. More posterior haemal
spines decrease in height, but expand anteriorly and posteriorly (Fig. 5a).

Each segment of gastralia is formed by two pairs of slender, tapering rods,
as in all non-avian theropods, *Protarchaeopteryx* and early birds.[8,9,13,14]

The paired sternals are similar to those of dromaeosaurids and oviraptorids. *Confuciusornis* had a relatively larger, unkeeled sternum. Some

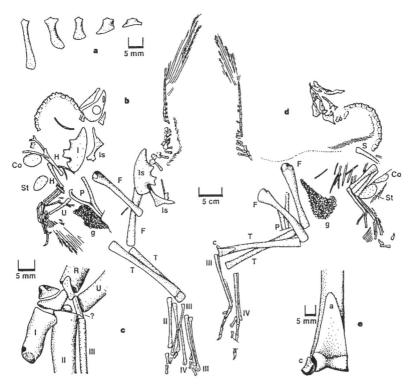

Figure 5 a, Haemal spines from the fourth, sixth, eighth, eleventh and thirteenth caudal vertebrae (from left to right) of NGMC 97-4-A in left lateral view. **b**, Drawing of the specimen shown in Fig. 4a. **c**, Wrist of NGMC 97-4-A. **d**, Drawing of NGMC 97-9-A. **e**, Proximal tarsals of NGMC 97-9-A. Abbreviations: a, astragalus; c, calcaneum; Co, coracoid; F, femur; g, gastroliths; H, humerus; I, ilium; Is, ischium; P, pubis; R, radius; S, scapula; St, sternal plate; T, tibia; U, ulna;?, possibly fragment of gastralia. Roman numerals represent digit numbers.

short bones with slight expansions at each end are found near the sternal plates of NGMC 97-9-A, and may be sternal ribs.

The scapula is longer than the humerus, whereas the scapula-to-humerus ratio is less than 1.0 in flying birds (Table 2) because of humerus elongation. The clavicles are fused into a broad, U-shaped furcula in NGMC 97-9-A as in *Archaeopteryx*, *Confuciusornis* and many non-avian theropods.

Compared to the humerus, forearm length is similar to that in oviraptorosaurs (Table 2), *Archaeopteryx* and *Protarchaeopteryx*. In more advanced birds,[15,16] the radius is longer than the humerus. The external surface of the ulna, as in *Archaeopteryx*,[8] lacks any evidence of quill nodes.

There are three carpals preserved in NGMC 97-4-A, including a large semi-lunate one that caps metacarpals I and II as in dromaeosaurids,

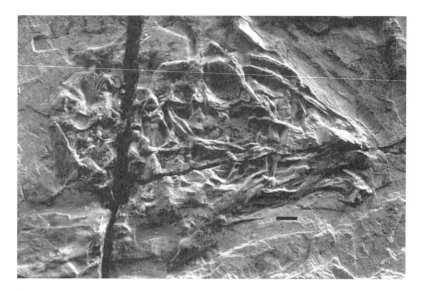

Figure 6 *Caudipteryx zoui*, skull of NGMC 97-9-A in right lateral view. Scale bar, 1 cm.

oviraptorids, troodontids, *Archaeopteryx*, *Confuciusornis* and other birds. Four carpals have been recognized in *Archaeopteryx*.[12] A large triangular radiale sits between the semi-lunate and the radius. A small carpal articulates with the third metacarpal. A thin wedge of bone at the end of the ulna is probably a fragment of gastralia.

The unfused metacarpals and digits of both specimens are well preserved. The third metacarpal is almost as long as the second, but is more slender. The hand has the normal theropod phalangeal formula of 2-3-4-x-x. The manus is longer than either the humerus or the radius, which is a primitive characteristic shared with most non-avian coelurosaurs, *Archaeopteryx*[9] and *Confuciusornis*. In contrast with *Archaeopteryx*, *Confuciusornis*, *Protarchaeopteryx* and many non-avian theropods (ornithomimids, troodontids, dromaeosaurids and oviraptorids), the manus is relatively short compared with the femur.

The curved second manual ungual is about two-thirds the size of the same element in *Protarchaeopteryx*, and is less than 70% the length of the penultimate phalanx.

Pelvic elements are unfused, as they are in all non-avian theropods (except some ceratosaurs) and the most primitive birds.[16] The acetabulum is large, comprising almost a quarter of the length of the ilium (the ratio of acetabulum-to-ilium length is less than 0.11 in birds[17]). It has a deeper, shorter, more squared-off preacetabular region than that of *Protarchaeopteryx*, and closely resembles the ilium of dromaeosaurids.[18]

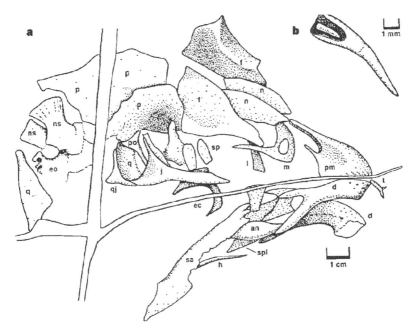

Figure 7 *Caudipteryx zoui*. **a**, Sketch of skull shown in Fig. 6. **b**, Premaxillary tooth of NGMC 97-4-A, showing resorption pit and germ tooth. Abbreviations: an, angular; d, dentary; ec, ectopterygoid; eo, exoccipital; f, frontal; h, hyoid; j, jugal; l, lacrymal; m, maxilla; n, nasal; ns, neural spine; p, parietal; pm, premaxilla; po, postorbital; q, quadrate; qj, quadratojugal; sa, surangular; sp, scleral plate; spl, splenial; t, premaxillary teeth.

The tapering postacetabular region is lower and longer than the preacetabular. The pubic penduncle is anteroposteriorly elongated, and has a notch (Figs 4 and 5b) in the ventral margin that divides the suture into two surfaces. This notch and the deep pubic penduncle of the ischium are characteristic of opisthopubic pelves. The ischium has no dorsal process such as that found in *Archaeopteryx* and *Confuciusornis*, and the shaft curves down and back. A well-developed ventromedial flange is present, perhaps indicating contact between elements. In general appearance, the ischium most closely resembles those of non-avian coelurosaurs.

The ratio of hindlimb-to-forelimb length is higher than in other coelurosaurs (Table 2) except alvarezsaurids,[19] which had exceptionally short arms. The greater trochanter is separated from the lesser trochanter of the femur by a shallow notch, and forms a raised, semi-lunate rim that is similar to the trochanter femoris of birds, troodontids and avimimids.

None of the fibulae is complete, but NGMC 97-9-A has a socket for the distal end of the fibula formed by the calcaneum, astragalus and tibia. The astragalus is not fused to the tibia. The ascending process of NGMC

97-9-A (Fig. 5e) extends 22% of the distance up the front surface of the tibia, compared with 12% in *Archaeopteryx*.[12] As in *Archaeopteryx*,[8] *Confuciusornis*[10] and most non-avian theropods, the calcaneum is retained as a separate, disk-like element. Two distal tarsals are positioned over the third and fourth metatarsals, as in *Archaeopteryx*, *Buluochia*[20] and all non-avian theropods that lack fused tarsometatarsals.

The metatarsals of *Caudipteryx* are not fused; this is the plesiomophic condition expressed in most non-avian theropods. Metatarsal I is centred about a quarter of the way up the posteromedial corner of the second metatarsal. The third is the longest of the metatarsals, and in anterior view completely separates the second and fourth metatarsals, unlike in the arctometatarsalian condition of many theropods.[21] Nevertheless, at midshaft the third metatarsal is thin anteroposteriorly and is triangular in cross-section. The pedal unguals are triangular in cross-section and are about the same size as the manual unguals.

At least fourteen remiges are attached to the second metacarpal, phalanx II-1, and the base of phalanx II-2 of NGMC 97-4-A (Fig. 8a). Each remex has a well-preserved rachis and vane. The most distal remex is less than 30 mm long. The second most distal remex is 63.5 mm long, is symmetrical, and has 6.5-mm-long barbs on either side of the rachis. The fourth most distal primary remex is 95 mm long and is longer than the humerus. Unfortunately, the distal ends of the remaining remiges are not preserved. In flying birds (even *Archaeopteryx*[12]), each remex is longer than they are in *Caudipteryx*, and the most distal remiges are the longest. For example, the remiges of *Archaeopteryx*[22] are more than double the length of the femur. The barbs on either side of the rachis are symmetrical, contrasting with *Archaeopteryx* and modern flying birds.[23]

The holotype preserves ten complete and two partial rectrices. Eleven are attached to the left side of the tail, and were probably paired with another eleven feathers on the right side (only the terminal feather is preserved). Two rectrices are attached to each side of the last five or six caudal vertebrae, but not to more anterior ones. NGMC 97-9-A preserves most of nine

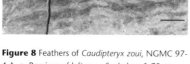

Figure 8 Feathers of *Caudipteryx zoui*, NGMC 97-4-A. **a**, Remiges of left arm. Scale bar, 1.75 cm. **b**, Rectrices, showing colour banding. Scale bar, 1 cm.

retrices. In *Archaeopteryx*, rectrices are associated with all but the first five or six caudals.[12,22] Each rachis has a basal diameter of 0.74 mm and tapers distally. All the feathers appear to be symmetrical (Fig. 8b), although in most cases the tips of the barbs of adjacent feathers overlap. The vane of the sixth feather is 6 mm wide on either side of the rachis.

The body of NGMC 97-4-A, especially the hips and the base of the tail, is covered by small, plumulaceous feathers of up to 14 mm long.

Both specimens have concentrations of small polished and rounded pebbles in the stomach region. These gastroliths are up to 4.5 mm in diameter, although most are considerably less than 4 mm wide.

Phylogenetic Analysis

We examined the systematic positions of *Protarchaeopteryx* and *Caudipteryx* by coding these specimens for the 90 characters used in an analysis of avialan phylogeny (for a matrix of these characters, see Supplementary Information). Characters were unordered, and a tree was produced using the branch-and-bound option of PAUP.[25] We rooted the tree with Velociraptorinae.[26,27] A single tree resulted with a length of 110 steps, a retention index of 0.849 and a consistency index of 0.855. Analysis shows *Caudipteryx* to be the sister group to the Avialae, and *Protarchaeopteryx* to be unresolved from the Velociraptorinae root (Fig. 9). The placement of *Protarchaeopteryx* as the sister group to *Caudipteryx* + Avialae, as the sister group to Velociraptorinae, or as the sister group to Velociraptorinae + (*Caudipteryx* + Avialae) are equally well supported by the data. Characters that define the *Caudipteryx* + Avialae clade in the shortest tree include unambiguous (uninfluenced by missing data or optimization) characters 2 and 12 and several more ambiguous ones (characters 4, 5, 10, 11, 15, 19, 24, 37, 85 and 86). *Caudipteryx* is separated from the Avialae by three unambiguous characters (7, 8 and 71) and additional ambiguous ones (characters 5, 6, 9, 10, 11, 18, 24, 39, 40, 56 and 69). The important characteristic of this phylogeny is that the Avialae (not including *Protarchaeopteryx* and *Caudipteryx*) is monophyletic; this placement is supported by the unequivocal presence of a quadratojugal that is joined to the quadrate by a ligament[17] (character 7), the absence

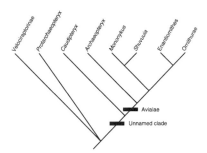

Figure 9 Cladogram of proposed relationships of *Protoarchaeopteryx* and *Caudipteryx*. This tree is based on 90 characters and has a length of 110 steps.

of a quadratojugal squamosal contact (character 8) and a reduced or absent process of the ischium (character 71).

As characters dealing with feathers cannot be scored relative to outgroup conditions, they were not used in the phylogenetic analysis. However, our analysis indicates that feathers can no longer be used in the diagnosis of the Avialae.

Discussion

The three *Protarchaeopteryx* and *Caudipteryx* individuals were close to maturity at the time of death. The neural spines seem to be fused to cervical and dorsal centra in *Protarchaeopteryx*. Sternal plates ossify late in the ontogeny of non-avian theropods, and are present in both *Caudipteryx* and *Protarchaeopteryx*. Well-ossified sternal ribs, wrist bones and ankle bones in *Caudipteryx* also indicate the maturity of the specimens.

The remiges of *Caudipteryx* and the rectrices of both *Protarchaeopteryx* and *Caudipteryx* have symmetrical vanes, whereas even those of *Archaeopteryx* are asymmetrical. Birds with asymmetrical feathers are generally considered to be capable of flight,[23] but it is possible that an animal with symmetrical feathers could also fly. Relative arm length of *Protarchaeopteryx* is shorter than that of *Archaeopteryx*, but is longer than in non-avian coelurosaurs. The arms of *Caudipteryx*, in contrast, are shorter than those of most non-avian coelurosaurs; the remiges are only slightly longer than the humerus; and the distal remiges are shorter than more proximal ones. It seems unlikely that this animal was capable of active flight. The relatively long legs of *Protarchaeopteryx* and *Caudipteryx*, both of which have the hallux positioned high and orientated anteromedially, indicate that they were ground-dwelling runners.

Paired rectrices of *Protarchaeopteryx* and *Caudipteryx* are restricted to the end of the tail, whereas in *Archaeopteryx* they extend over more than two-thirds the length of the tail.[12] Wherever preservation made it possible, we found semi-plumes and down-like feathers around the periphery of the bodies, suggesting that most of the bodies were feather-covered, possibly like *Archaeopteryx*.[28] Feathers found with *Otogornis*[29] were also apparently plumulaceous. Plumulaceous and downy feathers cover the bodies of *Protarchaeopteryx* and *Caudipteryx*, and possibly that of *Sinosauropteryx*[7] as well. This suggests that the original function of feathers was insulation.

Phylogenetic analysis shows that both *Caudipteryx* and *Protarchaeopteryx* lie outside Avialae and are non-avian coelurosaurs. This indicates that feathers are irrelevant in the diagnosis of birds. It can no longer be certain that isolated down and semi-plume feathers[30–33] discovered in Mesozoic rocks belonged to birds rather than to non-avian dinosaurs. Furthermore, the presence of feathers on flightless theropods suggests that the hypothesis that feathers and flight evolved together is incorrect. Finally, the presence of remiges, rectrices and plumulaceous feathers on non-avian theropods provides unambiguous evidence supporting the theory that birds are the direct descendants of theropod dinosaurs.

References

1. Ji, Q. *et al.* On the sequence and age of the protobird bearing deposits in the Sihetun-Jianshangou area, Beipaio, western Liaoning. *Prof. Pap. Strat. Paleo.* (in the press).

2. Hou, L.-H., Zhou, Z.-H., Martin, L. D. & Feduccia, A. A beaked bird from the Jurassic of China. *Nature* **377**, 616–618 (1995).

3. Smith, P. E. *et al.* Dates and rates in ancient lakes: 40Ar- 39Ar evidence for an Early Cretaceous age for the Jehol Group, northeast China. *Can. J. Earth Sci.* **32**, 1426–1431 (1995).

4. Smith, J. B., Hailu, Y. & Dodson, P. in The Dinofest Symposium, Abstracts (eds Wolberg, D. L. *et al.*) 55 (Academy of Natural Sciences, Philadelphia, 1998).

5. Colbert, E. H. The Triassic dinosaur *Coelophysis*. *Bull. Mus. N. Arizona* **57**, 1–160 (1989).

6. Ostrom, J. H. The osteology of *Compsognathus longpipes* Wagner. *Zitteliana* **4**, 73–118 (1978).

7. Chen, P.-j., Dong, Z.-m. & Zhen, S.-n. An exceptionally well-preserved theropod dinosaur from the Yixian Formation of China. *Nature* **391**, 147–152 (1998).

8. Wellnhofer, P. A new specimen of *Archaeopteryx* from the Solnhofen Limestone. *Nat. Hist. Mus. Los Angeles County Sci. Ser.* **36**, 3–23 (1992).

9. Sereno, P. C. & Rao, C. G. Early evolution of avian flight and perching: new evidence from the Lower Cretaceous of China. *Science* **255**, 845–848 (1992).

10. Zhou, Z. H. in Sixth Symposium on Mesozoic Terrestrial Ecosystems and Biota, Short Papers (eds Sun, A. & Wang, Y.) 209–214 (China Ocean, Beijing, 1995).

11. Ji, Q. & Ji, S. A. Protarchaeopterygid bird (*Protarchaeopteryx* gen. nov.)—fossil remains of archaeopterygids from China. *Chinese Geol.* **238**, 38–41 (1997).

12. Wellnhofer, P. Das fünfte skelettexemplar von *Archaeopteryx*. *Palaeontogr.* A**147**, 169–216 (1974).

13. Wellnhofer, P. Das siebte Examplar von *Archaeopteryx* aus den Solnhofener Schichten. *Archaeopteryx* **11**, 1–48 (1993).

14. Hou, L. H. A carinate bird from the Upper Jurassic of western Liaoning, China. *Chinese Sci. Bull.* **42**, 413–416 (1997).

15. Dong, Z. M. A lower Cretaceous enantiornithine bird from the Ordos Basin of Inner Mongolia, People's Republic of China. *Can. J. Earth Sci.* **30**, 2177–2179 (1993).

16. Forster, C. A., Sampson, S. D., Chiappe, L. M. & Krause, D. W. The theropod ancestry of birds: new evidence from the Late Cretaceous of Madagascar. *Science* **279**, 1915–1919 (1998).

17. Chiappe, L. M., Norell, M. A. & Clark, J. M. The skull of a relative of the stem-group bird *Mononykus*. *Nature* **392**, 275–278 (1998).

18. Norell, M. A. & Makovicky, P. Important features of the dromaeosaur skeleton: information from a new specimen. *Am. Mus. Novit.* **3215**, 1–28 (1997).

19. Perle, A., Chiappe, L. M., Barsbold, R., Clark, J. M. & Norell, M. Skeletal morphology of *Mononykus olecranus* (Theropoda: Avialae) from the Late Cretaceous of Mongolia. *Am. Mus. Novit.* **3105**, 1–29 (1994).

20. Zhou, Z. H. The discovery of Early Cretaceous birds in China. *Courier Forschungsinstitut Senckenberg* **181**, 9–22 (1995).

21. Holtz, T. R. The phylogenetic position of the Tyrannosauridae: implications for theropod systematics. *J. Paleontol.* **68**, 1100–1117 (1994). deBeer, G. *Archaeopteryx lithographi.*
22. deBeer, G. *Archaeopteryx lithographica. Br. Mus. Nat. Hist.* **244**, 1–68 (1954).
23. Feduccia, A. & Tordoff, H. B. Feathers of *Archaeopteryx*: asymmetric vanes indicate aerodynamic function. *Science* **203**, 1021–1022 (1979).
24. Chiappe, L. M. in *The Encyclopedia of Dinosaurs* (eds Currie, P. J. & Padian, K.) 32–38 (Academic, San Diego, 1997).
25. Swofford, D. & Begle, D. P. *Phylogenetic Analysis Using Parsimony. Version 3.1.1.* (Smithsonian Institution, Washington DC, 1993).
26. Gauthier, J. in *The Origin of Birds and the Evolution of Flight* (ed. Padian, K.) 1–55 (Calif. Acad. Sci., San Francisco, 1986).
27. Holtz, T. R. Jr Phylogenetic taxonomy of the Coelurosauria (Dinosauria: Theropoda). *J. Paleontol.* **70**, 536–538 (1996).
28. Owen, R. On the *Archaeopteryx* of von Meyer, with a description of the fossil remains of a long-tailed species, from the Lithographic Stone of Solenhofen. *Phil. Trans., Lond.* **153**, 33–47 (1863).
29. Hou, L. H. A late Mesozoic bird from Inner Mongolia. *Vert. PalAsiatica* **32**, 258–266 (1994).
30. Kurochkin, E. N. A true carinate bird from Lower Cretaceous deposits in Mongolia and other evidence of Early Cretaceous birds in Asia. *Cretaceous Res.* **6**, 271–278 (1985).
31. Sanz, J. L., Bonapart, J. F. & Lacasa, A. Unusual Early Cretaceous birds from Spain. *Nature* **331**, 433–435 (1988).
32. Kellner, A. W. A., Maisey, J. G. & Campos, D. A. Fossil down feather from the Lower Cretaceous of Brazil. *Palaeontol.* **37**, 489–492 (1994).
33. Grimaldi, D. & Case, G. R. A feather in amber from the Upper Cretaceous of New Jersey. *Am. Mus. Novit.* **3126**, 1–6 (1995).

Acknowledgements

We thank A. Brush, B. Creisler, M. Ellison, W.-D. Heinrich, N. Jacobsen, E. and R. Koppelhus, P. Makovicky, A. Milner, G. Olshevsky, J. Ostrom and H.-P. Schultze for advice, access to collections and logistic support; and the National Geographic Society, National Science Foundation (USA), the American Museum of Natural History, National Natural Science Foundation of China and the Ministry of Geology for support. Photographs were taken by O. L. Mazzatenta and K. Aulenback; the latter was also responsible for preliminary preparation of the *Caudipteryx* specimens. Line drawings are by P.J.C.

14

When Is a Bird Not a Bird?

Kevin Padian

Birds were once thought to have a large number of features exclusive to the group. One by one those features have also been identified in fossils of certain theropod dinosaurs. Now feathers join the list.

Among all living creatures, only birds have feathers. So when the fossil of a single isolated feather was found in the Late Jurassic rocks of Bavaria shortly after Darwin published *On the Origin of Species* in 1859, it was enough to demonstrate that, astonishingly, birds must have existed in those remote times (now known to be some 150 million years ago). When the skeleton of *Archaeopteryx* was discovered in the same area in 1861, it fulfilled that plumose prediction. But the skeleton was recognized as a bird's mainly because it had feathers on its wings and tail. Nothing else except an odd, boomerang-shaped wishbone seemed to ally it to living birds. In fact, some other specimens were assigned to *Archaeopteryx* only belatedly, because no feathers were preserved in the fossils concerned. Those specimens had instead been taken for small carnivorous (theropod) dinosaurs.

On page 753 of this issue, Ji et al.[1] show unequivocally that clothes don't make the bird. They describe two small theropod dinosaurs from geological beds in the Liaoning province of China.[1-4] The age of the beds is disputed; although initial reports suggested they are Late Jurassic, radiometric dates and other evidence now point to the Early Cretaceous[4]— that is, around 145 million years or maybe later. These theropods have both plumulaceous (down-like) and vaned, barbed feathers on the body, arms, legs and tail. But these animals were clearly not birds, and they were clearly not capable of flight.

One form, *Protarchaeopteryx*, has been briefly described in a Chinese journal.[2] It has down-like feathers on its body and tail, and vaned, barbed, symmetrical feathers along at least the end of the tail, in a fan-like pattern. The second, *Caudipteryx*, has remiges (primary feathers) attached to the second (longest) finger of the hand, though the arms are much shorter than in birds. These feathers are also vaned and barbed, and down-like feathers are also preserved. Were preservation more complete, we would have an even fuller idea of the plumage of these animals. This is the most that can be said at the moment, but new specimens continue to emerge from this locality.

If these are true feathers—and it can scarcely be doubted—what does this do to our conception of birds? The bottom line is that it simply forces us to revise our idea of the association of feathers with the animals we call birds. This is why.

Systematists *define* the names of organisms (taxa) by their ancestry; in this case, birds (Aves) consist of *Archaeopteryx* plus living birds (Neornithes) and all the descendants of their most recent common ancestor (Fig. 1). But we *diagnose* (recognize) birds by unique features that only they possess; these are inherited from that most recent common ancestor. (Ji *et al.*[1] give three such features, none of which relates to feathers.) So the diagnostic characteristics follow from our definition of ancestry. Why worry if feathers turn out to be shared by a wider group than birds alone? We still define birds as *Archaeopteryx* and its later relatives. *Protarchaeopteryx* and *Caudipteryx* may have feathers but they're not birds, because they're not members of that ancestral club of *Archaeopteryx* and living birds. Ji *et al.*[1] show that these animals belong to a group of dinosaurs known as the maniraptoran coelurosaurs, which include the small theropods most closely related to birds. However, their analysis of evolutionary relationships does not encompass a broader group of coelurosaurs, which might eventually help to pin down the position of *Protarchaeopteryx*.

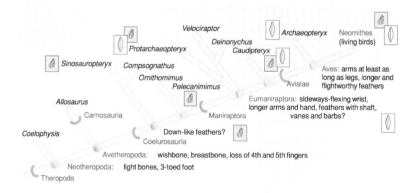

Figure 1 Distribution of known feather types and other features in selected groups of theropod dinosaurs, including birds (Aves).

By admitting that plumage did not first spring full-blown on the wings of *Archaeopteryx*, we are free to examine how feathers evolved in the first place. At the 1996 meeting of the Society of Vertebrate Paleontology, Dr Chen Pei-ji disclosed photographs of an amazing specimen—a small theropod dinosaur, also from the Liaoning beds, with a fringe of hair-like or down-like structures along its neck and backbone. This animal, dubbed *Sinosauropteryx* by Ji and Ji,[3,4] turns out to be closely related to the basal (relatively 'primitive') coelurosaur *Compsognathus*, a chicken-sized, short-armed form found in the same deposits as *Archaeopteryx*.[5] Without even seeing the specimens, critics of the theropod-bird connection tried to pass off these remains as artefacts of preservation. They claimed that the structures were frayed internal collagen fibres, not external or epidermal in origin at all, and that similar structures in sea snakes suggested that these theropods lived in an aquatic environment.[6] Those doubts can now be put to rest.[3,4]

The evolution of carnivorous dinosaurs through basal coelurosaurs into birds shows some unmistakable trends.[7] As their evolutionary history has become better known, we have seen wishbones, breastbones, hollow bones, long arms and hands, sideways-flexing wrists, nesting behaviour and rapid growth rates all disappear from the avian catalogue of exclusive features. Like feathers, as Ji and colleagues' work[1] shows, these features first evolved in coelurosaurs for purposes completely unrelated to flight.[7] The proto-feathery fringe of *Sinosauropteryx*, now known to extend to its flanks as well as along its midline,[4] was obviously not made for flight and it is questionable whether it could have served any kind of function in insulation. Camouflage, display and species recognition come to mind as other possibilities.[8]

Caudipteryx and *Protarchaeopteryx* go one better, evolving long feathers with a central rachis (shaft). Were these feathers airworthy? Their vanes are symmetrical and very even, suggesting interlocking barbs, although most flying birds have asymmetrical feathers. However, the arms of *Caudipteryx* are only 60% as long as the legs, and they are only two-thirds as long in *Protarchaeopteryx* (in *Archaeopteryx* they are of more or less equal length). Evidently, the arms and feathers were not large enough for flight (the plumage is not yet known well enough to say if it was effective as insulation, or what other functions it may have served). The feathers of *Archaeopteryx* seem to be a natural extension of this trend, so to speak, but they are not much different qualitatively. So the available evidence suggests that structurally airworthy feathers may have evolved before they were long enough, or their possessors able to use them, for flight.

The work of Ji *et al.*[1] should lay to rest any remaining doubts that birds evolved from small coelurosaurian dinosaurs. These new discoveries will excite the public and scientists alike by showing that down-like and later vaned body feathers evolved before flight feathers, and that a full complement of feathers was present in coelurosaurs before birds were invented.

References

1. Ji, Q., Currie, P. J., Norell, M. A. & Ji, S.-A. *Nature* **393**, 753–761 (1998).
2. Ji, Q. & Ji, S.-A. *Chinese Geol.* **238**, 38–41 (1997).
3. Ji, Q. & Ji, S.-A. *Chinese Geol.* **233**, 30–33 (1996).
4. Chen, P.-j., Dong, Z.-m. & Zhen, S.-n. *Nature* **391**, 147–152 (1998).
5. Ostrom, J. H. *Zitteliana* **4**, 73–118 (1978).
6. Geist, N. R., Jones, T. D. & Ruben, J. A. *J. Vert. Paleont.* **17**, 48A (1997).
7. Padian, K. & Chiappe, L. M. *Biol. Rev.* **73**, 1–42 (1998).
8. Cowen, R. & Lipps, J. H. *Proc. 3rd N. Am. Paleont. Conv. (Montreal)* 109–112 (1985).
9. Griffiths, P. J. *Archaeopteryx* **14**, 1–26 (1994).
10. Pérez-Moreno, B. P. et al. *Nature* **370**, 363–367 (1994).
11. Briggs, D. E. G. et al. *J. Geol. Soc. Lond.* **154**, 587–588 (1997).

A Diapsid Skull in a New Species of the Primitive Bird *Confuciusornis*

Lian-Hai Hou, Larry D. Martin, Zhong-He Zhou, Alan Feduccia, and Fu-Cheng Zhang

Since the description of *Confuciusornis* (the oldest beaked bird) in 1995, based on three partial specimens, large numbers of complete skeletons have been recovered.[1,2] Most new material of *Confuciusornis*[3,4] can be assigned to a single sexually dimorphic species, *C. sanctus*. Here we report a new species based on a remarkably well preserved skeleton with feathers and, for the first time in the Mesozoic record, direct evidence of the shape of a horny beak. It has a complete and large preserved postorbital that has a broad contact with the jugal bone. This character is presently only known in *Confuciusornis*, and may confirm previous suggestions of a postorbital in *Archaeopteryx*.[5] The squamosal is in tight contact with the postorbital. These two bones form an arch dividing the upper and lower temporal fenestrae, as in other diapsid reptiles.[6] The presence of a typical diapsid cheek region with two openings in *Confuciusornis* may preclude the presence of prokinesis (upper jaw mobility against the braincase and orbital area), a feeding adaptation found in most modern birds. The presence of a horny beak, characteristic of modern birds, coupled with a primitive temporal region provides new evidence for a mosaic pattern in the early evolution of birds.

Aves Linnaeus 1758
Sauriurae Haeckel 1866
Confuciusornithiformes Hou *et al.* 1995
Confuciusornithidae Hou *et al.* 1995
Confuciusornis Hou *et al.* 1995
Confuciusornis dui sp. nov.

ETYMOLOGY.   The species name is dedicated to Mr. Wenya Du, who collected and donated the specimen to the Institute of Vertebrate Paleontology and Paleoanthropology (IVPP) for scientific research.

HOLOTYPE.   A nearly complete skeleton. IVPP Collection Number V 11553.

PARATYPE.   IVPP 11521, a partial skeleton consisting of a sternum, ribs, vertebrae, pelvis, femora and tail.

HORIZON AND LOCALITY.   A two-metre thick interval within the Yixian Formation (Late Jurassic-Early Cretaceous); Libalanggou, Zhang-jiying, Beipiao, Liaoning, northeast China.

DIAGNOSIS.   The holotype, a presumed male, is about 15% smaller than the holotype of *C. sanctus* (a small individual and presumed female). Large male individuals of *C. sanctus* are about 30% larger than the new species. The mandible is more slender anteriorly, without the distinctive anteroventral expansion of the dentary found in *C. sanctus*. The upper jaw is also more pointed anteriorly than that of *C. sanctus*. The claw of the alular digit is not enlarged as in *C. sanctus*. The sternum is more elongate with an anterior notch and a pair of short lateral processes. The tarsometatarsus is relatively shorter than in *C. sanctus*, and is shorter than the pygostyle.

MEASUREMENTS OF HOLOTYPE (mm).   Wing chord, 201; length of lower jaw 40; length of humerus, 42; length of ulna, 39; length of femur, 35; length of tibiotarsus, 41; length of tarsometatarsus, 19.5; length of pygostyle, 23; length of carpometacarpus, 19.

DESCRIPTION.   This is a smaller, more gracile species than *C. sanctus*. The tarsometatarsus is fused proximally, and the semilunate bone is fused to the major metacarpal, indicating that this is an adult specimen. Like other *Confuciusornis*, it lacks teeth. Impressions of a single pair of long tail feathers (Fig. 1) indicate that IVPP V11553 was a male.

1 cm

Figure 1 Cast of the elongate tail feathers of *C. dui* (IVPP Collection V 11553) whitened with ammonium chloride, showing that it was probably male.

The skull (Fig. 2a, b) is well preserved and shows some critical features of *Confuciusornis*. The impression of the horny rhamphotheca is preserved (Fig. 2a, b). The horny bill extends in front of the bony core with a long and pointed tip. The distal end of the beak curves dorsally and the bill was not raptorial. It is likely that *Confuciusornis* was an herbivore. The nasal processes of the premaxillae separate the nasals and overlap the anterior end of the frontals. The lacrimal joins the nasal above a small slit-shaped antorbital fenestra.

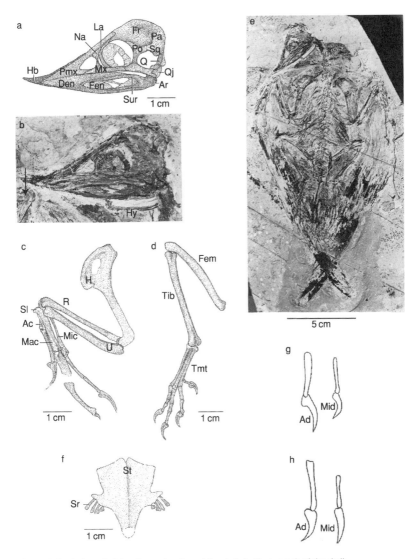

Figure 2 *Confuciusornis dui.* **a**, Reconstruction of the skull. **b**, Photograph of the skull (IVPP Collection V 11553); the arrow indicates the horny beak. **c**, Forelimb (IVPP Collection V 11553). **d**, Hindlimb. **e**, Photograph of the holotype (IVPP Collection V 11553). **f**, Reconstruction of the sternum and sternal ribs (based on IVPP Collection V 11521). **g**, Alular and minor manual digits of *C. sanctus.* **h**, Alular and minor manual digits of *C. dui.* Abbreviations: Ac, alular metacarpal; Ad, alular manual digit; Ar, articular; Den, dentary; Fen, fenestra of the lower jaw; Fem, femur; Fr, frontal; H, humerus; Hb, horny beak; Hy, hyoid bone; J. jugal; La, lacrimal; Mac, major metacarpal; Mic, minor metacarpal; Mid, distal minor manual digit; Mx, maxilla; Na, nasal; Pa, parietal; Po, postorbital; Q, quadrate; Qj, quadratojugal; R, radius; Sl, semilunate bone (fused with major metacarpal); Sq, squamosal; Sr, sternal ribs; St, sternum; Sur, surangular; Tib, tibiotarsus; Tmt, tarsometatarsus; U, ulna.

The postorbital is large and 'Y' shaped; it contacts the jugal, squamosal and frontal. The contact on the jugal for the postorbital appears to be a rounded process. The quadrate is similar to that of *Archaeopteryx* and lacks an orbital process. The quadratojugal seems to be small and 'L' shaped, as in *Archaeopteryx*. The maxillary process of the premaxilla tapers posteriorly and overlaps the maxilla below the middle of the narial opening. The maxilla has a dorsal ascending process in contact with the nasal, forming the anterior margin of the antorbital fenestra. There is a 'V'-shaped groove formed by backwards-extending processes of the dentary surrounding a large fenestra in the surangular.

The neck is relatively short. The exact number of cervical vertebrae in the new species is unclear, but it is probably fewer than eight. The vertebrae are short with deep pleurocoels. The pygostyle is longer than the tarsometatarsus and is well fused. The exact number of vertebrae that are fused into the pygostyle is unknown in the new species, but we have counted between 8 and 10 in *C. sanctus*. The sternal plate, which is visible on the holotype but best seen in the referred specimen (Fig. 2f), is longer than it is wide. It is relatively flat and lacks an obvious keel, as in *C. sanctus*. There are a pair of short lateral processes that serve as attachment sites for four short sternal ribs (Fig. 2f). Gastralia are present. The flight feathers (Fig. 2e) are very asymmetric with a wing chord of about 201 mm. The twin tail feathers are clearly visible in low-incidence lighting and resemble those of other *Confuciusornis* (Fig. 1).

The humerus is typical of *Confuciusornis* in having a very expanded proximal end with an oval depression (it is not clear whether any of the many specimens of *Confuciusornis* shown with a humeral perforation had that character before preparation). The humerus (Fig. 2c) is slightly longer than the ulna, as in *Archaeopteryx* but different from later birds where the ulna is usually longer than the humerus. The ulna is nearly twice as wide as the radius. The semilunate bone is fused to the major metacarpal, forming a carpometacarpus, but not to the alular metacarpal, indicating that the semilunate bone in *Confuciusornis*, as in modern birds, is a single distal carpal. The major and minor metacarpals are of nearly equal length and the major metacarpal is more robust than the other two. The minor metacarpal is significantly narrower proximally than distally, as in *C. sanctus*. There are three clawed fingers and the claw of the alular digit is about the same size as that of the minor digit, rather than about one-third larger as in *C. sanctus* (Fig. 2g, h). The middle manual claw is reduced, as in other *Confuciusornis*, although it is not completely preserved. The first phalanx of the alular digit is almost as long as the major and minor metacarpals. As is typical of *Confuciusornis*, the first phalanx of the minor digit is short (the first and second are short in *Archaeopteryx*); the others of the same digit are long, and the second is longer than the penultimate phalanx.

The pelvis is narrow and opisthopubic. The femur (Fig. 2d) is almost as long as the tibiotarsus. The tarsometatarsus is fused only proximally, as in *C. sanctus* and other saururine birds. The foot has highly recurved claws and a reversed hallux.

DISCUSSION.   The evolution of the skull in early birds is less known than that of the postcranial skeleton. This is particularly true of the temporal region. The cheek region in modern birds is reduced, owing to an expanded brain and enlarged orbit. Fossils of the oldest bird, *Archaeopteryx*, have unfortunately not preserved the postorbital and provide little information about the transition from the typical diapsid skull to the situation in modern birds.

Confuciusornis dui preserves a large 'Y'-shaped postorbital with a broad contact with the jugal bone. The postorbital process of the jugal does not lie behind the postorbital in contrast to that of *Dromaeosaurus*.[7] The postorbital has not been found in *Archaeopteryx*, but its presence has been extrapolated on the basis of the postorbital process on the jugal bone and the shape of the squamosal.[5,8] The jugal bone in *Confuciusornis* is strong and relatively short compared with that of modern birds, where it is slender and fused with the quadratojugal bone, forming a rod-like bar. The postorbital was probably absent in the Chinese Early Cretaceous enantiornithine *Cathayornis*.[9–11] The only other Mesozoic bird that has retained this structure is an Early Cretaceous bird from Spain.[12] However, in this bird the postorbital is much more reduced and has lost its contact with the jugal bar.

The squamosal is in tight contact with the postorbital. They form an intact intertemporal arch separating the upper and lower temporal fenestrae. The unreduced postorbital and its contacts with the jugal bone and the squamosal might prevent prokinesis; this may be confirmed by the fact that the quadrate has no orbital process and the quadratojugal is reduced so it cannot participate in a prokinetic push-rod. The presence of two derived quadrate systems indicates that a streptostylic quadrate with an orbital process would not have been a feature of the first bird.

Although *Confuciusornis* retains an intact diapsid temporal region, it shows several cranial modifications also found in more advanced birds. The enlargement of the frontal and reduction of the parietal make the frontal the major contributor to the skull internal to the upper arch, whereas in typical diapsid animals the parietal extends further anteriorly. The dorsal process of the jugal bone is low and has a rounded contact with the postorbital. The quadrate of *Confuciusornis* is single-headed; it lacks the prominent orbital process that is present in modern birds but is also lacking in the Early Cretaceous enantiornithine *Cathayornis*.[11]

Although the sternum has no obvious keel, the shape of the wing and flight feathers are highly adapted for powered flight. *Confuciusornis* probably lived in large aggregations along the margin of a freshwater lake, as shown by hundreds of individuals found in a limited area. There is abundant evidence for lush forested conditions and the freshwater mudstones contain a variety of conchostracans, insects, fishes, amphibians, reptiles and mammals. The long tails on the presumed males seem to preclude much activity on the ground or the water surface and, along with the highly recurved pedal claws and reversed hallux, indicate that *Confuciusornis* was a perching bird. None of the many specimens preserve stomach contents but it seems likely that *Confuciusornis* was herbivorous.

The Early Cretaceous saururine birds from younger geological sections are smaller than the Late Jurassic *Archaeopteryx* and the Late Jurassic-Early Cretaceous *Confuciusornis*. The new species of *Confuciusornis* is smaller than *C. sanctus*. It has been proposed that size reduction was important in the early evolution of avian flight.[13] The early ornithurine birds and Early Cretaceous enantiornithines are smaller than *Archaeopteryx* and *Confuciusornis*. In addition, the new species has a more elongated sternum.

The findings of different species of *Confuciusornis* show that the oldest known beaked bird was not only abundant but had also diversified, as had *Archaeopteryx*, with at least two species in both genera.[5] A recently found enanthiornithine bird[14] from the same locality as *Confuciusornis* and *Liaoningornis* indicates that Enantiornithes had a longer history than previously known. Thus, the origin and early diversification of birds might be earlier and more significant than expected.[15,16]

All known Late Jurassic-Early Cretaceous ornithurines such as *Chaoyangia* and enantiornithines have retained teeth in the jaw,[1] and only *Confuciusornis* lost its teeth completely in favour of a horny beak. It provides an unusual mosaic of primitive and derived features with the anterior skull more like that of a modern bird than in any other Late Jurassic-Early Cretaceous bird, but with a temporal region nearly unchanged from its remote archosaurian ancestors. Like the early evolution of mammals, the original diversification of birds was probably also a complicated bush with many extinct lines that may at one time have been more advanced in some features than their ultimately more successful contemporaries.[17] The combination of distinctively advanced and primitive features found in the skull provides new evidence for a mosaic pattern in the early evolution of birds. *Confuciusornis* is not the progenitor of either modern birds or later enantiornithines, but must be regarded as an early twig in a bush-like radiation of birds.

References

1. Hou, L., Martin, L. D., Zhou, Z. & Feduccia, A. Early adaptive radiation of birds: evidence from fossils from northeastern China. *Science* **274**, 1164–1167 (1996).
2. Peters, D. S. Ein nahezu vollständiges Skellette eines urtümlichen Vogels aus China. *Natur und Museum* **126**, 298–302 (1996).
3. Hou, L., Zhou, Z., Gu, Y. & Zhang, H. *Confuciusornis sanctus*, a new Late Jurassic sauriurine bird from China. *Chin. Sci. Bull.* **40**, 1545–1551 (1995).
4. Hou, L., Zhou, Z., Martin, L. D. & Feduccia, A. A beaked bird from the Jurassic of China. *Nature* **377**, 616–618 (1995).
5. Elzanowski, A. & Wellnhofer, P. Cranial morphology of *Archaeopteryx*: evidence from the seventh skeleton. *J. Vert. Paleontol.* **16**, 81–94 (1996).
6. Reisz, R. R. A diapsid reptile from the Pennsylvanian of Kansas. *Special Publ. Nat. Hist. Mus. Univ. Kansas* **7**, 1–74 (1981).
7. Colbert, E. H. & Russell, D. A. The small Cretaceous dinosaur *Dromaeosaurus*. *Amer. Mus. Novit.* **2380**, 1–49 (1969).
8. Wellnhofer, P. Das fünfte Skelettexemplar von *Archaeopteryx*. *Palaeontographica A* **147**, 169–216 (1974).
9. Zhou, Z., Jin, F. & Zhang, J. Preliminary report on a Mesozoic bird from Liaoning, China. *Chin. Sci. Bull.* **37**, 1365–1368 (1992).
10. Zhou, Z. The discovery of Early Cretaceous birds in China. *Cour. Forchungsinst. Senckenb.* **181**, 9–22 (1995).
11. Martin, L. D. & Zhou, Z. *Archaeopteryx*-like skull in enantiornithine bird. *Nature* **389**, 556 (1997).
12. Sanz, J. L. *et al.* An Early Cretaceous bird from Spain and its implication for the evolution of avian flight. *Science* **276**, 1543–1546 (1997).
13. Zhou, Z. & Hou, L. *Confuciusornis* and the early evolution of birds. *Vertebr. PalAsiat* **36**, 136–146 (1998).
14. Hou, L., Martin, L. D., Zhou, Z. & Feduccia, A. *Archaeopteryx* to opposite birds— missing link from the Mesozoic of China. *Vertebr. PalAsiat.* **37** (in the press).
15. Feduccia, A. *The Origin and Evolution of Birds* (Yale Univ. Press, New Haven, 1996).
16. Chatterjee, S. *The Rise of Birds* (John Hopkins University Press, Baltimore, 1997).
17. Martin, L. D. & Miao, D. in *Short Papers of the Sixth Symposium on Mesozoic Terrestrial Ecosystems and Biota* 217–219 (China Ocean Press, Beijing, 1995).

Acknowledgements

We thank L. Witmer and S. Chatterjee for critical and helpful comments and reviews, and D. Miao for assistance and suggestions. The Chinese Natural Science Foundation, the National Geographic Society (U.S.) and the Grand Project of the Chinese Academy of Sciences supported fieldwork in Liaoning Province, northeast China. M. Tanner did the drawings and J. Chorn the photographs. D. Miao and J. Chorn critically read the manuscript.

A Chinese Triconodont Mammal and Mosaic Evolution of the Mammalian Skeleton

Qiang Ji, Zhe-Xi Luo, and Shu-An Ji

Here we describe a new triconodont mammal from the Late Jurassic/Early Cretaceous period of Liaoning, China. This new mammal is represented by the best-preserved skeleton known so far for triconodonts which form one of the earliest Mesozoic mammalian groups with high diversity. The postcranial skeleton of this new triconodont shows a mosaic of characters, including a primitive pelvic girdle and hindlimb but a very derived pectoral girdle that is closely comparable to those of derived therians. Given the basal position of this taxon in mammalian phylogeny, its derived pectoral girdle indicates that homoplasies (similarities resulting from independent evolution among unrelated lineages) are as common in the postcranial skeleton as they are in the skull and dentition in the evolution of Mesozoic mammals. Limb structures of the new triconodont indicate that it was probably a ground-dwelling animal.

Class Mammalia
Infraclass Triconodonta (McKenna and Bell 1997)
Order Eutriconodonta (Kermack *et al.* 1973)
Family *Incertae sedis*
Jeholodens jenkinsi gen. et sp. nov.

ETYMOLOGY. Jehol: an ancient geographic name for the western part of the Liaoning Province, China; the namesake of the Jehol fauna in the Yixian Formation that yielded the holotype; *odens* (Latin): tooth; *jenkinsi* (Latin): in honour of F. A. Jenkins Jr for his pioneer studies of the evolutionary morphology of the mammalian post-cranial skeleton.

HOLOTYPE. GMV2139 a, b, a nearly complete skeleton consisting of a partial skull and all of the postcranial skeleton preserved as two counterparts (Fig. 1a; a reconstruction of the specimen is shown in Fig. 1b).

LOCALITY AND HORIZON. The Sihetun site (at roughly 41° 40′ 12″ N, 120° 47′ 36″ E), about 32 km east of Chaoyang City, Liaoning Province, China. The holotype is from the lacustrine shales that are intercalated with neutrobasic volcanic beds in the Yixian Formation.

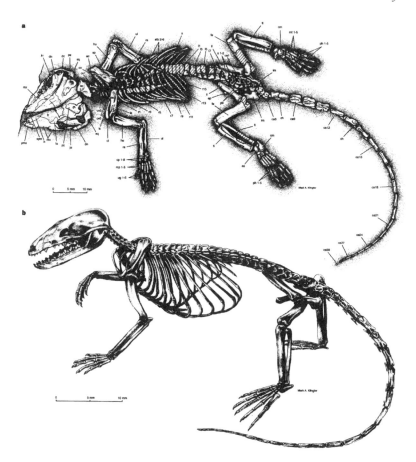

Figure 1 *Jeholodens jenkinsi* (National Geological Museum of China, holotype GMV 2139a). **a**, Dorsal view of the dorsoventrally compressed skeleton (dashed lines indicate impressions, for which the bone structures are almost completely preserved on the counterpart GMV 2139b, not shown). **b**, Reconstruction of *J. jenkinsi* as a ground-dwelling animal that had a plantigrade gait, sprawling hindlimbs and a mobile pectoral girdle with relatively wide range of excursion of the scapula but a sprawling elbow. Abbreviations: ac, acromion of scapula; as, astragalus; c, canine; c1–c7, cervical vertebrae 1–7; ca–ca30, caudal vertebrae 1–30; ch, chevron (hemal arch); cl, clavicle; cm, calcaneum; co, coracoid process of scapula (or the unfused coracoid in cynodonts); cp1–cp9, carpals 1–9; dc, dentary condyle; dn, dentary; ep, epipubis; fe, femur; fi, fibula; fr, frontal; gl, glenoid of the scapula; hu, humerus; ic, interclavicle; il, ilium; im, ischium; is, infraspinous fossa (of scapula); i1–i4, incisors 1–4; i2r, replacement incisor 2; ju, jugal; la, lacrimal; L1–L7, lumbar vertebrae 1–7; ma, metacromion (on the spine of the scapula); mf, mandibular foramen; mg, meckelian groove; m1–4, molars 1–4; mp1–5, metacarpals 1–5; mt1–mt5, metatarsals 1–5; mx, maxillary; n, nasal; pb, pubic; pc, pars cochlearis of petrosal; pcd, coronoid process of dentary; pe, petrosal, preserved in the dorsal/endocranial view; pf, pterygoid fossa; ph1–5; phalanges 1–5; pmx, premaxillary; ps, pterygoid shelf on the ventral border of the mandible; p1, p2, premolars 1, 2; ra, radius; r1–r15, thoracic ribs 1–15; sc, scapula; sn, scapular notch; so, supraoccipital; sp, spine of scapula; sq, squamosal; ss, supraspinous fossa of scapula; stb2–6, sternebrae 2–6 (the interclavicle, the sternal manubrium and sternebra 1 are preserved and exposed by preparation, but are not shown in **a**); sym, mandibula symphysis; s1, s2, sacral vertebrae 1, 2; ti, tibia; tm, teres major fossa (on scapula); t1–t15, thoracic vertebrae 1–15; ug1–5, ungual phalanges 1–5; ul, ulna; x, xiphoid process of sternum.

GEOLOGICAL AGE AND FAUNA.   Correlation of the Yixian Formation is equivocal. It has been suggested to be Late Jurassic,[1] near the Jurassic–Cretaceous boundary,[2–4] or Early Cretaceous.[5] The associated fauna includes the theropods *Sinosauropteryx*,[1,4] *Protarchaeopteryx*,[5] *Caudipteryx*,[5] the birds *Confuciusornis* and *Liaoningornis*,[2] the pterosaur *Eosipterus*, the mammal *Zhangheotherium*[3], and diverse fossil fishes, invertebrates and plants.

DIAGNOSIS.   Dental formula 4.1.2.3/4.1.2.4 (incisors, canine, pre-molars, molars); linguobuccally compressed molars with three main cusps in a straight alignment (Fig. 2a, b); uniquely derived among triconodonts in possessing spoon-shaped incisors (Fig. 2c). *Jeholodens jenkinsi* differs from morganucodontids in lacking the molar-interlocking mechanism (by cingular cuspule e and cusp b) found in morganucodontids,[6–8] in having weak labial cingula of the upper molars (Fig. 2a), in lacking lower cingular cuspules e, f and g (kuhneocone), and in lacking the angular process and the postdentary trough on the mandible (Fig. 2c). In *J. jenkinsi*, lower molar cusp a occludes into the valley-groove between cusps A and B of the opposite upper molar, thus differing from amphilestids and gobiconodontids, [9,10] in which lower cusp a occludes into the embrasure anterior to upper cusp B. *J. jenkinsi* also lacks cingular cuspules e and f in the lower molars of the latter groups. The lower molars are interlocked, with a crescent-shaped distal cusp d of the preceding molar fitted into the concave mesial margin of cusp b of the succeeding molar, which is diagnostic of the Triconodontidae.[11–14] The main cusp, a, of the new taxon is much higher than cusps b and c, a primitive character that is absent in triconodontids.[11–15] *J. jenkinsi* differs from most triconodontids, other than *Alticonodon*[11] and *Ichthyoconodon*,[14] in lacking the continuous lingual cingulum on the lower molars; it differs from *Alticonodon* and *Ichthyoconodon* in having a shelf-like cusp d.

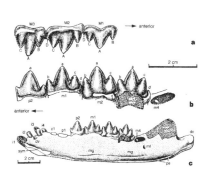

Figure 2 Dentition and mandible of *J. jenkinsi*. **a**, Upper molars M[1–3] (right side labial view). **b**, Lower postcanines p$_2$–m$_4$ (right side, lingual view; m$_4$ is in the process of eruption). **c**, Mandible (right side, lingual view, corrected for slight distortion in the original specimen). Upper molars are labelled in the figure as M1–M3. Lower teeth are labelled as m1–m4; p1, p2, i1–i4 and i2r. We follow ref. 7 for the designation of cusps A–D on the upper molars and cusps a–d on the lowers. For other abbreviations see Fig. legend. An analysis of mandibular and dental characters is found in Supplementary Information.

Lower incisor i$_2$ is about to be replaced by an erupting tooth (labelled i2r in Fig. 2c) and m$_4$ is erupting in the holotype of

Jeholodens jenkinsi. This dental replacement indicates that the individual probably had not reached a fully adult stage. We interpret the cusp pattern and the occlusion of laterally compressed molars in *J. jenkinsi* to be indicative of an insectivorous diet.

A noteworthy skeletal feature of this new taxon is a highly derived scapula (Figs 1a, 3a), confirming an earlier observation of a triconodontid from the Cretaceous of Montana.[12] It has a robust, peg-like acromion and a low elevation on the spine that resembles the metacromion of *Didelphis* (Fig. 3f); the supraspinous fossa is fully developed. The dorsal part of the scapula has a prominent triangular area (Figs 1 and 3), similar to the large attachment area for the teres major muscle in monotremes,[16] a condition also present in the archaic therian *Zhangheotherium.*[3] The clavicle is curved. Its lateral end has a tapering point with a reduced contact to the acromion, whereas in living monotremes the two structures have a rigid and broad articulation. This suggests that *Jeholodens* had a mobile scapuloclavicular articulation. Medially, the clavicle has a limited overlap with the lateral process of the interclavicle, and lacks the rigid claviculointerclavicle articulation found in tritylodont reptiles[16] and monotremes.[17] We interpret this joint to have had at least some degree of mobility, like the conditions in multi-tuberculates[18-20] and in archaic[3] and living[21] therians.

The distal end of the humerus bears the radial and ulnar condyles

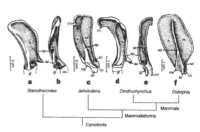

Figure 3 Mosaic evolution of the scapular characters in non-mammalian cynodonts and major mammalian clades. **a**, **b**, The tritylodontid *Bienotheroides* (lateral and anterior views of right scapula, after ref.16, representing the outgroup condition of cynodonts.[22] **c**, *J. jenkinsi* (lateral view, right scapula, composite reconstruction from both left and right scapulae of holotype GMV 2139 a). **d**, **e**, The monotreme *Ornithorhynchus* (Carnegie Museum specimen CM 1478; lateral and anterior views of right scapula). **f**, The marsupial *Didelphis* (lateral view, right scapula). For abbreviations, see Fig. 1 legend. Given the phylogeny in Fig. 5 and those of other references[3,27] the following derived characters of the pectoral girdle and forelimb of *J. jenkinsi* and therian mammals (including *Zhangheotherium*) would be best interpreted as convergences: a full supraspinous fossa; a protruding and laterally positioned acromion; coracoid fused to scapula; a ventral and uniformly concave scapular glenoid; mobile claviculoscapular and clavicleinterclavicle joints; a greatly reduced interclavicle; reduced ectepicondyles and entepicondyles of humerus; an incipient trochlea of humerus for the ulna. Alternatively, if the features of *J. jenkinsi* and therians are regarded as primitive for mammaliaforms as a whole, then those similarities between *Ornithorhynchus* and the outgroup tritylodonts would have to be interpreted as atavistic reversals in *Ornithorhynchus* to those of cynodonts; such features include: weak supraspinous fossa on the anterior aspect of scapula; acromion on the anterior margin of scapula; a lateral and saddle-shaped glenoid; and rigid and broad articulations of the clavicle–interclavicle and the scapula–clavicle. In either scenario, homoplasies are prominent in features of the pectoral girdles and forelimbs.[8,19,22] Clade ranks following ref. 27; see also the alternatives given in refs 8, 29.

on its anterior aspect, an incipient ulnar trochlea on its poster-oventral aspect, and reduced ectepicondyles and entepicondyles (on the counterpart slab; not shown), like the condition in *Zhangheotherium*.[3] These features differ from the primitive condition in cynodonts,[22] morganucodontids,[23] multituberculates[18–20] and living monotremes, in which the humerus has no ulnar trochlea. The epiphyseal suture is not present in the long bones in the holotype of *J. jenkinsi*.

In contrast to the derived forelimb and pectoral girdle of *Jeholodens jenkinsi*, its pelvic girdle, hindlimb and pes share many plesiomorphic characters with morganucodontids,[23] tritylodontids[16] and other cynodonts.[22] The epipubis is present. The patellar groove on the distal femur is far less developed than in monotremes, multituberculates and therians. The calcaneum (Fig. 4b) has a very short tubercle, a broad anterolateral ('peroneal') shelf, and extensive fibular and tibial facets, all of which are identical to those of morganucodontids[23] and non-mammalian cynodonts,[22] but different from the distinct peroneal process in monotremes, multituberculates[18,19] and therians[24] and from the elongate tubercle of multituberculates and therians. The astragalus has a broad and uniformly convex tibial facet, but a weakly developed neck and navicular facet. The astragalus and calcaneum contact each other in juxtaposition; this condition is

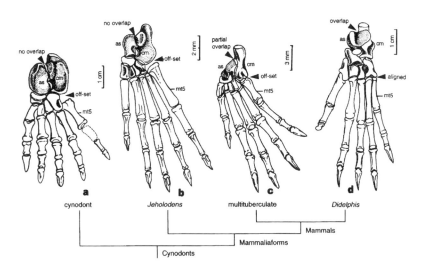

Figure 4 Pedes of mammals and outgroup cynodonts. **a**, A generalized cynodont condition (after refs 22, 28). **b**, *J. jenkinsi* (composite reconstruction of holotype GMV 2139 a,b). **c**, The multituberculate *Eucosmodon* (after refs 18, 19). **d**, The therian *Didelphis*. *J. jenkinsi* has little or no overlap of the astragalus and calcaneum, a primitive feature of cynodonts, contrasting with the partial overlap of these characters in multituberculates and the complete overlap in therians. Clade ranks follow ref. 27; see also the alternatives given in refs 8, 29.

similar to that in non-mammalian cynodonts[22] and morganucodontids,[23] but different from the partial overlap of these two ankle bones in multituberculates and the complete overlap of these bones in archaic[24] and living therians (Fig. 4). Metatarsal 5 is offset from the cuboid (Fig. 4), allowing a wide range of abduction of the lateral pedal digits, as in morganucodontids,[23] monotremes and multituberculates.[19] The manual and pedal ungual phalanges are slightly curved dorsally and concave on both sides, similar to those of most ground-dwelling small mammals.[19,25] A flexor tubercle on the midpoint in most of the ungual phalanges indicates a certain degree of ability to flex the claws. *J. jenkinsi* lacks the sesamoid ossicle(s) of the monotremes, multituberculates and therians. We interpret *J. jenkinsi* to have been a ground-dwelling small mammal with a plantigrade gait (Fig. 1b) and some capability for climbing on uneven substrates,[23] but not an arboreal mammal.

The eutriconodonts first appeared in the Middle Jurassic[15] and were quite diverse, with a worldwide distribution, in the Cretaceous.[9–14] Although abundantly represented by teeth[9–15] and some cranial[12,15,26] and postcranial[9] materials, no fully articulated skeleton was known, until now, for this diverse group. In traditional classifications of early mammals that were based primarily on dentition,[6,12,15] eutriconodonts and morganucodontids belonged to the order Triconodonta. The traditional grouping of 'triconodonts' (morganucodontids + eutriconodonts) is considered to be a grade, rather than a monophyletic group, according to more recent phylogenetic analyses.[26–28] Studies of some dental and cranial characters indicate that eutriconodonts may also be a heterogeneous group.[10,28]

It has been proposed that the postcranial features of cynodonts and early mammals were subjected to functional constraints of locomotion and therefore susceptible to homoplasies.[19,22,29] Several cladistic studies of cynodonts and early mammals that incorporated postcranial characters[27,30] indicated that dental characters are highly homoplastic, whereas the postcranial characters can be very informative for higher-level phylogeny. Because of the paucity of the relatively complete postcrania of the basal mammals, it is unknown whether (or to what degree) homoplasies exist among different postcranial skeletal parts (such as the forelimb versus the hindlimb) in early mammalian evolution.

The discovery of the first fully articulated triconodont skeleton offers an unprecedented opportunity for a more comprehensive assessment of the relationships of triconodonts, combining all evidence of the dentition, basicranium and postcranium, and will also allow us to elucidate the pattern of postcranial evolution in early mammals. Our phylogenetic analysis has placed *J. jenkinsi* in the basal part of the mammalian phylogenetic tree, outside the crown group of extant Mammalia (Fig. 5). Available evidence indicates that *J. jenkinsi* is a eutriconodont and far more derived than morganucodontids. Within eutriconodonts, the new taxon is more closely related to the triconodontid clade than it is to amphilestids and gobiconodontids (see Supplementary Information). These results are consistent with the ideas that triconodonts are paraphyletic[26–28] and that the triconodont-like molar cusp and occlusal patterns are characters for a functional grade.

Jeholodens jenkinsi shows a mosaic of derived, therian-like characters for most parts of the pectoral girdle (Fig. 3) and the humerus, but very primitive characters for the vertebral column, pelvic girdle, hindlimb and pes (Fig. 4). Our phylogenetic analysis indicates that many apomorphies of the pectoral girdle and forelimb in *J. jenkinsi* are independently derived and convergent with those of therians. Such apomorphies include a fully developed supraspinous fossa, the acromion and metacromion on the scapular spine, the incipient ulnar trochlea and the reduced epicondyles of the humerus. Given the phylogeny shown in Fig. 5, the mobile clavicle and scapula in *J. jenkinsi* would represent convergences with those of the multituberculate–therian clade.[3,20,27] Thus these pectoral characters, which allow a greater range of excursion of the shoulder joint in the locomotion of multituberculates[3,19,20] and living therians,[21] have evolved at least twice among the Mesozoic mammals.

Alternatively, the mobile joints between the clavicle, interclavicle and scapula could be ancestral conditions shared by *J. jenkinsi* and the more derived mammals. If so, then the rigid clavicle–interclavicle articulation and relatively immobile scapula of monotremes (Fig. 3) would have to be regarded as atavistic reversals to the primitive conditions in the more distantly related non-mammalian cynodonts.[16,22] For either evolutionary scenario, we must conclude that the pectoral girdles and

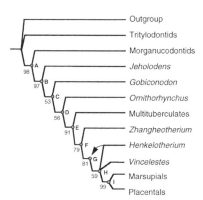

Figure 5 Evolutionary relationships of *J. jenkinsi*, gen. et sp. nov. The derived characters related to a mobile pectoral girdle occur separately on nodes B and E. The 'primitive' characters related to an immobile pectoral girdle occur in tritylodontids and the cynodont outgroup, and separately in *Ornithorhynchus* (node D). This indicates that there are many homoplasies of pectoral girdles and forelimbs (in contrast to the few or no convergences in the pelvic girdle and hindlimb), given the same tree topology of major clades of mammals. The arrow indicates an alternative placement of *Henkelotherium* at node G. For details of phylogenetic analysis see Methods and Supplementary Information.

The figure shows the cladogram with taxa: Outgroup, Tritylodontids, Morganucodontids, *Jeholodens*, *Gobiconodon*, *Ornithorhynchus*, Multituberculates, *Zhangheotherium*, *Henkelotherium*, *Vincelestes*, Marsupials, Placentals, with nodes A (98), B (97), C (53), D (56), E (91), F (79), G (81), H (59), I (99).

forelimbs of early mammals underwent extensive convergent evolution, not only by comparison with the dental and cranial features, but also in relation to more conservative features of the pelvis and hindlimbs.

Methods

Phylogeny of mammals (Fig. 5) is based on a strict consensus of two equally parsimonious trees (tree length = 210; consistency index = 0.638; retention index = 0.724) from PAUP analysis (3.1.1. Branch and Bound search) of 101 dental, cranial and postcranial characters that can be scored for the 12 major clades of mammals (see Supplementary Information). Most of the characters are preserved in the holotype of *Jeholodens jenkinsi*. The two most pasimonious trees differ only in the alternative placements of *Henkelotherium*, which either is at node G (represented by an arrow in Fig. 5) or switches positions with *Vincelestes*. These alternative placements of *Henkelotherium* do not alter the positions of any other clades, including *J. jenkinsi*. Numbers on branches represent the percentage of bootstrap values in 1,000 bootstrap replicas for a 50% majority bootstrap consensus tree that has identical topology to one of the two most parsimonious trees (that in which *Henkelotherium* is positioned at node G).

References

1. Ji, Q. & Ji, S.-A. Discovery of the earliest bird fossils in China and the origin of birds. *Chinese Geol.* **1996**, 30–33 (1996).
2. Hou, L., Martin, L. D., Zhou, Z. & Feduccia, A. Early adaptive radiation of birds: evidence from fossils from Northeastern China. *Science* **274**, 1164–1165 (1996).
3. Hu, Y., Wang, Y., Luo, Z. & Li, C. A new symmetrodont mammal from China and its implications for mammalian evolution. *Nature* **390**, 137–142 (1997).
4. Chen, P.-J., Dong, Z.-M. & Zhen, S.-N. An exceptionally well-preserved theropod dinosaur from the Yixian Formation of China. *Nature* **391**, 147–152 (1998).
5. Ji, Q., Currie, P. J., Norell, M. A. & Ji, S.-A. Two feathered dinosaurs from northeastern China. *Nature* **393**, 753–761 (1998).
6. Kermack, K. A., Mussett, F. & Rigney, H. W. The lower jaw of *Morganucodon*. *Zool. J. Linn. Soc.* **53**, 87–175 (1973).
7. Crompton, A. W. The dentitions and relationships of the southern African Triassic mammals *Erythrotherium parringtoni* and *Megazostrodon rudnerae*. *Bull. Br. Mus. Nat. Hist.* **24**, 399–437 (1974).
8. Luo, Z. in *In the Shadow of Dinosaurs—Early Mesozoic Tetrapods* (eds Fraser, N. C. & Sues, H.-D.) 980–128 (Cambridge Univ. Press, Cambridge, 1994).
9. Jenkins, F. A. Jr & Schaff, C. R. The Early Cretaceous mammal *Gobiconodon* (Mammalia, Triconodonta) from the Cloverly Formation in Montana. *J. Vert. Paleontol.* **6**, 1–24 (1988).
10. Kielan-Jaworowska, Z. & Dashzeveg, D. New Early Cretaceous amphilestid ('triconodont') mammals from Mongolia. *Acta Palaeont. Polonica.* **43**, 413–438 (1998).
11. Fox, R. C. Additions to the mammalian local fauna from the upper Milk River Formation (Upper Cretaceous), Alberta. *Can. J. Earth Sci.* **13**, 1105–1118 (1976).
12. Jenkins, F. A. Jr & Crompton, A. W. in *Mesozoic Mammals: The First Two-thirds of Mammalian History* (eds Lillegraven, J. A., Kielan-Jaworowska, Z. & Clemens, W. A.) 74–90 (Univ. Calif. Press, Berkeley, 1979).
13. Cifelli, R. L., Wible, J. R. & Jenkins, F. A. Jr Triconodont mammals from the Cloverly Formation (Lower Cretaceous), Montana and Wyoming. *J. Vert. Paleontol.* **16**, 237–241 (1998).
14. Sigogneau-Russell, D. Two possibly aquatic triconodont mammals from the Early Cretaceous of Morocco. *Acta Palaeont. Polonica* **40**, 149–162 (1995).

15. Simpson, G. G. *A catalogue of the Mesozoic Mammalia in the Geological Department of the British Museum* (Oxford Univ. Press, London, 1928).

16. Sun, A. & Li, Y. The postcranial skeleton of the late tritylodont *Bienotheroides. Vert. PalAsiat.* **23**, 136–151 (1985).

17. Klima, M. Die Frühentwicklung des Schültergürtels und des Brustbeins bei den Monotremen (Mammalia: Prototheria). *Adv. Anat. Embryol. Cell Biol.* **47**, 1–80 (1973).

18. Krause, D. W. & Jenkins, F. A. Jr The postcranial skeleton of North American multituberculates. *Bull. Mus. Comp. Zool.* **150**, 199–246 (1983).

19. Kielan-Jaworowska, Z. & Gambaryan, P. P. Postcranial anatomy and habits of Asian multituberculate mammals. *Fossils Strata* **36**, 1–92 (1994).

20. Sereno, P. & McKenna, M. C. Cretaceous multituberculate skeleton and the early evolution of the mammalian shoulder girdle. *Nature* **377**, 144–147 (1995).

21. Jenkins, F. A. Jr & Weijs, W. A. The functional anatomy of the shoulder in the Virginia opossum (*Didelphis virginiana*). *J. Zool.* **188**, 379–410 (1979).

22. Jenkins, F. A. Jr The postcranial skeleton of African cynodonts. *Bull. Peabody Mus. Nat. Hist. Yale Univ.* **36**, 1–216 (1971).

23. Jenkins, F. A. Jr & Parrington, F. R. Postcranial skeleton of the Triassic mammals *Eozostrodon, Megazostrodon,* and *Erythrotherium. Phil. Trans. R. Soc. Lond. B* **273**, 387–431 (1976).

24. Rougier, G. W. *Vincelestes neuquenianus* Bonaparte (Mammalia, Theria), un primitivo mammifero del Cretacico Inferior de la Cuenca Neuqina. Thesis, Univ. Nacional de Buenos Aires (1993).

25. McLeod, N. & Rose, K. D. Inferring locomotory behavior in Paleogene mammals via eigenshape analysis. *Am. J. Sci.* **293**, 300–355 (1993).

26. Rougier, G. W., Wible, J. R. & Hopson, J. A. Basicranial anatomy of *Priacodon fruitaensis* (Triconodontidae, Mammalia) from the Late Jurassic of Colorado, and a reappraisal of mammalia-form interrelationships. *Am. Mus. Novit.* **3183**, 1–28 (1996).

27. Rowe, T. Definition, diagnosis, and origin of Mammalia. *J. Vert. Paleontol.* **8**, 241–264 (1988).

28. Kielan-Jaworowska, Z. Characters of multituberculates neglected in phylogenetic analyses of early mammals. *Lethaia* **29**, 249–255 (1997).

29. Hopson, J. A. in *Major Features of Vertebrate Evolution* (eds Prothero, D. R. & Schoch, R. M.) 190–219 (Short Courses in Paleontol. No. 7, Paleontol. Soc. Knoxville, Tennessee, 1994).

30. Kemp, T. S. The relationships of mammals. *Zool. J. Linn. Soc.* **77**, 353–384 (1983).

For Supplementary information see page 200.

Acknowledgements

We thank K. C. Beard, R. L. Cifelli, W. A. Clemens, A. W. Crompton, M. R. Dawson, J. A. Hopson, F. A. Jenkins, Z. Kielan-Jaworowska, J. Meng, T. Rowe, D. Sigogneau-Russell, J. R. Wible and X.-c. Wu for discussions and for reviews of the manuscript; A. Henrici for preparation of the specimen; M. Klingler for preparing Fig. 1; and N. Wuethele for assistance. This research was supported by funding from the Ministry of Geology and Mineral Resources of China and the National Natural Science Foundation of China (to J.Q.), and the US National Science Foundation, the National Geographic Society, and the M. Graham Netting Fund of the Carnegie Museum (to Z.L.).

17

At the Roots of the Mammalian Family Tree
Timothy Rowe

An exceptionally complete skeleton, dating back roughly 140–150 million years, offers our closest look yet at the last common ancestor of modern mammals.

Mesozoic mammals are among the most challenging and prized entries on a fossil hunter's list of discoveries. The essence of the challenge is size—the Mesozoic history of mammals was played out by tiny animals. Few specimens survived the destructive agencies of fossilization, and those that did are supremely difficult to find and collect. Most are fragmentary, and most named species are based on isolated teeth and jaws. In nearly two centuries of searching, only a few precious complete specimens have been recovered and, without better fossils, long stretches of mammalian history have remained in the dark. But, on page 326 of this issue, Ji, Luo and Ji[1] describe one of the most complete and exquisitely preserved specimens ever found. It comes from the same Late Jurassic/Early Cretaceous deposit of Liaoning, China, that recently yielded spectacular feathered dinosaurs[2] and one other complete mammal skeleton.[3] The latest discovery helps to fill a wide gap in the fossil record, and brings new information to classic problems on the origin and interrelationships of early mammals.

The first Mesozoic mammals were discovered in 1812 by a mason in a tilestone quarry near Headington, England. These specimens came from the Middle Jurassic (roughly 165-Myr-old) Stonesfield Slate, and consisted of two isolated lower jaws, each belonging to a different species. Today, the Stonesfield jaws are still the oldest known fossils of the 'crown clade' Mammalia[4]—the lineage founded by the last common ancestor of living mammals (Fig. 1).

The tiny jaws quickly found their way to the University of Oxford where, in 1818, Baron Georges Cuvier examined them while on a sojourn in England. Renowned for his ability to judge the nature and affinities of an extinct animal from a part or even a single fragment of a skeleton,

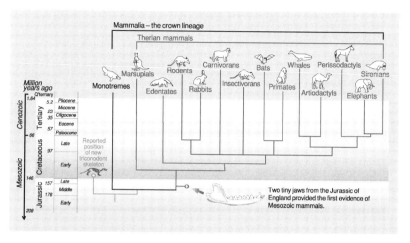

Figure 1 Timescale of mammalian evolution. The new skeleton discovered by Ji *et al.* 1 has been placed near the base of the crown lineage, Mammalia.

Cuvier pronounced the Stonesfield specimens to be mammalian.[5] He lived up to his reputation, and his identification was the first of many violations of what had been considered a very general rule—that mammals did not live during the Age of Reptiles.

A century and a half later and hundreds more Mesozoic mammal fossils had been discovered, yet the Stonesfield jaws remained among the most complete specimens known. Screen-washing techniques pioneered by Claude Hibbard in the 1940s offered the first clues to an unsuspected diversity of Mesozoic mammalian species.[6] Many tonnes of Mesozoic sediments were sieved through a series of screens designed to trap even the smallest fossils. But most recovered specimens consisted only of teeth and broken jaws, and the emerging view of early mammalian history became overly focused. In 1968, Alfred Sherwood Romer[7] admonished: "So great has been this concentration on dentitions that I often accuse my 'mammalian' colleagues, not without some degree of justice, of conceiving of mammals as consisting solely of molar teeth and of considering that mammalian evolution consisted of parent molar teeth giving birth to filial molar teeth and so on down through the ages."

The first great advance towards a more complete knowledge of the structure and relationships of Mesozoic mammals came in the 1960s, when the Polish Academy of Sciences sent a series of expeditions into central Asia. Dozens of Late Cretaceous mammal skeletons representing several different lineages were collected.[6] Throughout the 1990s, Asian expeditions led by the American Museum have been collecting hundreds more Late Cretaceous specimens that document, in even greater detail, the initial diversification of therian mammals (Fig. 1).[8,9]

Computer-assisted cladistic analyses of data from these more complete specimens[4,8,9] profoundly altered the picture of Late Cretaceous mammalian diversity that was painted in Romer's time.[10] For example, Romer's generation accepted that Mammalia arose in the Triassic (which immediately preceded the Jurassic), whereas the new analyses indicate that the last common ancestor of living mammals probably lived in the Early or Middle Jurassic. In other words, Mammalia is 20–40 Myr younger than once believed. Until very recently, however, the earliest details of mammalian history were obscured owing to the lack of complete fossils.

The remarkable specimen described by Ji and colleagues[1] is, along with a primitive therian mammal announced last year from the same locality,[3] by far the most complete and informative fossil discovered from a roughly 20-Myr or longer segment of Jurassic and Early Cretaceous time. Ji *et al.* present an analysis of evolutionary relationships, including dental evidence and data from throughout the skeleton, that places the new find very near the base of the mammalian crown clade (Fig. 1). Their analysis indicates that Triconodontidae—a group once believed to contain the direct ancestors of modern mammals—is not a natural group. Originally founded on the dental attributes that inspired its name, some triconodonts seem to be closer to mammals than to other so-called triconodonts.

But Ji and colleagues' specimen also highlights the presence of homoplasy—the independent evolution of similar features—in skeletal characters, corroborating an earlier finding that no region of the skeleton is immune to homoplasy.[11] Still, the data to be gleaned from the skeleton are strong enough to overthrow the apparent dental resemblance of the triconodonts to one another. And it is the skeletal characteristics that largely support the sister-group relationship of multituberculates (a long-lived and long-enigmatic lineage of extinct mammals) with therian mammals, once again contradicting hypotheses derived from dental evidence.

This beautiful specimen also offers new insight into what the ancestor of modern mammals was like. Working out when mammals first moved into the trees, and whether this happened more than once, has been problematic. Ji and colleagues' find indicates that mammals arose as terrestrial forms, and that only later did their therian descendants take to the trees.

Even with this spectacular new find, long gaps still punctuate our Mesozoic record of mammals and their extinct relatives. But this exciting Chinese locality has now produced so many exquisite tetrapod fossils that additional complete specimens of early mammals are likely to be unearthed. We can then expect rapid increases in the resolution of what was once the most fragmented segment of our early history.

References

1. Ji, Q., Luo, Z. & Ji, S.-A. *Nature* **398**, 326–330 (1999).
2. Ji, Q., Currie, P. J., Norell, M. A. & Ji, S.-A. *Nature* **393**, 753–761 (1998).
3. Hu, Y., Wang, Y., Luo, Z. & Li, C. *Nature* **390**, 137–142 (1997).
4. Rowe, T. *J. Vert. Paleontol.* **8**, 241–264 (1988).
5. Owen, R. *Geol. Trans* (2nd series) **6**, 47–65 (1839).
6. Lillegraven, J. A., Kielan-Jaworowska, Z. & Clemens, W. A. *Mesozoic Mammals— The First Two-thirds of Mammalian History* (Univ. California Press, Berkeley, 1979).
7. Romer, A. S. *Notes and Comments on Vertebrate Paleontology* (Univ. Chicago Press, 1968).
8. Novacek, M. J. *et al. Nature* **389**, 483–486 (1997).
9. Rougier, G. W., Wible, J. R. & Novacek, M. J. *Nature* **396**, 459–463 (1998).
10. Dingus, L. & Rowe, T. *The Mistaken Extinction—Dinosaur Evolution and the Origin of Birds* (Freeman, New York, 1997).
11. Rowe, T. in *Mammalian Phylogeny* (eds Szalay, F. S., Novacek, M. J. & McKenna, M. C.) 129–145 (Springer, New York, 1993).

18

A New Symmetrodont Mammal from China and Its Implications for Mammalian Evolution

Yao-Ming Hu, Yuan-Qing Wang, Zhe-Xi Luo, and Chuan-Kui Li

A new symmetrodont mammal has been discovered in the Mesozoic era (Late Jurassic or Early Cretaceous period) of Liaoning Province, China. Archaic therian mammals, including symmetrodonts, are extinct relatives of the living marsupial and placental therians. However, these archaic therians have been mostly documented by fragmentary fossils. This new fossil taxon, represented by a nearly complete postcranial skeleton and a partial skull with dentition, is the best-preserved symmetrodont mammal yet discovered. It provides a new insight into the relationships of the major lineages of mammals and the evolution of the mammalian skeleton. Our analysis suggests that this new taxon represents a part of the early therian radiation before the divergence of living marsupials and placentals; that therians and multituberculates are more closely related to each other than either group is to other mammalian lineages; that archaic therians lacked the more parasagittal posture of the forelimb of most living therian mammals; and that archaic therians, such as symmetrodonts, retained the primitive feature of a finger-like promontorium (possibly with a straight cochlea) of the non-therian mammals. The fully coiled cochlea evolved later in more derived therian mammals, and is therefore convergent to the partially coiled cochlea of monotremes.

Systematic Palaeontology

Class Mammalia
Subclass Theria
Order Symmetrodonta
Family Spalacotheriidae
Zhangheotherium quinquecuspidens gen. et sp. nov.

HOLOTYPE. IVPP V7466 (Institute of Vertebrate Paleontology and Paleoanthropology, Chinese Academy of Sciences, Beijing, China), a well-preserved skeleton (Fig. 1) consisting of a partial skull, and most of the postcranial skeleton, including all cervical and thoracic vertebrae, forelimbs and pectoral girdle, as well as the excellent impressions of the

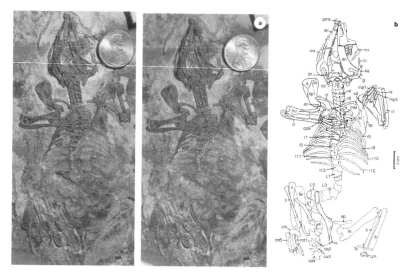

Figure 1 *Zhangheotherium quinquecuspidens* (IVPP V7466, holotype). Stereo-photographs (**a**) and outline (**b**) of the skeleton in ventral view of the dorsoventrally compressed specimen (broken lines indicate the morphology preserved in impressions that can be examined on the silicon rubber mould of the impressions). Vertical scale bar in **b** represents 1 cm. Abbreviations: ac, acromion of scapula; c1–c7, cervical vertebrae 1 to 7; ca1, canine; ca2–ca4, caudal vertebrae 2 to 4 (caudal vertebrae are incomplete); cd, coracoid process of scapula; cl, clavicle; cm, calcaneum; co? coronoid fossa?; cp 1–9, carpals 1 to 9; csp, crista parotica of petrosal; ctp, caudal tympanic recess of petrosal; dn, dentary; ep, epipubis; er, epitympanic recess; fc, fenestra cochlearis (round window); fe, femur; ff?, facial nerve foramen?; fi, fibula; frs, foramen for ramus superior; fst, fossa for stapedial muscle; fv, fenestra vestibuli (oval window); g, glenoid of the scapula; gj, groove for jugal (on the squamosal); gl, glenoid fossa of squamosal; hu, humerus; ic, interclavicle; if, infraspinous fossa; il, ilium; is, ischium; i1–i3, incisors 1 to 3; ju, jugal; L1–L6, lumbar vertebrae 1 to 6 (impressions only); lm, lambdoidal crest; mas, mastoid exposure of petrosal; mc, metacoracoid; mf, mandibular foramen; mg, Meckelian groove; mp1–mp5, metacarpals I–V; mt1–mt5, metacarpals I–V; mst, manubrium of sternum; mx, maxillary; oc, occipital condyle; of, obturator foramen of pectoral girdle; on, odontoid notch for dens of atlas; pc, procoracoid; pcd, condylar process of dentary; pcl, preclavicle (a homologue of procoracoid, *sensu* ref. 50); pcr, coronoid process of dentary; pf, pterygoid fossa; pgc, postglenoid crest; pgd, postglenoid depression; pmx, premaxillary; pp, paroccipital process of petrosal; ppe, paroccipital process of exoccipital (incomplete); pr, promontorium of petrosal; prs, prootic sinus canal; ptc, post-temporal canal; pts, post-tympanic sinus; p1 and p2, premolars 1 and 2; ra, radius; r1–r13, thoracic ribs 1 to 13 (posterior thoracic ribs preserved only in impressions); sc, scapula; sl?, fossa for splenial? on the dentary; sm, stylomastoid notch; sp, spine of scapula; sq, squamosal; ss, supraspinous fossa of scapula; stb1–stb6, sternebra 1 to sternebra 6; stl, embryonic sternal bands; sym, mandibular symphysis; s1–s4, sacral vertebrae 1 to 4 (represented mostly by impression); th, attachment site for tympanohyal; ti, tibia; ts, lateral tarsal spur of ankle; t1–t13, thoracic vertebrae 1 to 13; ul, ulna; x, xiphoid process of sternum.

hindlimbs and pelvic girdle, and ribs. Lumbar and sacral vertebrae are represented by poor impressions. Some tarsals and most caudal vertebrae are missing (for measurements, see Tables 1 and 2).

ETYMOLOGY.   Zhanghe, in honour of Zhang He, who collected and donated the holotype specimen to the Institute of Vertebrate Paleontology and Paleoanthropology; therium, beast (Greek); quinque, five (Latin);

cuspis, point (Latin); dens, tooth (Latin), for the three main cusps plus two large accessory cuspules on the lower molars.

LOCALITY.   Jianshangou Valley (approximately 41° 41′ 01″ N, 120° 59′ 30″ E), about 32 km east of Chaoyang City, Liaoning Province, northeastern China.[1]

HORIZON.   The Jianshangou Beds, consisting primarily of shales, are the lowest lacustrine intercalation in the neutro-basic volcanic beds of the Yixian Formation.[1–3]

ASSOCIATED FAUNA.   The Jianshangou Beds have yielded diverse fossil fish,[4] the birds *Confuciusornis, Liaoningornis*[5,6] and *Protarchaeopteryx*,[7] the theropod *Sinosauropteryx*,[8] and diverse gastropods, bivalves, ostracods, conchostracans and insects.[2]

AGE.   The age of the Yixian Formation is equivocal. Vertebrate faunal correlation[1,4,5] and previous radiometric dates[4] suggest that the Jianshangou Beds are either of the latest Jurassic age, or near the Jurassic–Cretaceous transition.[6] An Early Cretaceous age was also suggested by invertebrate faunal correlation,[3] and supported by a recent radiometric date.[9] According to our most recent field investigation, this date[9] should be regarded as an upper age limit for the Jianshangou Beds.

DIAGNOSIS.   A spalacotheriid symmetrodont (Fig. 2) with the dental formula 3·1·2·5/3·1·2·6; differing from all other known symmetrodonts in having a hypertrophied cusp B′ between the main cusp A and the stylocone (cusp B) (Fig. 2); distinguishable from other spalacotheriids[10–12] in the conical shape of the main cusps and the lack of lingual and labial cingulids on the lower molars;[11] distinctive from *Spalacotherium*[11] in having more robust and rounded main cusps that lack connecting cristae; unique among early mammals[13–18] in having fused sternebrae, and a posteriorly expanded xiphoid process.

Description and Comparison

Zhangheotherium resembles other known spalacotheriid symmetrodonts[10–12,19,20] in that the central cusp (a) and the two accessory cuspules (b, d) of the lower molar form an acute triangle. The associated upper

Table 1 Skeletal measurements of *Zhangheotherium quinquecuspidens*

Bone	Length (mm)
Dentary	30.2
Axis (C2)	5.0
Cervical vertebrae 3–7	10.0 (total length)
Thoracic vertebra 8	2.9
Caudal vertebra 2	3.5
Clavicle	11.6
Scapula (posterior border to glenoid)	17.2
Humerus	22.1
Ulna	21.5
Radius	17.1
Metacarpal III	6.5
Ilium	20.1
Ischium	8.5
Epipubis	9.1
Femur	22.0
Tibia	23.5
Fibula	23.5
Calcaneum	6.6
Matatarsal III	8.6

and lower teeth of *Zhangheotherium* could occlude into the embrasures of the opposing tooth row, a pattern unique to spalacotheriids among archaic therians.[10] This suggests that the teeth were used more for crushing and puncturing than shearing.[20]

Zhangheotherium is unique among spalacotheriids in having short and weak cingula in the upper molars, and a relatively low trigonid and two large accessory cuspules (cusps d and e) on the lower molars.[10–12,19,20] Upper molars have a wide labial shelf and their cusp pattern resembles that of *Peralestes*, a Late Jurassic symmetrodont from England.[10–12] Lower incisors are procumbent, with the first incisor I_1 being enlarged. C_1 is small and single-rooted, similar to $I_{2–3}$.

The dentary has a dorsally curved condylar process and a posteriorly tilted coronoid process. Like other symmetrodonts, it lacks an angular process (Fig. 2). The mandibular symphysis is oval, unfused and was probably mobile in life. The meckelian groove is narrow, and becomes shallow and faint anteriorly. The posterior part of the groove is separated from the mandibular foramen in the pterygoid fossa. The topographic relationships of these mandibular features (Fig. 2) are very similar to those of more derived therian mammals ('eupantotheres'),[10,16] but differs from that of the earliest known therian *Kuehneotherium* (Late Triassic), and from those of the non-therian mammals *Morganucodon*[21] and *Sinoconodon*.[22] A slightly rugose area along the anterior meckelian groove suggests the presence of the splenial, and a rough area near the base of the coronoid process is probably for a poorly developed coronoid, as occurs in the more derived therian *Henkelotherium* from the Upper Jurassic of Portugal.[16] No other postdentary bones are preserved.

The squamosal has an anteroposteriorly elongate glenoid fossa flanked by a low postglenoid crest (Fig. 3). The postglenoid region has a constricted neck between the glenoid and the broad cranial moiety that forms a wall lateral to the epitympanic recess. *Zhangheotherium* is similar in the latter two features to derived therians,[17,23,24] but is different from non-therian mammals

Table 2 Dental measurements of *Zhangheotherium quinquecuspidens*

Tooth	Length (mm)	Height (mm)
Lower dentition		
I1	1.2	
I2	0.5	
I3	0.5	
C	0.8	
P1	1.2	
P2	1.3	
M1	1.5	1.0
M2	1.6	(1.5)
M3	1.8	1.7
M4	1.7	1.8
M5	1.5	1.8
M6	1.1	1.0
Upper dentition		
M2		1.5
M3	(1.6)	1.5
M4	(1.6)	1.6
M5	1.8	1.5

Parentheses indicate approximate estimates.

which have a narrow cranial moiety of the squamosal but no squamosal wall for the epitympanic recess.[25,26] The postglenoid area of the squamosal has a posterolateral depression that resembles the broad external auditory meatus of *Vincelestes*,[17] the marsupial *Pucadelphys*[24] and placentals, in contrast to non-therian mammals in which the squamosal has no postglenoid region or the external auditory meatus.[21,22,27–29]

The petrosal bones that contained the inner ear are complete but fractured and distorted in preservation. The promontorium that houses the cochlea has a cylindrical and finger-like shape (Fig. 3). Its structure is very similar to those of *Sinoconodon*,[30] morganucodontids,[21,22,31] triconodonts[22,26] and multituberculates.[32–34] As documented for a wide range of early mammals,[30,31,33–35] a cylindrical and finger-like promontorium is closely correlated with either a straight or slightly curved (but uncoiled) cochlea. We therefore infer from the slender promontorium of *Zhangheotherium* (Fig. 3) that it has a more or less straight (definitely uncoiled) cochlea. *Zhangheotherium* is very different from derived therian mammals, which have oval-shaped and more bulbous promontoria with cochleae coiled for at least 270 degrees.[17,23,33,36]

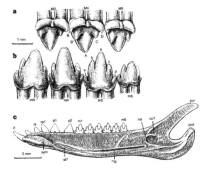

Figure 2 Dentition and mandible of *Zhangheotherium quinquecuspidens*. **a**, Upper molars M;[3–5] **b**, lower molars M$_{3-6}$ (left, labial views); **c**, mandible (right side, reconstruction in medial view). *Zhangheotherium* closely resembles *Spalacotherium* in lower molars, and *Peralestes* in upper molars. *Spalacotherium* and *Peralestes* were established on some dissociated lower and upper teeth, respectively, from the same fauna in the Late Jurassic of England.[10–12] The associated upper and lower dentitions of *Zhangheotherium* suggest that *Spalacotherium* and *Peralestes* are the lower and upper teeth of the same species, and that *Peralestes* should be synonymized with *Spalacotherium*.[10–12] Cusp B' is too large and positioned too far lingually to be the stylocone (cusp B). It could be either a homologue to the much smaller intermediate cusp in the comparable position in other symmetrodonts, or a neomorphic cusp. We follow ref. 45 for the designation of cusps A, B, C, D and E on the upper molars, and cusps a, b, c, d and e on the lower molars. For abbreviations, see Fig. 1 legend.

The petrosal has a prominent paroccipital process that is similar to those of morganucodontids,[21,22] triconodonts,[22,26] multituberculates and ornithorhynchids.[37] The posttemporal canal is positioned between the paroccipital region of the petrosal and lambdoidal crest of the squamosal, also a primitive character of mammals. *Zhangheotherium* has a large post-tympanic recess, and a broad and shallow epitympanic recess bound medially by a low crista parotica, and laterally by the squamosal. Both these features are characteristic of more derived therians.

Zhangheotherium has 7 cervical and 13 thoracic vertebrae. The postaxial cervical ribs remain unfused in adults, as in morganucodontids[13] and multituberculates.[38] All thoracic ribs are robust and compressed anteroposteriorly. Six lumbar vertebrae and three of four sacral vertebrae are represented by impressions. The sacral vertebrae and three preserved caudal vertebrae have wide transverse processes (Fig. 1).

The interclavicle is V-shaped. It is clearly articulated with, and overlaps the ventral side of the anterior part of, the sternal manubrium. Its sharp posterior process is continuous with the sternal keel (Fig. 4). The manubrium, although fused to other sternebrae, is still discernible by its facet for the first thoracic rib. The short and broad lateral process of the interclavicle has a loose and mobile articulation with the clavicle. *Zhangheotherium* and multituberculates share strong similarities in the articulations of the interclavicle to the clavicle and to the sternum.[18] The lateral end of the clavicle has a spiral articular surface for the acromion of the scapula, allowing some mobility in the clavicle–scapular joint. The triangular scapular blade is demarcated from the glenoid by a scapular notch. The supraspinous fossa is developed along the entire length of the blade, but is much narrower than the infraspinous fossa. The spine is very high, with a prominent acromion. The coracoid process is hook-like. The glenoid faces ventrally, and is much narrower than the spherical humeral head. *Zhangheotherium* has mobile articulations of the interclavicle, clavicle and scapula that allow the clavicle to move and act as a pivotal strut[39] for a wider rotation of the scapula, as observed during the locomotion of the opossum.[40]

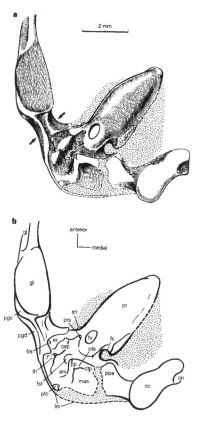

Figure 3 Reconstruction of the partial basicranium of *Zhangheotherium quiquecuspidens* (IVPP V7466). Right side, ventral view. Composite reconstruction (**a**) and outline with labels (**b**). Hatched pattern indicates the areas of damage or matrix. Arrows in **a** indicate the constricted neck of the squamosal. For abbreviations, see Fig. 1 legend.

The robust humerus of *Zhangheotherium* has a long deltopectoral crest. The intertubercular groove is narrow and deep, and separates the greater and lesser tubercles, the former being slightly wider than the latter. Torsion of the proximal end of humerus relative to the distal end is about 30 degrees, about the same as in *Henkelotherium*,[16] and close to the 40 degree torsion in *Vincelestes*.[17] Distally, the humerus has an incipient trochlea for the ulna, a therian apomorphy[16,17,41] that is absent in multituberculates.[14,18,42] The humerus also has a weakly developed ulnar condyle, resembling those of non-therian mammals.[13,14,38,41] The radial condyle is prominent and spherical, a primitive feature of non-therian mammals. Nine carpals are present. The pisiform is very large.

The pelvic bones are slender. The ilium is long and rod-like, and three times as long as the ischium. The acetabulum is partially preserved, showing a dorsal rim. Although the ischiopubic plate is not well preserved in IVPP V7466, its impressions indicate a shallow pelvis, differing from the deep pelvis of multituberculates.[38] The epipubis is as long as the ischiopubic plate. As in multituberculates,[14,38] the spherical femoral head is set off from the shaft by a well-defined neck, and the greater trochanter is directed dorsally. The calcaneum has an elongate tubercle and a well-defined peroneal tubercle. Its astragalar process contacted the fibula, a plesiomorphy for therians.[27] As in living monotremes and *Gobiconodon*,[15] *Zhangheotherium* has an external pedal spur. This spur is associated with a poisonous gland in the modern male platypus.

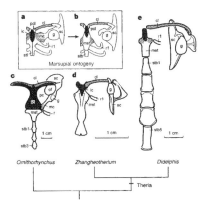

Figure 4 Comparison of the sternal apparatus and pectoral girdle of *Zhangheotherium* and living mammals. **a, b,** Embryonic (**a**) and adult (**b**) stages of the sternum and pectoral girdle in marsupials (modified from ref. 50). **c–e,** The adult sternal and girdle structures of *Ornithorhynchus* (modified from ref. 49) (**c**), *Zhangheotherium* (**d**) and *Didelphis* (CMNH c45, and several uncatalogued specimens at Carnegie Museum) (**e**). The chondral element of the interclavicle and the medial part(s) of the embryonic coracoscapular plate are considered to be incorporated into the sternal manubrium in adult marsupials.[49,50] *Zhangheotherium* has retained a separate interclavicle, a primitive character of non-therian mammals, but a more derived character of a mobile clavicle-interclavicle joint which is present in multituberculates and in a modified form in living therians. For abbreviations, see Fig. 1 legend.

Phylogenetic Implications for Early Mammals

The relationships of therians to extinct non-therian mammalian lineages have received much attention, as hypotheses of therian relationships are central to the understanding of early mammalian evolution.[27,43] Therians have been hypothesized to be the sister taxon of either multituberculates[18,26,27] or monotremes on the basis of some derived molar characters.[44] Basicranial studies[34,37] have suggested that monotremes and multituberculates were sister taxa to the exclusion of therians. There are few relatively complete fossils of archaic therians.[16,17] The discovery of the skeleton of *Zhangheotherium* offers the first opportunity to assess the phylogenetic relationships of early therians using nearly all dental, basicranial and postcranial characters.

Symmetrodont mammals were previously known only from teeth and jaws. Owing to the lack of better evidence, the affinities of this lineage to the more derived therians have been based solely on dental evidence.[12,43–46] It has been argued that dental characters are as homoplasic as non-dental characters,[27,43,47,48] and the reliability of dental characters for inferring the relationships of major lineages of mammals has been questioned. *Zhangheotherium* has provided more extensive basicranial and postcranial evidence to corroborate the traditional hypothesis[12,43–46] that symmetrodonts represent a part of the basal therian radiation. Our analysis (Fig. 5) suggests that *Zhangheotherium*, as the best-preserved taxon of all symmetrodonts, is a basal clade in therian phylogeny. It is more primitive than *Henkelotherium*[16] and *Vincelestes*[17] in retaining the interclavicle in its pectoral girdle. It also has a primitive dentition in which the lower molars have no distinct talonid, and the upper and lower molars occlude in the embrasures of the opposing tooth rows (Fig. 2), in contrast to more derived therians in which the lingual part of the upper molar occludes with the talonid of the lower molar.[16,45]

The sister-taxon relationship between therians and multituberculates[18,26,27] is strongly supported by the evidence from *Zhangheotherium*. Additional derived characters from *Zhangheotherium*, such as the clavicle–interclavicle joint and some features of the femur, corroborate the therian affinities of multituberculates (Fig. 5).

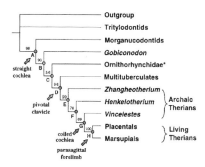

Figure 5 Phylogenetic relationships of *Zhangheotherium quinquecuspidens*. Asterisk: given this phylogeny, the coiled membranous labyrinth without corresponding coiling of the bony labyrinth in monotremes[28,33] is considered to be a convergence to the coiled cochlea of derived therians. See Methods for details of nodes A-G.

Implications for Anatomical Evolution

A large and reptile-like interclavicle is present in adult monotremes.[49] In living marsupials (Fig. 4), it first appears embryonically as a separate ossification(s), but later becomes incorporated, at least in part, within the sternal manubrium in the adult.[50] *Zhangheotherium* shows that the interclavicle was still present in at least some archaic therians. The loss of the interclavicle, or its incorporation into the sternal manubrium, occurred only in more derived therians. The reduced interclavicle and the mobile clavicle–interclavicle articulation in *Zhangheotherium* are far more derived than the large interclavicle and the rigid clavicle–interclavicle articulation in living monotremes (Fig. 4).

In contrast to the sprawling posture of living monotremes, most living therian mammals have a more parasagittal posture (*sensu* ref. 42), with the elbows positioned close to the thorax.[39–41] Such a forelimb posture has been hypothesized to have evolved only once in the common ancestry of multituberculates, living therians and their extinct relatives,[18] although this view has been contested.[42] Several osteological characters are considered to be crucial to the parasagittal posture of the forelimb of therians:[18,38,39,41] a mobile joint between the clavicle and sternal apparatus (including the interclavicle); a greater tubercle much wider than the lesser tubercle; a narrow intertubercular groove;[42] a humeral trochlea that constrains the movement of the ulna;[41] and possibly the lack of torsion of the humerus[18] (but see ref. 42). The skeleton of *Zhangheotherium* provides additional information on the phylogenetic distribution of these characters.

Zhangheotherium, *Vincelestes*, modern therians, and multituberculates all have a mobile joint between the clavicle and the sternal apparatus. This allows the clavicle to function as a pivot for the shoulder joint, and allows a greater range of rotation of the scapula during locomotion[39,40] than is possible in monotremes. However, torsion of the humerus and the large lesser tubercle relative to the greater tubercle in *Zhangheotherium*, *Henkelotherium* and *Vincelestes*[17] suggest that the humeri of these archaic therians were more abducted than those of most living therians. The humerus of *Zhangheotherium* has an incipient trochlea but also retains a vestigial ulnar condyle. The trochlea is also incipient in the Late Jurassic *Henkelotherium*,[16] but more developed in the Early Cretaceous *Vincelestes*.[17] This differs from the fully functioning (but more primitive) ulnar condyle of multituberculates.[14,18,38,42] The incipient trochlea in such archaic therians as *Zhangheotherium* and *Henkelotherium* represents an intermediate condition towards the less abducted posture of the forelimb in living therians.

These new data suggest that the mobility of the clavicle and scapula has a more ancient origin than the more parasagittal posture of the fore-limbs (Fig. 5). The mobile and pivotal clavicle evolved before the divergence of multituberculates and therians. The parasagittal forelimb posture of living therians was not present in such archaic therians as *Zhangheotherium, Henkelotherium*[16] and *Vincelestes*,[17] and it is best considered to have arisen later in therian evolution.

Although the cochleae of living monotremes and therian mammals all have some degree of coiling, it remains unclear whether a coiled cochlea evolved only once or in parallel among monotremes and therians. In early marsupials and placentals,[33,36] the promontorium that houses the cochlea is inflated and bulbous as a result of the cochlear coiling. Within the cochlear canal, the osseous spiral laminae of the bony labyrinth are well developed to support the membranous labyrinth.[36] In contrast, the cochleae of monotremes lack such internal bony structures.[28,33]

Because the finger-like promontorium is closely correlated to an un-coiled cochlea in diverse early mammals,[30,31,33-35] its presence in *Zhangheotherium* suggests that the coiled cochlea and the inflated promontorium of more derived therians[17,33,36] are not present in sym-metrodonts. Within the framework of mammalian phylogeny supported by many independent characters (Fig. 5), the coiled cochlea of therians would best be considered to have evolved later in such derived mammals as *Vincelestes*,[17] marsupials and placentals, independently of the partially coiled cochlea of monotremes (Fig. 5).

Conclusions

The nearly complete skeleton of *Zhangheotherium quinquecuspidens* has yielded new and more comprehensive anatomical information about the early therian mammals. New evidence on basicranial and postcranial anatomy from *Zhangheotherium* corroborates the hypothesis that sym-metrodonts are a part of the basal therian radiation. Postcranial features of this new therian mammal support a sister-group relationship between multituberculates and therian mammals. A mobile clavicle–interclavicle joint that allows a wide range of movement of the forelimb has an ancient origin in mammalian phylogeny. The abducted forelimb inferred for *Zhangheotherium* and other archaic therians suggests that early therian mammals lacked the more parasagittal forelimb posture of most living therians. The presence of a finger-like promontorium in *Zhangheotherium* indicates, albeit indirectly, that an uncoiled cochlea was present in sym-metrodonts and that the coiled cochlea was a development later in therian evolution.

Methods

PHYLOGENY OF MAMMALS. Based on a most parsimonious tree from paup (3.1.1. Exhaustive search) from 66 dental, cranial and postcranial characters that can be examined in *Zhangheotherium* (see Supplementary information for the character list and matrix). The most parsimonious tree has: tree length = 5 108, consistency index = 5 0.759, retention index = 0.840. Numbers on the branches represent the bootstrap value of 100 bootstrap replicas (the 50% majority bootstrap consensus tree and the most parsimonious tree have identical topology).

ABBREVIATED APOMORPHY LIST. **Node A** (Mammalia of ref. 43 or Mammaliaformes of ref. 27): prezygapophysis absent on axis; shallow patellar groove present on femur; notches for quadrate and quadratojugal absent in squamosal; promontorium present, unilateral occlusion of lower jaw; rotation of lower jaw during occlusion; differentiation of postcanine crowns into premolars and molars. Equivocal apomorphies: proatlas neural arch absent in adults; number of divided postcanine roots is not more than three. **Node B** (*Gobiconodon* plus the crown group of mammals): atlas ribs fused in adults; postaxial ribs fused in adults; sesamoid bone present in the pedal flexor tendon; a distinctive mandibular foramen fully formed; meckelian groove is vestigial or absent in adult; glenoid of scapula concave (instead of saddle shaped) and facing posteroventrally. Equivocal apomorphies: acetabular dorsal emargination closed; presence of external pedal (tarsal) spur; foramen for superior ramus of stapedial artery enclosed. **Node C** (the crown group of mammals, or the Mammalia of ref. 27): presence of well-developed patellar groove on femur; greater rotation of the mandible during occlusion; main cusps of the molars arranged in reversed triangle; more transverse orientation of protocristid. Equivocal apomorphies: absence of meckelian groove on the mandible (reversed in *Zhangheotherium* and *Henkelotherium*); main cusps of molar arranged in reverse triangle (lost in multituberculates). **Node D** (Theriiformes of ref. 27): mobile joint between clavicle and sterno-interclavicular apparatus; supraspinous fossa present on scapula; strong acromion extending below the level of glenoid of scapula; coracoid reduced to a process and fused to scapula; glenoid of scapula narrower than humeral head and facing posteroventrally; humeral and femoral heads spherical and strongly inflected; entepicondyle and ectepicondyle of humerus more reduced; styloid process of radius better developed; greater trochanter of femur directed dorsally; calcaneal tubercle elongate; presence of a separate peroneal tubercle; incipient superposition of astragalus over calcaneum; sesamoid bones in pedal flexor tendon paired; epitympanic recess in petrosal fully developed. Equivocal apomorphies: anterior sternal element (procoracoid) either fused with the manubrium sterni or lost; lesser trochantor of femur oriented ventromedially or ventrally; distinct tibial malleolus and fibular styloid process. **Node E** ('holotheres' of ref. 43): supraspinous fossa developed to the full length of scapula; flat medial surface of scapula; greater tubercle of humerus larger than the lesser; intertubercular groove very deep and narrow; squamosal with a postglenoid depression; cranial moiety of squamosal broad; post-tympanic recess present; craniomandibular joint anterior to fenestra vestibuli; crista interfenestralis limited to promontorium; squamosal wall flanking the epitympanic recess. Equivocal apomorphies: distal humerus forming a trochlea for ulnar articulation. **Node F**: talonid present on lower molar; wear facet present on talonid; protocristid of lower molar very transverse. Equivocal apomorphies: interclavicle absent in adults. **Node G** (Tribosphenida of ref. 43): upper molar with protocone; coronoid absent in adult; meckelian groove absent in adults. **Node H** (crown group of marsupials and placentals, or the Theria of ref. 27): manubria sterni small; the torsion of humerus weak; ulnar condyle on the humerus lost; fibula lost contact to calcaneum; cochlea elongate and coiled at least 360 degrees; lower molar talonid fully developed with a basin. Equivocal apomorphies: fusion of atlas neural arch. *Gobiconodon*: equivocal apomorphy: ulnar articulation on distal humerus forming a trochlea (convergent to therians). Ornithorhynchidae: equivocal apomorphies: postcanines with multiple roots (convergent to tritylodontids). Autapomorphy: lower molars with interoposteriorly compressed talonid. Multituberculates: equivocal apomorphy: meckelian groove absent in adults. Autapomorphy: deep peroneal groove.

References

1. Wang, Y.-Q., Hu, Y.-M., Zhou, M.-Z. & Li, C.-K. in *Sixth Symposium on Mesozoic Terrestrial Ecosystems and Biota, Short Papers* (eds Sun, A.-L & Wang, Y.-Q.) 221–227 (China Ocean Press, Beijing, 1995).
2. Chen, P.-J. *et al.* Studies on the Late Mesozoic continental formations of western Liaoning. *Bull. Nanjing Inst. Geol. Palaeontol. Acad. Sin.* 1, 22–55 (1980). (In Chinese.)
3. Chen, P.-J. & Chang, Z.-L. Nonmarine Cretaceous stratigraphy in eastern China. *Cretaceous Res.* 15, 245–257 (1994).
4. Jin, F. New advances in the Late Mesozoic stratigraphic research of Western Liaoning, China. *Vert. PalAsiat.* 34, 102–122 (1996).
5. Hou, L.-H., Zhou, Z.-H., Martin, L. D. & Feduccia, A. A beaked bird from the Jurassic of China. *Nature* 377, 616–618 (1995).
6. Hou, L.-H., Martin, L. D., Zhou, Z.-H. & Feduccia, A. Early adaptive radiation of birds: evidence from fossils from Northeastern China. *Science* 274, 1164–1165 (1996).
7. Ji, Q. & Ji, S.-A. *Protarchaeopteryx*, a new genus of Archaeopteridae of China. *Chin. Geol.* 1997 (3), 38–41 (1997). (In Chinese.)
8. Ji, Q. & Ji, S.-A. Discovery of the earliest bird fossils in China and the origin of birds. *Chin. Geol.* 1996 (10), 30–33 (1996). (In Chinese.)
9. Smith, P. E. *et al.* Date and rates in ancient lakes: ^{40}Ar–^{39}Ar Evidence for an early Cretaceous age for the Jehol group, Northeast China. *Can. J. Earth Sci.* 32, 1426–1431 (1995).
10. Simpson, G. G. *A Catalogue of the Mesozoic Mammalia in the Geological Department of the British Museum* (Oxford Univ. Press, London, 1928).
11. Clemens, W. A. Late Jurassic mammalian fossils in the Sedgwick Museum, Cambridge. *Palaeontology* 6, 373–377 (1963).
12. Cassiliano, M. L. & Clemens, W. A. Jr in *Mesozoic Mammals: the First Two-thirds of Mammalian History* (eds Lillegraven, J. A., Kielan-Jaworowska, Z. & Clemens, W. A.) 150–161 (Univ. California Press, Berkeley, 1979).
13. Jenkins, F. A. Jr & Parrington, F. R. Postcranial skeleton of the Triassic mammals *Eozostrodon*, *Megazostrodon*, and *Erythrotherium*. *Phil. Trans. R. Soc. Lond. B* 273, 387–431 (1976).
14. Krause, D. W. & Jenkins, F. A. Jr The postcranial skeleton of North American multituberculates. *Bull. Mus. Comp. Zool.* 150, 199–246 (1983).
15. Jenkins, F. A. Jr & Schaff, C. R. The Early Cretaceous mammal *Gobiconodon* (Mammalia Triconodonta) from the Cloverly Formation in Montana. *J. Vert. Paleontol.* 6, 1–24 (1988).
16. Krebs, B. Das Skelett von *Henkelotherium guimarotae* gen. et sp. nov. (Eupantotheria Mammalia) aus dem Oberen Jura von Portugal. *Berlin. Geowiss. Abh. A* 133, 1–110 (1991).
17. Rougier, G. W. *Vincelestes neuquenianus* Bonaparte (Mammalia, Theria), un primitivo mammifero del Cretacico Inferior de la Cuenca Neuqina. Thesis, Univ. Nacional de Buenos Aires (1993).
18. Sereno, P. & McKenna, M. C. Cretaceous multituberculate skeleton and the early evolution of the mammalian shoulder girdle. *Nature* 377, 144–147 (1995).
19. Fox, R. C. Upper molar structure in the Late Cretaceous symmetrodont *Symmetrodontoides* Fox, and a classification of the Symmetrodonta (Mammalia). *J. Paleontol.* 59, 21–26 (1985).
20. Cifelli, R. Cretaceous mammals of southern Utah. III. Therian mammals from the Turonian (Early Late Cretaceous). *J. Vert. Paleontol.* 10, 332–345 (1990).
21. Kermack, K. A., Mussett, F. & Rigney, H. W. The skull of *Morganucodon*. *Zool. J. Linn. Soc.* 71, 1–158 (1981).
22. Crompton, A. W. & Luo, Z. in *Mammal Phylogeny* (eds Szalay, F. S., Novacek, M. J. & McKenna, M. C.) 30–44 (Springer, New York, 1993).

23. Rougier, G. W., Wible, J. R. & Hopson, J. A. Reconstruction of the cranial vessels in the Early Cretaceous mammal *Vincelestes neuquenianus*: implications for the evolution of the mammalian cranial vascular system. *J. Vert. Paleontol.* **12**, 188–216 (1992).

24. Marshall, L. G. & Muizon, C. de in *Pucadelphys andinus (Marsupialia, Mammalia) from the early Paleocene of Bolivia* (ed. Muizon, C. de) *Mém. Mus. Natl Hist. Nat. Paris* **165**, 21–90 (1995).

25. Luo, Z. & Crompton, A. W. Transformations of the quadrate (incus) through the transition from nonmammalian cynodonts to mammals. *J. Vert. Paleontol.* **14**, 341–374 (1994).

26. Rougier, G. W., Wible, J. R. & Hopson, J. A. Basicranial anatomy of *Priacodon fruitaensis* (Triconodontidae, Mammalia) from the Late Jurassic of Colorado, and a reappraisal of mammaliaform interrelationships. *Am. Mus. Novit.* **3183**, 1–28 (1996).

27. Rowe, T. 1988. Definition, diagnosis, and origin of Mammalia. *J. Vert. Paleontol.* **8**, 241–264 (1988).

28. Zeller, U. Die Entwicklung und Morphologie des Schädels von *Ornithorhynchus anatinus* (Mammalia: Prototheria: Monotremata). *Abh. Senckenb. Naturf. Gesel.* **545**, 1–188 (1989).

29. Wible, J. R. Origin of Mammalia: the craniodental evidence reexamined. *J. Vert. Paleontol.* **11**, 1–28 (1991).

30. Luo, Z., Crompton, A. W. & Lucas, S. G. Evolutionary origins of the mammalian promontorium and cochlea. *J. Vert. Paleontol.* **15**, 113–121 (1995).

31. Graybeal, A., Rosowski, J., Ketten, D. R. & Crompton, A. W. Inner ear structure in *Morganucodon*, an early Jurassic mammal. *Zool. J. Linn. Soc.* **96**, 107–117 (1989).

32. Kielan-Jaworowska, Z., Presley, R. & Poplin, C. The cranial vascular system in taeniolabidoid multituberculate mammals. *Phil. Trans. R. Soc. Lond. B* **313**, 525–602 (1986).

33. Luo, Z. & Ketten, D. R. CT scanning and computerized reconstructions of the inner ear structure of multituberculate mammals. *J. Vert. Paleontol.* **11**, 220–228 (1991).

34. Meng, J. & Wyss, A. Monotreme affinities and low-frequency hearing suggested by multituberculate ear. *Nature* **377**, 141–144 (1995).

35. Lillegraven, J. A. & Krusat, G. Cranio-mandibular anatomy of *Haldanodon exspectatus* (Docondontia; Mammalia) from the Late Jurassic of Portugal and its implications to the evolution of mammalian characters. *Contrib. Geol.* **28**, 39–138 (1991).

36. Meng, J. & Fox, R. C. Therian petrosals from the Oldman and Milk River Formations (Late Cretaceous), Alberta, Canada. *J. Vert. Paleontol.* **15**, 122–130 (1995).

37. Wible, J. R. & Hopson, J. A. in *Mammal Phylogeny* Vol. 1 (eds Szalay, F. S., Novacek, M. J. & McKenna, M. C.) 45–62 (Springer, New York, 1993).

38. Kielan-Jaworowska, Z. & Gambaryan, P. P. Postcranial anatomy and habits of Asian multituberculate mammals. *Fossils & Strata* **36**, 1–92 (1994).

39. Jenkins, F. A. Jr The movement of the shoulder in claviculate and aclaviculate mammals. *J. Morphol.* **144**, 71–84 (1974).

40. Jenkins, F. A. Jr & Weijs, W. A. The functional anatomy of the shoulder in the Virginia opossum (*Didelphis virginiana*). *J. Zool.* **188**, 379–410 (1979).

41. Jenkins, F. A. Jr The functional anatomy and evolution of the mammalian humero-ulnar joint. *Am. J. Anat.* **137**, 281–298 (1973).

42. Gambaryan, P. P. & Kielan-Jaworowska, Z. Sprawling versus parasagittal stance in multituberculate mammals. *Acta Palaeontol. Pol.* **42**, 13–44 (1997).

43. Hopson, J. A. in *Major Features of Vertebrate Evolution* (eds Prothero, D. R. & Schoch, R. M.) 190–219 (Paleontological Society Short Courses, Knoxville, TN, 1994).

44. Kielan-Jaworowska, Z., Crompton, A. W. & Jenkins, F. A. Jr The origin of egg-laying mammals. *Nature* **326**, 871–873 (1987).

45. Crompton, A. W. in *Early Mammals* (eds Kermack, D. M. & Kermack, K. A.) 65–87 (Academic, London, 1971).

46. Crompton, A. W. & Jenkins, F. A. Jr in *Mesozoic Mammals: the First Two-thirds of Mammalian History* (eds Lillegraven, J. A., Kielan-Jaworowska, Z. & Clemens, W. A.) 59–72 (Univ. California Press, Berkeley, 1979).

47. Kemp, T. S. The relationships of mammals. *Zool. J. Linn. Soc.* **77**, 353–384 (1983).

48. Rowe, T. in *Mammal Phylogeny* Vol. 1 (eds Szalay, F. S., Novacek, M. J. & McKenna, M. C.) 129–145 (Springer, New York, 1993).

49. Klima, M. Die Frühentwicklung des Schültergürtels und des Brustbeins bei den Monotremen (Mammalia: Prototheria). *Adv. Anat. Embryol. Cell Biol.* **47**, 1–80 (1973).

50. Klima, M. Early development of the shoulder girdle and sternum in marsupials (Mammalia: Metatheria). *Adv. Anat. Embryol. Cell Biol.* **109**, 1–91 (1987).

For supplementary information see page 236.

Acknowledgements

We thank K. C. Beard, M. Dawson, R. Fox, Z. Kielan-Jaworowska, B. Krebs, T. Martin, J. Meng, R. Presley, G. Rougier, P. Sereno, A. Sun and J. Wible for suggestions on the manuscript; L.-H. Hou for field assistance; G.-H. Cui for photography; and H. Zhang for preparation. Research was supported by National Natural Science Foundation of China (to C.L.), the Rea Postdoctoral Fellowship of Carnegie Museum (to Y. W.), National Science Foundation of USA, National Geographic Society, and the M. Graham Netting Fund of Carnegie Museum (to Z.L.)

19

Biostratigraphy of New Pterosaurs from China

Shu-An Ji, Qiang Ji, and Kevin Padian

Pterosaurs are represented in China by five genera and some isolated bones rang-
ing in age from the Middle Jurassic to the Late Cretaceous period.[1,2] Four of these
genera belong to the derived monophyletic subgroup Pterodactyloidea; only the
Middle Jurassic *Angustinaripterus* from Dashanpu, Sichuan, is a non-pterodactyloid
(traditionally 'rhamphorhynchoid', a paraphyletic taxon). Two further pterosaurs[1,2]
(Fig. 1) from the Chaomidianzi Formation of the Beipiao area, western Liaoning
Province, occur in the Liaoning beds, several metres higher than the compsognathid coelurosaur *Sinosauropteryx* and the basal bird *Confuciusornis*. Our analysis
of these two fossils and other components of the fauna suggest a Late Jurassic biostratigraphic age for the Liaoning beds, which are important in the study of avian
origins.

One new pterosaur, *Dendrorhynchus curvidentatus*,[1] is morphometrically
most similar to the Late Jurassic (Tithonian) Solnhofen form *Rhamphorhynchus* but does not fit within that taxon. Principal components
analysis (by K. I. Warheit and K. P.) (Fig. 2) of the long bones of various
pterosaurs indicates that *Dendrorhynchus* clusters most closely with
Rhamphorhynchus and has no obvious unique features, but its proportions differ from those of the ontogenetic trajectories of *Rhamphorhynchus*. Instead, like *Scaphognathus*, it is distinct at the generic level.
The generic name *Dendrorhynchus* is preoccupied by a nemertine,[3] so we
propose replacing it with the name *Dendrorhynchoides*.

The second new fossil, *Eosipterus yangi*[2] (Fig. 1c), is a large pterodactylid with a wingspan of about 1.25 metres. The incomplete fusion of
the carpals and tarsals indicates that this specimen, although large for a
pterodactylid, was not fully adult. A plot of the second and third principal
components (the first was size) of the preserved long bones of *Eosipterus*
(Fig. 2) groups *Eosipterus* with the Late Jurassic Solnhofen pterodactylids
Pterodactylus kochi, *P. antiquus* and *Germanodactylus*, which have been

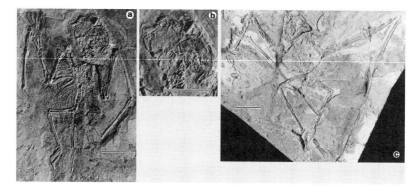

Figure 1 The new Liaoning pterosaurs. **a**, *Dendrorhynchoides ("Dendrorhynchus") curvidentatus* n. g. (Ji & Ji, 1998) (holotype, National Geological Museum of China GMV2128). Scale bar, 20 mm. **b**, Detail of *D. curvidentatus* skull. Scale bar, 20 mm. **c**, *Pterodactylus ("Eosipterus") yangi* Ji & Ji, 1997 (holotype, GMV2117). Scale bar, 50 mm.

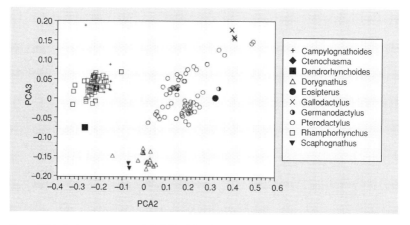

Figure 2 Principal components analysis (PCA). PCA plot of second and third major axes, comparing radius, wing metacarpal, first two wing phalanges, and tibia in *Campylognathoides, Ctenochasma, Dendrorhynchoides,*[1] *Dorygnathus, "Eosipterus,"*[2] *Gallodactylus, Germanodactylus, Pterodactylus, Rhamphorhynchus* and *Scaphognathus*. *Dendrorhynchoides* differs significantly from *Rhamphorhynchus* along the third major axis; "*Eosipterus*" falls within the cluster of points of the congenerics *Pterodactylus, Germanodactylus* and Ctenochasma[4] for pairwise comparisons of PCAs 1–3, and so is probably a large specimen of *Pterodactylus* (including *Germanodactylus*).

viewed as part of a single ontogenetic series[4] of *P. antiquus*, with which *Eosipterus* may be synonymous.

The age of the Liaoning beds has been argued to be Late Jurassic or Early Cretaceous, and radiometric dates give conflicting results.[5] The biostratigraphy of the fauna through the entire section of the Yixian and Chaomidianzi formations is ambiguous. At Sihetun (in the Beipiao area), fish, frogs, turtles, lizards and mammals have been found, as well as theropod and sauropod (saurischian) and psittacosaurid (ornithischian) dinosaurs.[6,7] Among the theropods, *Sinosauropteryx*, *Protarchaeopteryx*, *Caudipteryx* and *Confuciusornis*[8,10] are of primary importance to studies of the origins of birds and avian features.

When exact faunal equivalents are not available, next-of-kin taxa can provide at least minimal divergence times. Psittacosaurid dinosaurs have so far been known only from the Early Cretaceous,[7] but the separation of these marginocephalians from other ornithischian dinosaurs was no later than Early Jurassic, perhaps much earlier.[11] *Sinosauropteryx* is the most closely related of the Liaoning coelurosaurs to the Late Jurassic basal coelurosaur *Compsognathus* from the Solnhofen limestones of Germany.[8] *Protarchaeopteryx*[3] is a basal maniraptoran, as are avians such as the Solnhofen form *Archaeopteryx*, and the Liaoning form *Caudipteryx* has been linked basally to these taxa.[3] *Confuciusornis* is the next most basal bird known after *Archaeopteryx*. The coelurosaurian lineages therefore provide a biostratigraphic signal of sister-taxa rooted in the Late Jurassic. None of these specific genera is known from the Early Cretaceous, although related lineages persisted.

The two new Liaoning pterosaurs provide a similar signal by clustering with Late Jurassic relatives. Pterodactyloids are known from both the Late Jurassic and the Cretaceous, but non-pterodactyloid pterosaurs are not reliably known from any Cretaceous deposits.[12] The Jurassic-Cretaceous transition among pterosaurs may therefore be sharper than in some other tetrapod faunal components, and the available data from coelurosaurs and particularly pterosaurs suggest a Late Jurassic age for the beds in which they are found. In contrast, the earliest Cretaceous faunas, such as that of the Wealden, bear little resemblance to the Liaoning and Solnhofen faunas.[13] But the most basal Cretaceous faunas must be better known before any biostratigraphic hypothesis can be considered iron-clad.

References

1. Ji, S.-A. & Ji, Q. *Jiangsu Geol.* **22**, 199–206 (1998).
2. Ji, S.-A. & Ji, Q. *Acta Geol. Sin.* **71**, 115–121 (1997).
3. Yin, Z. & Zeng, F. *Oceanol. Limnol. Sin.* **16**, 323–335 (1995).
4. Bennett, S. C. *J. Vert. Paleontol.* **16**, 432–444 (1996).
5. Ji, Q. *et al. Prof. Pap. Stratigr. Paleontol.* (in the press).
6. Wang, X.-L. *et al. Vert. PalAsiat.* **36**, 81–101 (1998).
7. Xu, X. & Wang, Y. *Vert. PalAsiat.* **36**, 147–158 (1998).
8. Chen, P.-J., Dong, Z.-M. & Zhen, S.-N. *Nature* **391**, 147–152 (1998).
9. Ji, Q., Currie, P. J., Norell, M. A. & Ji, S.-A. *Nature* **393**, 753–761 (1998).
10. Hou, L.-H., Martin, L. D., Zhou, Z.-H. & Feduccia, A. *Science* **274**, 1164–1167 (1996).
11. Padian, K. in *Encyclopedia of Dinosaurs* (eds Currie, P. J. & Padian, K.) 549 (Academic, San Diego, 1997).
12. Padian, K. *Modern Geol.* **23**, 57–68 (1998).
13. Norman, D. B. in *Encyclopedia of Dinosaurs* (eds Currie, P. J. & Padian, K.) 783–785 (Academic, San Diego, 1997).

20

Palaeobiology: A Refugium for Relicts

Zhe-Xi Luo

The transition between the Late Jurassic and Early Cretaceous, 157–100 million years (Myr) ago, is a defining time in the history of terrestrial biodiversity. It witnessed the descent of birds from two-legged, meat-eating dinosaurs,[1,2] and their early diversification;[3-5] the rapid evolution of non-bird dinosaurs; diversification of the major mammalian groups;[6,7] the origins of flowering plants (angiosperms);[8] and diversification of the nectar-feeding flies—the finest early examples of plant–insect coevolution.[9,10] These emerging lineages came to dominate the world's terrestrial biotas, and some are still thriving today.

A window with a grand view of this evolutionary spectacle is the Yixian Formation in China's Liaoning Province, one of the richest and most important sources of fossils from the Mesozoic era (245–65 Myr ago). On page 58 of this issue,[11] Carl Swisher, Yuan-qing Wang and their colleagues show that the Yixian beds are 124 Myr old, placing them firmly within the Early Cretaceous. Such precisely dated rich fossil assemblages are rare in the Early Cretaceous,[12] making the new work a welcome step towards a better temporal calibration of the Mesozoic terrestrial biotas.

A cornucopia of beautiful fossils has been unearthed, in unprecedented quantities, from the Yixian Formation. The discovery of such exceptionally complete fossils has filled the gaps in our knowledge about lineages that were previously represented only by sparse, fragmentary fossils. Although the Yixian beds have yielded fossil fish, plants and arthropods since the 1920s, most of these species are restricted to eastern and central Asia. Likewise, the abundant dinosaurs, such as the horned psittacosaurs and enigmatic therizinosaurs found in the Yixian Formation, were also endemic to Asia during the earliest Cretaceous.[13] Without fossils from outside Asia, it has been difficult to correlate the Yixian biota with the worldwide geological timescale.

Ideally, to date fossil beds we need 'index' fossils that are abundant, broadly distributed and have rapid evolution. Age correlation of terrestrial biotas—as in the case of the Yixian—can be more difficult, because the preservation of terrestrial fossils is rarer, less complete and more uneven geographically than in marine environments. So, for decades, divergent opinions have waxed and waned about the age of the Yixian biota, which was suggested to be either Late Jurassic or Early Cretaceous—or anywhere in between.

The controversy has implications for both temporal and geographical patterns in the evolution of Mesozoic plants and animals, as many fossils from the Yixian beds occupy critical positions on their respective family trees. For instance, *Sinosauropteryx*[1] is the most primitive coelurosaurian dinosaur with downy feathers. *Protarchaeopteryx* and *Caudipteryx*[2] are basal in the maniraptoran dinosaur lineage that also includes *Archaeopteryx* and other birds, suggesting that true feathers evolved before the origin of avian flight. Other examples include the symmetrodont mammal *Zhangheotherium*,[6] which is basal to the diverse spalacotheriid tree,[14] and *Jeholodens*, a close relative of the triconodontids[7] that were found around the world in the Late Jurassic and Early Cretaceous. Then there are *Confuciusornis* and *Changchengornis*, the primitive beaked birds capable of powered flight, which are relatives of the enantiornithine birds that dominated the global avian faunas in the Cretaceous.[3,15]

Similar examples abound in plants and invertebrates. The primitive angiosperm *Archaefructus*, for instance, has a typical angiosperm carpel (the female reproductive structure) but no flower petals.[8] It tells us much about the early anatomical evolution of flowering plants. Angiosperm plants and the flies that pollinate them are thought to have first diversified in the temperate climate of East and Central Asia, before spreading elsewhere.[8,9] This would be consistent with a Late Jurassic age for the Yixian Formation, and could mean that northeastern China was a source for the emergence and early diversification of many evolutionary lineages that dominated the world's biota in the Cretaceous and the Cenozoic (65 Myr to present).

In harmony with this view, several vertebrates from the Yixian Formation have been shown to be closely clustered with European and North American relatives from the Late Jurassic (Fig. 1). For instance, the flying reptiles (pterosaurs) from Yixian closely resemble those of the Jurassic Solnhofen limestone in Germany.[16] Likewise, the Yixian Formation's *Sinosauropteryx* and Solnhofen's *Compsognathus* are very similar.[1,16] The two new mammals from Yixian, *Zhangheotherium*[6,14] and *Jeholodens*,[7] are both related to groups common in the Late Jurassic of Europe and

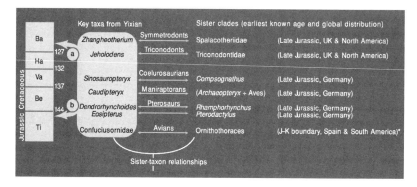

Figure 1 Correlation of taxa from the Yixian Formation with their close relatives from the Lower Jurassic or earliest Cretaceous in other continents.

North America. Such a parsimonious correlation by vertebrate sister taxa[17] on an intercontinental scale tends to suggest a Late Jurassic age for the Yixian fossil assemblage.

But Swisher *et al.*[11] have now found that the sediments that yielded the feathered dinosaurs and mammals in the Yixian Formation date back to the Early Cretaceous. They have based this figure on measurements taken from sanidine, a mineral that is known to be reliable for determining the $^{40}Ar/^{39}Ar$ isotope age. The authors collected samples from the volcanic tuffs that are a part of the original sediments that bear the fossils—so, the association of the dated samples and the fossils is unambiguous.

All of this means that the primitive organisms from the Yixian Formation are best interpreted as long-lived relics, extending from the Late Jurassic into the Early Cretaceous.[11] A corollary is that western Liaoning province and its neighbouring areas are a refugium—an isolated area in which a population survived much longer than elsewhere—for these relict lineages from a bygone era. In other words, *Sinosauropteryx*, the primitive pterosaurs and pre-flower angiosperm plants should be considered as 'living fossils' of their time. They survived into the Early Cretaceous only in Yixian, having previously been widely distributed around the world in the Late Jurassic. This agrees with the observation that non-theropod dinosaur faunas of eastern and central Asia were isolated from the rest of the world during the Late Jurassic and earliest Cretaceous.[13]

Primitive fossil taxa may not always be older than their more derived relatives. Terrestrial fossils are often incompletely preserved and, of those that are, only a fraction are ever discovered. Where the primitive 'stage of evolution' for a taxon does not match up with its apparently young age (compared with its more derived relatives), it is reasonable to infer that this taxon had a 'ghost lineage', or an earlier history that must have existed but is not yet documented by the fossil record.[18] So, an Early Cretaceous age for the Yixian Formation is not inconsistent with its seemingly archaic biota, in view of its restricted geographical distribution in China and Central Asia.

Had Swisher and colleagues come up with a Late Jurassic age, the Yixian Formation could have been viewed as a paradise for the origination and early diversification of many groups that are still thriving today. As it is, though, the Yixian beds are more likely to have been a geographic refugium for many relict lineages of the Jurassic.

References

1. Chen, P.-J., Dong, Z.-M. & Zhen, S.-N. *Nature* **391**, 147–152 (1998).
2. Ji, Q., Currie, P. J., Norell, M. A. & Ji, S.-A. *Nature* **393**, 753–761 (1998).
3. Hou, L., Zhou, Z., Martin, L. D. & Feduccia, A. *Nature* **377**, 616–618 (1995).
4. Hou, L., Martin, L. D., Zhou, Z. & Feduccia, A. *Science* **274**, 1164–1165 (1996).
5. Ji, Q., Chiappe, L. M. & Ji, S.-A. *J. Vert. Paleontol.* **19**, 1–7 (1999).
6. Hu, Y., Wang, Y., Luo, Z. & Li, C. *Nature* **390**, 137–142 (1997).
7. Ji, Q., Luo, Z. & Ji, S. *Nature* **398**, 326–330 (1999).
8. Sun, G., Dilcher, D. L., Zhen, S.-L. & Zhou, Z.-K. *Science* **274**, 1164–1167 (1998).
9. Ren, D. *Science* **280**, 85–88 (1998).
10. Labandeira, C. C. *Science* **280**, 57–58 (1998).
11. Swisher, C. C. II, Wang, Y.-Q., Wang, X.-L., Xu, X. & Wang, Y. *Nature* **400**, 58–61 (1999).
12. Gradstein, F. M. et al. in *Geochronology, Time Scales and Global Stratigraphic Correlation* (eds Berggren, W. A., Kent, D. V., Aubry, M. P. & Hardenbol, J.) 95–126 (SEPM, Spec. Publ. No. 54, 1995).
13. Russell, D. A. *Can. J. Earth Sci.* **30**, 2002–2012 (1993).
14. Cifelli, R. L. & Madsen, S. K. *Geodiversitas* (in the press).
15. Chiappe, L. M. *Nature* **378**, 349–355 (1995).
16. Ji, S.-A., Ji, Q. & Padian, K. *Nature* **398**, 573–574 (1999).
17. Padian, K., Lindberg, D. R. & Polly, D. P. *Annu. Rev. Earth Planet. Sci.* **22**, 63–91 (1995).
18. Norell, M. A. *Am. J. Sci.* **293A**, 407–417 (1993).

21

A Refugium for Relicts?

Makoto Manabe, Paul M. Barrett, and Shinji Isaji

Luo[1] suggests that the vertebrate fauna from the Yixian Formation (Liaoning Province, China) shows that this region of eastern Asia was a refugium, in which several typically Late Jurassic lineages (compsognathid theropod dinosaurs, 'rhamphorhynchoid' pterosaurs, primitive mammals) survived into the Early Cretaceous[1] (Fig. 1). Data from slightly older sediments in the Japanese Early Cretaceous, however, suggest that the faunal composition of this region can only be partly explained by the concept of a refugium.

The Kuwajima Formation of Ishikawa Prefecture, central Japan, is yielding an important Early Cretaceous vertebrate fauna. This unit is a lateral equivalent of the Okurodani Formation that outcrops in neighbouring Gifu Prefecture.[2] Stratigraphic, biostratigraphic and radiometric data show that the Okurodani Formation is basal Cretaceous (Valanginian or Hauterivian) in age.[3] The Kuwajima Formation has yielded more than one hundred isolated teeth of a new genus of tritylodontid synapsid.[4] Before these discoveries, tritylodontids were thought to have become extinct sometime in the Middle or early Late Jurassic, as the youngest-known tritylodontid (*Bienotheroides*) was recovered from late Middle Jurassic deposits. This discovery supports the concept of an East Asian refugium, but other evidence suggests that

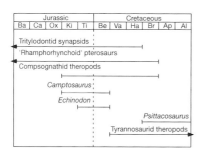

Figure 1 Stratigraphic ranges of clades that include taxa recovered from the Yixian Formation, China, and the Kuwajima and Itsuki Formations, Japan.[1,4,9] Data on *Camptosaurus* and *Echinodon* are from ref. 13. Arrows, lineage extends beyond the time range shown here; solid bars, first and last occurrences. Al, Albian; Ap, Aptian; Ba, Bathonian; Be, Berriasian; Br, Barremian; Ca, Callovian; Ha, Hauterivian; Ki, Kimmeridgian; Ox, Oxfordian; Ti, Tithonian; Va, Valanginian.

different factors may have had an equally strong influence on faunal composition.

A theropod dinosaur referable to the unnamed clade Oviraptorosauria + Therizinosauroidea[5] has also been found in the tritylodontid locality. This clade is best known from the Late Cretaceous of mainland Asia, although several taxa referable to this clade are known from the late Early Cretaceous of Liaoning (*Beipiaosaurus*[6] and *Caudipteryx*[7]), and possibly from the Early Jurassic of Yunnan Province, China.[8] The Japanese material, consisting of a single manual ungual (Fig. 2) with a pronounced posterodorsal lip (a feature synapomorphic of this group of theropods[5]), is one of the earliest representatives of this group. The Itsuki Formation of Fukui Prefecture, a lateral equivalent of the Okurodani and Kuwajima Formations,[2] has produced an isolated tyrannosaurid tooth, identifiable by its D-shaped cross-section—a synapomorphy of tyrannosaurids.[9]

These Japanese discoveries, combined with the presence of late Early Cretaceous taxa in the Yixian Formation (such as the ornithischian dinosaur *Psittacosaurus*[10]), suggest that several dinosaur clades (such as tyrannosaurids and psittacosaurids) may have originated and diversified in eastern Asia while a number of other lineages (tritylodontid synapsids, compsognathid dinosaurs and 'rhamphorhynchoid' pterosaurs) persisted in this region. Moreover, the presence of hypsilophodontid and iguanodontid ornithopod dinosaurs in the Japanese Early Cretaceous[11] suggests faunal connections with western Asia and Europe. The historical biogeography of this region appears to be much more complex than was thought previously.

Alternatively, the so-called relict taxa in eastern Asia may indicate that faunal turnover at the Jurassic–Cretaceous boundary was not as marked as has been suggested.[12] The presence of camptosaurid (*Camptosaurus*) and heterodontosaurid (*Echinodon*) ornithopods in European Early Cretaceous faunas[13] indicates faunal similarities to the Late Jurassic Morrison Formation of North America. The presence of 'Late Jurassic' taxa in eastern Asia may simply represent another example of this more gradual Jurassic–Cretaceous faunal transition (Fig. 1), although more evidence is needed to distinguish between these alternatives.

Figure 2 Manual ungual of a theropod dinosaur from the Kuwajima Formation (Valanginian or Hauterivian) of Shiramine, Ishikawa Prefecture, Japan. Note the prominent lip posterodorsal to the articular surface of the ungual, a synapomorphy of the clade Oviraptorosauria + Therizinosauroidea.[5] Scale bar, 5 mm.

References

1. Luo, Z. *Nature* **400**, 23–25 (1999).
2. Maeda, *S. J. Collect. Arts Sci. Chiba Univ.* **3**, 369–426 (1961).
3. Evans, S. E. *et al. New Mexico Mus. Nat. Hist. Sci. Bull.* **14**, 183–186 (1998).
4. Setoguchi, T., Matsuoka, H. & Matsuda, M. in *Proc. 7th Annu. Meet. Chinese Soc. Vert. Paleontol.* (eds Wang, Y.-Q. & Deng, T.) 117–124 (China Ocean, Beijing, 1999).
5. Makovicky, P. J. & Sues, H.-D. *Am. Mus. Novitates* **3240**, 1–27 (1998).
6. Xu, X., Tang, Z.-L. & Wang, X.-L. *Nature* **399**, 350–354 (1999).
7. Sereno, P. C. *Science* **284**, 2137–2147 (1999).
8. Zhao, X.-J. & Xu, X. *Nature* **394**, 234–235 (1998).
9. Manabe, M. *J. Paleontol.* **73**, 1176–1178 (1999).
10. Xu, X. & Wang, X.-L. *Vertebrata PalAsiatica* **36**, 147–158 (1998).
11. Manabe, M. & Hasegawa, Y. in *6th Symp. Mesozoic Terrest. Ecosyst. Biotas* (eds Sun, A.-L. & Wang, Y.-Q.) 179 (China Ocean, 1995).
12. Bakker, R. T. *Nature* **274**, 661–663 (1978).
13. Norman, D. B. & Barrett, P. M. *Spec. Pap. Palaeontol.* (in the press).

Cretaceous Age for the Feathered Dinosaurs of Liaoning, China

Carl C. Swisher III, Yuan-Qing Wang, Xiao-Lin Wang, Xing Xu, and Yuan Wang

The ancient lake beds of the lower part of the Yixian Formation, Liaoning Province, northeastern China, have yielded a wide range of well-preserved fossils: the 'feathered' dinosaurs *Sinosauropteryx*,[1] *Protarchaeopteryx* and *Caudipteryx*,[2] the primitive birds *Confuciusornis*[3] and *Liaoningornis*,[4] the mammal *Zhangheotherium*[5] and the reportedly oldest flowering plant, *Archaefructus*[6]. Equally well preserved in the lake beds are a wide range of fossil plants, insects, bivalves, conchostracans, ostracods, gastropods, fish, salamanders, turtles, lizards, the frog *Callobatrachus*[7] and the pterosaur *Eosipterus*.[1,8] This uniquely preserved assemblage of fossils is providing new insight into long-lived controversies over bird–dinosaur relationships,[1,2] the early diversification of birds[3,9,10] and the origin and evolution of flowering plants.[6] Despite the importance of this fossil assemblage, estimates of its geological age have varied widely from the Late Jurassic to the Early Cretaceous. Here we present the first [40]Ar/[39]Ar dates unambiguously associated with the main fossil horizons of the lower part of the Yixian Formation, and thus, for the first time, provide accurate age calibration of this important fauna. The results of this dating study indicate that the lower Yixian fossil horizons are not Jurassic but rather are at least 20 Myr younger, placing them within middle Early Cretaceous time.

The bulk of the fossils from the lower part of the Yixian Formation come from a small region surrounding the village of Sihetun, located about 20–25 km south of the city of Beipiao (Fig. 1). Fossils in this area occur primarily at two horizons separated stratigraphically by about 30–40 m (ref. 11). Most of the vertebrate fossils listed above come from an interval of a few metres, designated as Bed 6 in our Fig. 2. The angiosperm *Archaefructus* and the pterosaur *Eosipterus*, as well as other well preserved fossils, are known from the stratigraphically higher fossiliferous level, here referred to as Bed 8 (Fig. 2). Some confusion surrounds the reported occurrence of the dinosaurs *Protarchaeopteryx* and *Caudipteryx*. These dinosaurs were reported as coming from a 'Chaomidianzi' Formation in the

Sihetun area. This formally unpublished formation consists of layers that are also known as the 'Jianshangou' beds and are here considered to be equivalent to Beds 1–9 of the lower part of the Yixian Formation, following the terminology in ref. 11. The designation of a separate formation for these beds has not gained wide acceptance in China.

Palaeontologists focusing on various aspects of the fossil assemblage have considered the fauna to be Late Jurassic,[3,4,12–17] Late Jurassic to Early Cretaceous[1,2] or Early to late Early Cretaceous[18–20] in age. Unfortunately, many of these ages are based on comparisons of freshwater invertebrates from sites that are also poorly dated. The Jurassic age for the *Archaefructus* site was based on comparisons of fossil insects from sites in Siberia and Kazakhstan.[16] However the ages of these sites are equally poorly known, lacking any independent isotopic age control. Looking at the vertebrate fauna, comparison of *Sinosauropteryx* and *Confuciusornis* with *Compsognathus* and *Archaeopteryx*, respectively, from the Solnhofen Limestone of Europe,[3,21] led some workers to argue for a Late Jurassic (Tithonian) age or an age near the Jurassic/Cretaceous boundary. However, derived features noted in some of the bird fossils (for example, in comparison with *Archaeopteryx*, *Confuciusornis* lacks teeth, possesses a beak, has a much shorter tail and so on), and the occurrence of the ceratopsian dinosaur *Psittacosaurus*,[22] have led others to argue for a younger Cretaceous age. A Cretaceous Valanginian age has also been proposed, based on pollen and spores associated with the fossil birds *Sinornis*, *Cathyornis* and *Chaoyangia* from the overlying Jiufotang Formation.[10]

Isotopic dates on volcanics from the Yixian Formation have not helped to resolve the controversy. Proponents of a Jurassic age for the Lower Yixian Formation refer to a K–Ar date of 137 ± 7 Myr and a Rb–Sr date of 143 ± 4 Myr (refs 23, 24) (these uncertainties are 2 standard deviations (s.d.), all others in this report are 1 s.d.). These workers propose an age of around 135 Myr for the Jurassic/Cretaceous boundary, although most workers now accept an age for the boundary of around 144 Myr (ref. 25). Most of the younger age estimates for a Jurassic/ Cretaceous boundary are based on low-temperature glauconite dates which we consider to be unreliable. Although the boundary is still not

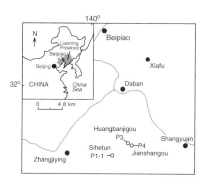

Figure 1 Map of the Sihetun area, Liaoning Province, northeastern China. Locations of stratigraphic sections and sites discussed in the text and in Fig. 2 are shown.

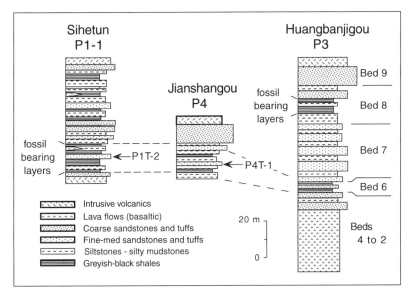

Figure 2 Stratigraphic correlation of the main fossil-bearing layers of the lower Yixian Formation. Arrows show stratigraphic locations of dated tuff samples P1T-2 and P4T-1. Precise stratigraphic relationship of key vertebrate fossils from Bed 6 to the dated tuff layers are reported in the text. The *Archaefructus* and *Eosipterus*-bearing layers of Bed 8 are located about 30 m stratigraphically above the dated tuff layers. The stratigraphic sections and correlations are modified slightly from those reported in ref. 11 as a result of additional geological field mapping in Autumn of 1998.

well calibrated, for the purposes of this study we defer to the discussion within ref. 25 and adopt their age of 144 Myr.

$^{40}Ar/^{39}Ar$ dates on a volcanic breccia and diabase from the upper part of the Yixian at Jingangshan have been used to argue for a Cretaceous age for the underlying fossils.[26] The average age of 121.1 ± 0.2 Myr for these dates suggests correlation with the Cretaceous Aptian. $^{40}Ar/^{39}Ar$ dates have also been obtained from volcanic rocks from the lower part of the Yixian Formation. An andesite from Daxinfangzi has been dated at 122.9 ± 0.3 Myr and a basalt from Beipiao at 121.2 ± 0.3 Myr (ref. 26). However, the relationship of these dated volcanics to fossil-bearing horizons has recently come into question. Additional data indicates that the basalt from Beipiao may be an intrusive sill rather than an interbedded flow within the lake beds of the Yixian Formation. Dated volcanics at Jingangshan and Daxinfangzi are geographically distant from the 'feathered' dinosaur and *Confuciusornis* sites and precise stratigraphic relationships are uncertain and require further investigation.

During stratigraphic studies of the Sihetun area in 1997–98, volcanic layers were found interbedded within the main fossiliferous horizons of Bed 6 (Fig. 2). To try and resolve the age of the fossils of the lower Yixian

Formation, two tuffs were collected for ^{40}Ar/^{39}Ar dating. Dates on these tuffs are potentially better than previous dates for two reasons. First, given the proximity of the tuffs to the fossiliferous layers, problems with correlation and provenance are obviated. Second, the tuffs are pristine water-lain tephra that contain abundant euhedral sanidine feldspars. Sanidine, because of its argon retentivity and potential reproducibility (as established from previous dating studies on replicate analyses of single crystals), is considered to be one of the most reliable materials for ^{40}Ar/^{39}Ar dating.[27,28]

The tuffs were collected from Bed 6 at two sites, about 4 km apart, in the main fossil-bearing region of Sihetun (Fig. 1, 2). Tuff P4T-1 is from the Jianshangou section (Fig. 2) and occurs about 50 cm above the type specimen of *Zhangheotherium quinquecuspidens*.[5,11] The second tuff (PIT-2) is from the Sihetun section (Fig. 2), located 3.40 m above the layer bearing the type specimen of *Confuciusornis sanctus*.[3,11] Bed 6 at Sihetun has also yielded well-preserved specimens of *Sinosauropteryx*, *Protarchaeopteryx* and *Liaoningornis*.[11] The Sihetun and Jianshangou tuffs are stratigraphically within a metre of each other, and may represent the same tuff, although changes in depositional facies preclude this determination. The reportedly oldest flowering plant, *Archaefructus*, comes from approximately 30 m above this level in Bed 8 at Huangbanjigou (Fig. 1, 2).

^{40}Ar/^{39}Ar dating of the P4T-1 and PIT-2 tuffs was done by replicate Ar-ion-laser total-fusion analyses of single crystals of sanidine and CO_2-laser incremental-heating of a bulk sanidine separate. Thirty-four single sanidine crystal dates obtained on Tuff P4T-1 (Fig. 3a) give a mean age of 124.6 ± 0.2 (s.d.) ± 0.03 Myr (s.e.) Thirty-five single sanidine crystal dates on Tuff PIT-2 (Fig. 3a) give an age of 124.6 ± 0.3 (s.d.) ± 0.04 Myr (s.e.), an age indistinguishable from that of P4T-1. The radiogenic ^{40}Ar yields from all of the sanidines were greater than 99%, indicating little alteration. No crystals of significantly older age that might suggest reworking or detrital contamination were found in either tuff. A few additional crystals that were dated gave lower radiogenic

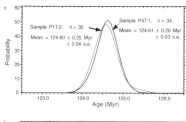

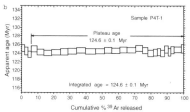

Figure 3 ^{40}Ar/^{39}Ar dating results. **a**, Age probability diagram for single sanidine crystal dates; **b**, incremental-heating spectrum on sanidine.

yields, but were determined subsequently to be plagioclase crystals based on their measured $^{37}Ar/^{39}Ar$ (Ca/K) ratios. These lower radiogenic plagioclase analyses gave a wider variance in age distribution (122.5 Myr to 125 Myr) than the sanidine, considered here to be a result of various alteration of the plagioclase. These analyses are omitted from the calculation of the age of the tuffs.

To address the possibility of argon loss and/or trapped excess argon, CO_2-laser incremental-heating of the sanidine of P4T-1 was also undertaken (Fig. 3b). The release profile (Fig. 3b) was relatively flat with only a slight deviation from a plateau occurring at the lowest three temperature increments. All of the 24 increments that formed the plateau yielded over 95% radiogenic ^{40}Ar and totalled more than 90% of the ^{39}Ar released from the sample. The calculated age for the plateau is 124.6 ± 0.1 Myr (Fig. 3b). The uniformly high radiogenic yields of all 24 increments resulted in a small cluster of points when plotted on an inverse ($^{36}Ar/^{40}Ar$ vs $^{39}Ar/^{40}Ar$) isochron, making any fit unreliable. If we include the three lower radiogenic steps that fell outside the plateau in the isochron, an age of 124.6 ± 0.1 Myr (MSWD (mean sum of weighted deviates) = 1.48) is obtained. The $^{40}Ar/^{36}Ar$ intercept of 286.2 ± 10.7 is not distinguishable from the air ratio. The incremental-heating and single-crystal experiments indicate little alteration of the sanidine, no discernible excess argon and no apparent detrital contamination of the sanidine.

The dates reported here are the first $^{40}Ar/^{39}Ar$ analyses on tuffs interbedded directly in the fossiliferous horizons of the lower Yixian Formation. Our $^{40}Ar/^{39}Ar$ ages for Tuffs P4T-1 and P1T-2 from the Bed 6 fossil-bearing levels are about 10% younger than the previously reported K/Ar and Rb/Sr dates.[23,24] Although we are unaware of the precise stratigraphic and geographic location of these dated samples, the dates are accompanied by large uncertainties and are generally not considered reliable owing to probably alterations and/or contamination. Our sanidine dates, although 2% older, are more consistent with the $^{40}Ar/^{39}Ar$ dates on andesites and basalts of the Yixian Formation reported in ref. 26. However we consider our dates to be a more accurate calibration of the fossils from the lower Yixian Formation. We base this on the superior argon retentivity of sanidine versus the whole rock and biotite[27,28] used in their dating study[26] and on the unambiguous relationships of our dated tuffs to fossiliferous levels. We also note the possibility of some ^{40}Ar loss and/or ^{39}Ar recoil during irradiation that accompanies some of the $^{40}Ar/^{39}Ar$ release spectra shown for those dates,[26] which may indicate that those samples are altered. At least a percentage of the age difference can be attributed to the age of Hb-3gr, the standard used in ref. 26. We can make a second-order comparison of our dates with those in ref. 26 using the interlaboratory standard MMHb which has been intercalibrated with both Hb-3gr[29] and our standard Fish Canyon Sanidine.[27] Compared with our age for Fish Canyon Sanidine, we obtain ages of 122–124 Myr for the dates in ref. 26, bringing them into close agreement with our sanidine dates for tuffs P4T-1 and P1T-2.

The new dates reported here, in conjunction with the results of ref. 26, indicate that the 'feathered' dinosaurs of Liaoning, although primitive in appearance, are not Late Jurassic or even earliest Cretaceous in age. Compared with the geologic timescale of ref. 25, the dates indicate a correlation with the middle Barremian (mid-Early Cretaceous), at least 20 Myr younger than *Archaeopteryx* from the Late Jurassic (Tithonian) Solnhofen Limestone of Europe. Given the similarities already noted with the Solnhofen and Liaoning fossils, it would appear that aspects of the terrestrial fauna were part of a long-lived chronofauna, persisting across the Jurassic–Cretaceous boundary. However the ages of many of these sites worldwide are poorly known and, in light of the new dates for the lower Yixian fauna reported here, their ages may need re-examination. A final outcome of the new dates is that *Archaefructus*, although remarkably well preserved, cannot be considered as early as originally thought.[5] Depending on the accuracy in dating of other fossil sites worldwide, *Archaefructus* appears to be comparable in age with early angiosperm evidence from the Barremian of China, Europe, Russia and eastern North America.[30]

Methods

[40]Ar/[39]Ar dating of the P4T-1 and P1T-2 tuffs follow procedures described in ref. 27 and references therein. Sanidine from the tuffs and multiple samples of the monitor mineral Fish Canyon Sanidine (to determine lateral flux gradients during irradiation) were co-irradiated for 20 h in the cadmium-lined core (CLICIT) facility of the Oregon State University TRIGA reactor. Nucleogenic interference corrections are those previously reported.[27] Mass discrimination was monitored before and after measurement of the sanidines using an on-line air pipette system. The P4T-1 and P1T-2 dates are referenced to the Fish Canyon Sanidine monitor mineral using the published age of 28.02 Myr (ref. 27). The ages reported are arithmetic means of the individual dates based on total fusion analyses of single sanidine crystals. We report one standard deviation (s.d.) and standard error of the mean (s.e.) uncertainties with the mean ages to document crystal-to-crystal age variation. Individual crystal dates are presented in the Supplementary Information, Table 1.

In the incremental-heating experiment, 15 handpicked euhedral sanidine crystals were laid flat on the bottom of a 2-cm-diameter well of a copper sample disk. The crystals were heated by a CO_2 laser beam directed through an integrator lens that produces relatively even and uniform heating. Twenty-seven discrete temperature increments based on a controlled stepwise increase in power of the CO_2 laser were obtained. The heating time for each increment was 60 s. Incremental heating data are supplied in the Supplementary Information, Table 2.

References

1. Chen, P.-J., Dong, Z.-M. & Zhen, S.-N. An exceptionally well-preserved theropod dinosaur from the Yixian Formation of China. *Nature* 391, 147–152 (1998).

2. Ji, Q., Currie, P. J., Norell, M. A. & Ji, S.-A. Two feathered dinosaurs from northeastern China. *Nature* 393, 753–761 (1998).

3. Hou, L.-H., Zhou, Z.-H., Martin, L. D. & Feduccia, A. A beaked bird from the Jurassic of China. *Nature* 377, 616–618 (1995).

4. Hou, L.-H. A carnate bird from the Upper Jurassic of western Liaoning, China. *Chinese Sci. Bull.* 41, 1861–1864 (1996).

5. Hu, Y.-M., Wang, Y.-Q., Luo, Z.-X. & Li, C.-K. A new symmetrodont mammal from China and its implications for mammal evolution. *Nature* 390, 137–142 (1997).

6. Sun, G., Dilcher, D. L., Zheng, S.-L. & Zhou, Z.-K. In search of the first flower: A Jurassic angiosperm, *Archaefructus*, from northeast China. *Science* 282, 1692–1695 (1998).

7. Wang, Y. & Gao, K.-Q. Earliest Asian discoglossid frog from western Liaoning. *Chinese Sci. Bull.* 44, 636–642 (1999).

8. Ji, S.-A. & Ji, Q. Discovery of a new pterosaur from western Liaoning, China. *Acta Geol. Sin.* 71, 1–5 (1997).

9. Hou, L.-H., Martin, L. D., Zhou, Z.-H. & Feduccia, A. Early adaptive radiation of birds: evidence from fossils from northeastern China. *Science* 274, 1164–1167 (1996).

10. Sereno, P. C. & Rao, C.-G. Early evolution of avian flight and perching: new evidence from the Lower Cretaceous of China. *Science* 255, 845–848 (1992).

11. Wang, X.-L. *et al.* Stratigraphic sequence and vertebrate-bearing beds of the lower part of the Yixian Formation, in Sihetun and neighboring area, western Liaoning, China. *Vert. PalAsiat.* 36, 81–96 (1998).

12. Chen, P.-J. Distribution and migration of the Jehol Fauna with reference to nonmarine Jurassic-Cretaceous boundary in China. *Acta Palaeont. Sin.* 27, 659–683 (1988).

13. Wang, S.-E. Origin, evolution and mechanism of the Jehol Fauna. *Acta Geol. Sin.* 63, 351–360 (1990).

14. Yu, J.-S., Dong, G.-Y. & Yao, P.-Y. in *Mesozoic Stratigraphy and Palaeontology of Western Liaoning* Vol. 3, 1–28 (Geological Publishing House, Beijing, 1987).

15. Zhang, L.-J. in *Mesozoic Stratigraphy and Palaeontology of Western Liaoning* Vol. 2, 1–212 (Geological Publishing House, Beijing, 1985).

16. Ren, D. Flower-associated *Brachycera* flies as fossil evidence for Jurassic angiosperm origins. *Science* 280, 85–88 (1998).

17. Wang, W.-L. in *Mesozoic Stratigraphy and Palaeontology of Western Liaoning* Vol. 3, 134–201 (Geological Publishing House, Beijing, 1987).

18. Hao, Y.-C., Su, D.-Y., Li, Y.-G. & Li, P.-X. Stratigraphical division of non-marine Cretaceous and the Juro-Cretaceous boundary in China. *Acta Geol. Sin.* 56, 187–199 (1982).

19. Li, P.-X., Su, D.-Y., Li, Y.-G. & Yu, J.-X. The chronostratigraphic status of the *Lycoptera*-bearing bed. *Acta Geol. Sin.* 68, 87–100 (1994).

20. Mao, S., Yu, J. & Lentin, J. K. Palynological interpretation of Early Cretaceous non-marine strata of northeast China. *Rev. Paleobot. Palynol.* 65, 115–118 (1990).

21. Wellnhoffer, P. Das siebte Exemplar von *Archaeopteryx* aux den Solnhofener Schichten. *Archaeopteryx* 11, 1–48 (1993).

22. Xu, X. & Wang, X.-L. New psittacosaur (Ornithischia, Ceratopsia) occurrence from the Yixian Formation of Liaoning, China, and its stratigraphical significance. *Vert. PalAsiat.* 36, 147–158 (1997).

23. Wang, D.-F. On the age of the Rehe Group in western Liaoning Province, China. *Bull. Chinese Acad. Geol. Sci.* 7, 65–82 (1983).

24. Wang, D.-F. & Diao, N.-C. in *Scientific Papers on Geology for International Exchange* Vol. 5, 1–12 (Geological Publishing House, Beijing, 1984).

25. Gradstein, F. M. *et al.* in *Geochronology, Time Scales and Global Stratigraphic Correlation* (eds Berggren, W. A, Kent, D. V., Aubry, M.-P. & Hardenbol, J.) 95–126 (S.E.P.M. Spec. Publ. No. 54, 1995).

26. Smith, P. et al. Dates and rates in ancient lakes: [40]Ar/[39]Ar evidence for an Early Cretaceous age for the Jehol Group, northeast China. Can. J. Earth Sci. 32, 1426–1431 (1995).

27. Renne, P. R. *et al.* Intercalibration of standards, absolute ages and uncertainties in [40]Ar/[39]Ar dating. *Chem. Geol.* **145**, 117–152 (1998).

28. McDougall, I. & Harrison, T. M. *Geochronology and Thermochronology* (Oxford Univ. Press, New York, 1988).

29. Roddick, J. C. High precision intercalibration of [40]Ar/[39]Ar standards. *Geochim. Cosmochim. Acta* **47**, 887–898 (1983).

30. Sun, G. & Dilcher, D. L. Early angiosperms from the Lower Cretaceous of Jixi, China and their significance for the study of the earliest occurrence of angiosperms in the world. *Palaeobotany* **45**, 393–399.

For supplementary information see page 248.

Acknowledgements

We thank all the members of the IVPP field crew, and M. M. Chang, Z. Luo and S. Kelley for discussion and comment on this study. This research was funded, in part, by the Chinese Academy of Sciences and the National Science Foundation, Earth Science Program.

Supplementary Information

Chapter 5   Lower Cambrian Vertebrates from South China
De-Gan Shu, Hui-Lin Luo, S. Conway Morris, Xin-Liang, Zhang, S.-X. Hu, Liang Chen, J. Han, Min Zhu, Y. Li, and L.-Z. Chen

I. Character list

1. Eye lens. 0 = absent (0); 1 = present.
2. Ribbon-shaped synaptic organelles in retina. 0 = absent; 1 = present.
3. Tabular muscles in 'tongue' musculature. 0 = absent; 1 = present.
4. Lateral-line neuromasts. 0 = absent; 1 = present.
5. Nerve control of heart. 0 = absent; 1 = present.
6. Heart response to catecholamine. 0 = absent; 1 = present.
7. Blood system. 0 = open; 1 = close.
8. Spleen. 0 = absent; 1 = present.
9. Concentrated exocrine pancreas. 0 = absent; 1 = present.
10. Typhlosole in intestine. 0 = absent; 1 = present.
11. Larval stage. 0 = present; 1 = absent.
12. Compartmentalized adenohypophysis. 0 = absent; 1 = present.
13. Pituitary control of melanophores. 0 = absent; 1 = present.
14. Kidney tubules with glomerulae. 0 = absent; 1 = present.
15. Osmoregulation. 0 = absent; 1 = present.
16. Granulocytes and neutrophils. 0 = absent; 1 = present.
17. Two types of giant Mauthner cells in the central nervous system. 0 = absent; 1 = present.
18. Electroreceptive cells. 0 = absent; 1 = present.
19. Sperm. 0 = not shed through coelomic cavity; 1 = shed through coelomic cavity.
20. Cartilaginous copula. 0 = absent; 1 = present.
21. Pancreas. 0 = absent; 1 = present.
22. Differentiated adenohypophysis. 0 = absent; 1 = present.
23. Skull. 0 = absent; 1 = present.
24. Cartilaginous or calcified braincase (neurocranium). 0 = absent; 1 = present.
25. Dorsally closed cartilaginous or calcified braincase (neurocranium). 0 = absent; 1 = present.
26. Olfactory organ. 0 = absent; 1 = present.
27. Terminal nasohypophysial opening (inhalant or not). 0 = absent; 1 = present.
28. Dorsal nasohypophysial opening (inhalant or not). 0 = absent; 1 = present.
29. Optic capsules (exclusive of the lens): absent (0), present (1).
30. Transversely biting teeth. 0 = absent; 1 = present.

31. Heart. 0 = absent; 1 = present.
32. Closed pericardium. 0 = absent; 1 = present.
33. Trunk and tail musculature with chevron-shaped muscle blocks. 0 = absent; 1 = present.
34. Zigmoid myomeres. 0 = absent; 1 = present.
35. Radials in fin. 0 = absent; 1 = present.
36. Numerous and closely set radials in unpaired fin. 0 = absent; 1 = present.
37. Radial muscles in fin. 0 = absent; 1 = present.
38. Dorsal fin. 0 = absent; 1 = present.
39. Separate dorsal fin. 0 = absent; 1 = present.
40. Preanal fin persisting in adult. 0 = absent; 1 = present.
41. Anal fin. 0 = absent; 1 = present.
42. Caudal fin with internal support. 0 = absent; 1 = present.
43. Tail. 0 = isoceral; 1 = hypoceral; 2 = epiceral.
44. Paired fin folds or fins. 0 = absent; 1 = present.
45. Paired fin folds or fins concentrated in the pectoral or epibranchial regions. 0 = absent; 1 = present.
46. Muscles in paired fins. 0 = absent; 1 = present.
47. Arcualia. 0 = absent; 1 = present.
48. Gill openings arranged in posteriorly slanting line. 0 = absent; 1 = present.
49. Number of gill units. 0 = more than ten; 1 = less than ten.
50. Olfactory tract. 0 = absent; 1 = present.
51. Olfactory organ unpaired or with closely set and confluent nasal sacs. 0 = absent; 1 = present.
52. Olfactory organ paired with entirely separated nasal sacs. 0 = absent; 1 = present.
53. Nasohypophysial duct serving branchial respiration. 0 = absent; 1 = present.
54. Nasohypophysial duct posteriorly closed and serving, only as a common "nostril". 0 = absent; 1 = present.
55. Extrinsic eye muscles. 0 = absent; 1 = present.
56. Photosensory pineal organ (or pineal foramen). 0 = absent; 1 = present.
57. Semicircular canals. 0 = absent; 1 = present.
58. Single semicircular canal. 0 = absent; 1 = present.
59. Two vertical semicircular canals. 0 = absent; 1 = present.
60. Two vertical semicircular canals forming distinct loops. 0 = absent; 1 = present.
61. Lateral lines enclosed in canals. 0 = absent; 1 = present.
62. Lateral-line grooves or canals. 0 = absent; 1 = present on head (1); 2 = present on head and body. (ORDERED)
63. Neuromasts. 0 = absent; 1 = isolated; 2 = in grooves; 3 = in tubes. (ORDERED)
64. Cerebellum. 0 = absent; 1 = present.
65. Large and paired cerebellum. 0 = absent; 1 = present.
66. Vagus and glossopharyngeus nerves included in occipital region. 0 = absent; 1 = present.
67. Closely set atrium and ventricle. 0 = absent; 1 = present.
68. Large dorsal jugular vein. 0 = absent; 1 = present.
69. Subaponeurotic vascular system. 0 = absent; 1 = present.
70. Calcified cartilage. 0 = absent; 1 = present.
71. Perichondral bone. 0 = absent; 1 = present.
72. Endoskeletal cranial roof. 0 = absent; 1 = present.
73. Occiput enclosing IX and X. 0 = absent; 1 = present.
74. Head endoskeleton expanded into a massive shield covering the gills. 0 = absent; 1 = present.
75. Endoskeletal scleral ossification or calcification. 0 = absent; 1 = present.
76. Orthodentine or metadentine. 0 = absent; 1 = present.
77. Mesodentine. 0 = absent; 1 = present.
78. Enameloid or enamel. 0 = absent; 1 = present.
79. Dermal skeleton. 0 = absent; 1 = present.

80. Acellular dermal bone. 0 = absent; 1 = present.
81. Cellular dermal bone. 0 = absent; 1 = present.
82. Honeycomb-like middle layer of exoskeleton. 0 = absent; 1 = present.
83. Three-layered dermal skeleton. 0 = absent; 1 = present.
84. Oakleaf-shaped tubercles or odontodes in ornamentation. 0 = absent; 1 = present.
85. Large median dorsal and ventral dermal shields in head. 0 = absent; 1 = present.
86. External opening of endolymphatic duct. 0 = absent; 1 = present.
87. Opercular flaps on external branchial openings. 0 = absent; 1 = present.
88. Preanal skin fold or scale ridge. 0 = absent; 1 = present.
89. Sclerotic ring. 0 = absent; 1 = present.
90. Scales made up by a single odontodes (microsquamose). 0 = absent; 1 = present.
91. Scales made up by several odontodes (macrosquamose) and diamond-shaped. 0 = absent; 1 = present.
92. Scales made up by several odontodes (macrosquamose) and rod-shaped. 0 = absent; 1 = present.
93. Pharyngeal dermal denticles. 0 = absent; 1 = present.
94. Elongate oral plate. 0 = absent; 1 = present.
95. Scales. 0 = absent; 1 = non-overlapping; 2 = overlapping. (ORDERED)
96. Head armour differentiated from trunk armour. 0 = absent; 1 = present.
97. Dorsal/ventral growing plates. 0 = absent; 1 = present.
98. Dorsal/ventral scutes. 0 = absent; 1 = present.
99. Scale-covered zones on tail. 0 = absent; 1 = present.
100. Horny teeth. 0 = absent; 1 = present.
101. Piston cartilage. 0 = absent; 1 = present.
102. Dentigerous cartilage. 0 = absent; 1 = present.
103. Tentacles strengthened by cartilage. 0 = absent; 1 = present.
104. Sucking disc with an annular cartilage around mouth. 0 = absent; 1 = present.
105. Mouth ventral. 0 = absent; 1 = present.
106. Tongue protractor/retractor musculature. 0 = absent; 1 = present.
107. Pouch-shaped gill. 0 = absent; 1 = present.
108. Trematic rings. 0 = absent; 1 = present.
109. Gill arterial supply. 0 = absent; 1 = pretrematic; 2 = pre- and post-trematic.
110. Gills ventral. 0 = absent; 1 = present.
111. Gills transversely elongated. 0 = absent; 1 = present.
112. Gills crowded posteriorly. 0 = absent; 1 = present.
113. Branchial basket. 0 = absent; 1 = present.
114. Visceral skeleton attached to head. 0 = absent; 1 = present.
115. Optic tectum. 0 = absent; 1 = present.
116. Eye dorsal. 0 = absent; 1 = present.

II. Data set with 116 characters for 16 taxa

Note: A majority of characters are compiled from Forey (1995) and Janvier (1996). 0 = plesiomorphic state; 1, 2, 3 = apomorphic states; ? = unavailable characters or logical impossibility; (01) = polymorphic states (0 and 1).

	1	1111111112	2222222223	3333333334
	1234567890	1234567890	1234567890	1234567890
anaspids	??????????	??????????	??1??1011?	??11102000
arandaspids	??????????	??????????	??1??1?01?	??1??????00
astraspids	??????????	??????????	??1????01?	???????000
cephalochordes	0000000000	0000000000	0000000000	0010000001

galeaspids	???1??????	??????????	??1111011?	??1?100?00
gnathostomes	1101111111	0111111100	1111110010	1111111110
hagfishes	0010000000	1000000011	0010011011	1011100001
Haikouichthys	??????????	??????????	??????????	1?1111?1?0
Heterostracans	???1??????	??????????	??1??1?01?	??1110?000
Jamoytius	??????????	??????????	??1??????1?	??1?11?1?0
Lampreys	1111111111	1111111111	1111010111	1111111110
Myllokunmingia	??????????	??????????	??????????	1?110001?0
Osteostracans	???1??????	??????????	??1111011?	1111111110
pituriaspids	??????????	??????????	??1111??1?	??????1??0
thelodonts	???1??????	??????????	??1??1101?	???110?110
tunicates	0000000001	0000000000	0000000000	0000000000

	4444444445	5555555556	6666666667	7777777778
	1234567890	1234567890	1234567890	1234567890
anaspids	1111?0?10?	?????1????	02????????	0??0000011
arandaspids	0??000?10?	?????1????	022???????	????100011
astraspids	0??000?01?	?????0????	0?2???????	????00111
cephalochordes	0000000000	0000000000	0000000000	0000000000
Galeaspids	?1?000?011	0110111011	123111?11?	1111000111
gnathostomes	1121111011	0101111011	1231111111	1100111110
hagfishes	0100000000	1010001100	0100000000	0000000000
Haikouichthys	???100?01?	??????????	?????????0	0??0000000
heterostracans	0110001011	01??011011	12311???10	0??010011
Jamoytius	1111?0??0?	??????????	?????????0	0??0000000
Iampreys	0110001110	1001111010	0011001111	0100000000
Myllokunmingia	0??100001?	??????????	?????????0	0??0000000
osteostracans	012111101?	1001111011	1231111111	1111101110
pituriaspids	???111?0??	??????????	???11?????	11?1????1?
thelodonts	11111??01?	01?0??????	123??????0	0??00(01)(01)011
tunicates	0000000000	0000000000	0000000000	0000000000

		1	1111111111	111111
	8888888889	9999999990	0000000001	111111
	1234567890	1234567890	1234567890	123456
anaspids	00000?0100	010010011?	??00001??0	?0???1
arandaspids	0111100010	01011110??	??00001??0	?00??0
astraspids	0010100?0	10?02001??	????0?1??0	10???0
cephalochordes	0000000000	0000000000	0000000000	000000
galeaspids	0000010101	00?021101?	??001?1??1	100110
gnathostomes	1010011111	1010210000	0000000020	000010
hagfishes	0000000000	0000000001	0110011110	000100
Haikouichthys	00000??200	000000000?	??0?0????0	001???
heterostracans	0111100000	100121111?	??000001??0	100??0
Jamoytius	00000??200	0000000000	??010?11?0	?01??0
Iampreys	0000000100	0000000001	1101011120	001110
Myllokunmingia	00000??200	000000000?	??0?0?1?0	000???
osteostracans	1010011110	100011010?	??00101?21	100111
pituriaspids	???00??1??	???0?10???	??000?1??1	?????1
thelodonts	00000?1?01	001020001?	??0?0?1??0	10???0
tunicates	0000000000	0000000000	0000000000	000000

III. Clades & supporting characters in Fig.3b

The category "uniquely shared derived characters" lists all characters with a C.I. equal to 1.

Node	Uniquely shared derived characters	Reversals	Homoplasies
Node 1	23, 26, 29, 31, 34, 42, 57, 109(1)		3, 11, 19, 20, 30, 35, 51, 62(1), 102, 106, 107, 108,114
Node 2	1, 2, 4, 5, 6, 7, 8, 9, 12, 13, 14, 15, 16, 17, 18, 21, 22, 24, 32, 43(1), 59, 64, 67, 68, 69, 72, 109(2), 115		10, 28, 38, 39, 44, 49, 55, 56, 63(1), 88
Node 3	47		37, 41
Node 4	101, 104, 113	62	36, 54
Node 5			48
Node 6	25, 50, 60, 65, 66, 79	3, 11, 19, 20, 30, 51, 102, 106, 108	52, 62(2), 63(2), 73, 80, 95(2), 99, 111
Node 7		38, 39, 55	98
Node 8	85	28, 41, 44, 88	83, 91
Node 9	82, 84, 94		96, 97
Node 10	86		45, 61, 63(3), 87, 90, 93
Node 11	43(2), 71		70, 74, 78, 96, 110
Node 12	46, 77, 81	80, 99	36, 54, 75, 83, 89, 91
Node 13		41, 52, 90, 93, 95(1)	51, 98, 116

Chapter 7 A Primitive Fossil Fish Sheds Light on the Origin of Bony Fishes   Min Zhu, Xiao-Bo Yu, and Philippe Janvier

I. Clades & supporting characters at major nodes in Fig. 4a

Note: Lower level nodes in taxa with asterisk (*) are not depicted in Fig.4.

Node	Clade	Supporting characters
Node 1.	(Acanthodes + Ctenacanthus) + (Psarolepis + Osteichthyes)	147
Node 2.	Psarolepis + Osteichthyes	149
Node 3.	Osteichthyes	No synapomorphy
Node 4.	Sarcopterygii	4, 18, 42, 49, 54, 63(2), 88, 105, 120, 128

Node 5.	Crossopterygii (sensu ref. 4)	No synapomorphy
Node 6.	*Powichthys* + {Porolepiformes + [(Actinistia + Onychodontida) + Tetrapodomorpha]}	No synapomorphy
Node 7.	Porolepiformes + [(Actinistia + Onychodontida) + Tetrapodomorpha]	89
Node 8.	(Actinistia + Onychodontida) + Tetrapodomorpha	137; reversals 2, 41, 43
Node 9.	TETRAPODOMORPHA* (= Choanata of ref. 4) Rhizodontida* (*Barameda, Strepsodus*) Osteolepidae* (*Beelarongia, Gyroptychius, Osteolepis*) *Eusthenopteron* Elpistostegalia* (*Elpistostege, Panderichthys*) Tetrapoda* (*Acanthostega, Crassigyrinus, Ichthyostega, Ventastega*)	44, 71, 119(2), 122
Node 10.	*Acanthodes* + *Ctenacanthus*	Reversal 141
Node 11.	Actinopterygii	6; reversals 52, 93(2->0), 110
Node 12.	DIPNOI* (*Diabolepis, Diphnorynchus, Dipterus, Speonesydrion, Uranolophus*)	17A, 30, 65, 80, 107
Node 13.	POROLEPIFORMES* (*Glyptolepis, Holoptychius, Porolepis*)	51, 55, 63(2->1), 75, 78A, 101
Node 14.	ACTINISTIA* (*Diplocercides kayseri, D. heiligenstockiensis*)	76, 93(2->1), 95, 97, 138
Node 15.	Actinistia + Onychodontida	No synapomorphy
Node 16.	ONYCHODONTIDA* (*Barameda, Strepsodus*)	41(2)

II. Clades & supporting characters at major nodes in Fig. 4b

Note: Lower level nodes in taxa with asterisk (*) are not depicted in Fig. 4.

Node	Clade	Supporting characters
Node 1.	(*Acanthodes* + *Ctenacanthus*) + ("Actinopterygii" + *Psarolepis* + Sarcopterygii)	147
Node 2.	"Actinopterygii" + *Psarolepis* + Sarcopterygii	149
Node 3.	*Cheirolepis* + [*Polypterus* + (*Psarolepis* + Sarcopterygii)]	No synapomorphy
Node 4.	*Polypterus* + (*Psarolepis* + Sarcopterygii)	46, 63(2), 69, 98, 134
Node 5.	*Psarolepis* + Sarcopterygii	4, 42, 49, 88, 93(2), 110, 128
Node 6.	Sarcopterygii	18, 54, 105, 120
Node 7.	Onychodontida + Rhipidistia (sensu ref. 14)	No synapomorphy

Node 8.	Rhipidistia (sensu ref. 14)	56
Node 9.	TETRAPODOMORPHA*	44, 71, 119(2), 122, reversal 2
	Rhizodontida* (Barameda, Strepsodus);	
	Osteolepidae* (Beelarongia,	
	Gyroptychius, Osteolepis);	
	Eusthenopteron;	
	Elpistostegalia* (Elpistostege,	
	Panderichthys) Tetrapoda*	
	(Acanthostega, Crassigyrinus,	
	Ichthyostego, Ventastega)	
Node 10.	Acanthodes + ctenacanthus	Reversal 141
Node 11.	Mimia + (Howqualepis + Moythomasia)	7
Node 12.	ACTINISTIA* (Diplocercides kayseri,	76, 93(2->1), 95, 97, 138
	D. heiligenstockiensis)	
Node 13.	ONYCHODONTIDA* (Barameda,	41(2)
	Strepsodus)	
Node 14.	POROLEPIFORMES* (Glyptolepis,	51, 55, 75, 101
	Holoptychius, Porolepis)	
Node 15.	Dipnomorpha (sensu ref. 14)	41, 43, 130
Node 16.	Dipnoiformes (sensu ref. 14)	77, 100, 119
Node 17.	Youngolepis + Dipnoi	17, 45
Node 18.	DIPNOI* (Diabolepis, Diphnorynchus,	30, 65, 80, 107
	Dipterus, Speonesydrion, Uranolophus)	

III. Characters and character states defining major nodes in Fig. 4a, b

Source: Ref. 14, except for characters 141–149.

---

2. Mesh canals. 0 = pore cavity with horizontal partition; 1 = pore cavity without horizontal cavity.
4. Peg on rhombic scale. 0 = narrow; 1 = broad.
6. Ganoine. 0 = absent; 1 = present.
7. Acrodin caps on teeth. 0 = absent; 1 = present.
17. Position of premaxilla 0 = marginal; 1 = ventral part turned in (17A uses codings from ref. 4).
18. Relation of premaxilla to the orbit. 0 = forming part of orbit; 1 = not forming part of orbit.
30. B-bone. 0 = absent; 1 = present.
41. Median extrascapular. 0 = overlapped by lateral extrascapular; 1 = overlapping lateral extrascapular; 2 = abutting lateral extrascapular.
42. Tectals. 0 = absent; 1 = present.
43. Number of tectals (not counting the "posterior tectal" of Jarvik). 0 = one; 1 = three or more.
44. Anterior and posterior nares. 0 = both present; 1 = only anterior naris present.
45. Position of posterior naris. 0 = external, far from jaw margin; 1 = external, close to jaw margin; 2 = palatal.
46. Position of posterior naris in relation to the orbit. 0 = associated with orbit; 1 = not associated with orbit.
49. Number of sclerotic plates. 0 = four or less; 1 = more than four.
51. Prespiracular. 0 = absent; 1 = present.
52. Dermohyal. 0 = present; 1 = absent.

54. Squamosal and preopercular. 0 = one bone ("preopercular");
    1 = two separate bones.
55. Subsquamosals. 0 = absent; 1 = present.
63. Number of branchiostegal rays per side. 0 = 10 or more; 1 = two to seven; 2 = one.
65. Width of submandibulars. 0 = narrow; 1 = broad.
69. Size of lateral gular. 0 = similar to branchiostegal rays; 1 = covering about half the intermandibular space.
71. Articulation of vomers. 0 = not articulating with each other; 1 = articulating with each other.
75. Buccohypophysial foramen of parasphenoid. 0 = single; 1 = double.
76. Rostral organ. 0 = absent; 1 = present.
77. Rostral tubuli. 0 = absent; 1 = present.
78. Fossa autopalatina. 0 = absent; 1 = present (78A uses codings from ref. 4).
80. Labial cavity. 0 = absent; 1 = present.
88. Condition of hyomandibular. 0 = with one proximal articular head; 1 = two proximal articular head.
89. Posttemporal fossa. 0 = absent; 1 = present.
93. Number of infradentaries. 0 = one; 1 = two; 2 = four.
95. Condition of most posterior coronoid. 0 = not distinctly differentiated from other coronoids; 1 = well developed and oriented vertically.
97. Articulation or symplectic with articular. 0 = absent; 1 = present.
98. Trajectory of supraorbital canal. 0 = passing between anterior & posterior nares;
    1 = passing anterior to both naries.
100. Relationship of infraorbital canal to premaxilla. 0 = entering Pmx; 1 = following dorsal margin of Pmx.
101. Trajectory of otic canal. 0 = not passing through growth center of postparietal;
    1 = passing through.
105. Preopercular canal. 0 = ending at dorsal margin of preopercular; 1 = not ending at dorsal margin of preopercular.
107. Position of infraorbital canal. 0 = ventral to anterior naris; 1 = dorsal to anterior naris.
110. Trajectory of mandibular canal. 0 = passing through dentary; 1 = not passing through dentary.
119. Proximal articular surface of humerus. 0 = concave; 1 = flat; 2 = convex.
120. Endoskeletal supports in pectoral fins. 0 = multiple elements articulating with girdle;
    1 = single element.
122. Deltoid and supinator processes. 0 = absent; 1 = present.
128. Basal plates in dorsal fin supports. 0 = absent; 1 = present.
130. Posterior branched radial complex in posterior dorsal fin. 0 = absent; 1 = present.
134. Epichordal lepidotrichia in tail. 0 = absent; 1 = present.
137. Supraneural spines. 0 = present on thoracic and abdominal vertebrae; 1 = restricted to a few anterior vertebrae or absent.
138. Condition of intercentra. 0 = ossified; 1 = not ossified.
141. Large dermal plates. 0 = absent; 1 = present.
142. Paired pectoral spines. 0 = absent; 1 = present.
143. Median fin spines. 0 = absent; 1 = present.
144. Denticulate postbranchial lamina of the cleithrum. 0 = absent; 1 = present.
145. Wide suborbital ledge. 0 = absent; 1 = present.
146. Eye stalk or unfinished area for similar structure. 0 = absent; 1 = present.
147. Ventral and otico-occipital fissures. 0 = absent; 1 = present.
148. Basipterygoid articulation. 0 = absent; 1 = present.
149. Endochondral bone. 0 = absent; 1 = present.

See Part IV for the expanded matrix. See ref. 14 for a complete list of the original 140 characters and character states; see ref. 4 for the changed codings for characters 10, 17, 78, 108.

IV. Expanded matrix with 149 characters for 37 taxa

Note: Characters 1–140 for taxa 1–33 are from ref. 14; modified codings at the end of the matrix (10A, 17A, 78A, 108A) are based on ref. 4; 0 = plesiomorphic state; 1, 2, 3 = apomorphic states; ? = unavailable character or logical impossibility; {01} = polymorphic states (0 and 1).

	1 1234567890	1111111112 1234567890	2222222223 1234567890	3333333334 1234567890	4444444445 1234567890
Acanthostega	0?0??00000	0111100101	?0?1100110	000101111?	?101??1110
Allenypterus	0?1?000000	000000011?	?000000120	0101011003	?1?0010011
Barameda	0?1?100000	011100010?	??00000110	0001101002	01????0??0
Beelarongia	1?010000??	????0???00	1???0001?0	000110?002	???10?0??0
Cheirolepis	0?0?010000	0000000{01}00	0000000100	01000{01}0001	?0?0000000
Crassigyrinus	0?0?000000	0?11100101	???1100110	000100111?	?101??01?0
Diabolepis	11???0001?	000000011??	?0??0001?1	2100?010??	???0111???
Diplocercides kaeseri	0?1?000000	000000011?	??00000120	0101012002	01?0010011
D. heligenstock- iensis	0?1?000000	0?00000011?	??00000120	0101012002	?1?0010?11
Dipnorynchus	1?0100?12?	000?01??1?	??000011?1	2000001003	?1102120??
Dipterus	111?00011?	000001??1?	??00011111	2100001002	?1?0212010
Elpistostege	0?010000??	0???100101	?000100110	0011011???	?101??1?70
Eusthenopteron	0?1?100000	0101000100	1010000110	0001001002	0101??0110
Glyptolepis	0?1?000000	1102000101	?000000100	0101112002	1110010110
Gyroptychius	1?01000000	0101000100	0010000110	0001101002	0101??0??0
Holoptychius	0?1?000000	1102000101	?000000120	0101112002	1110010110
Howqualepis	0?00011000	0000000000	01?00000?0	000?000001	?0?0000?00
Ichthyostega	0?1??00000	0111100101	?0?1100110	000101111?	?101??11?0
Miguashaia	0?1?000000	000000011?	?000000120	0101001002	01?0010?10
Mimia	0?00011000	0000000000	00?00000?0	000?000001	?0?0000000
Moythomasia	0?00011000	0000000000	01?00000?0	000?000000	?0?0000000
Onychodus	0?1?000000	1100000100	0000000120	0101111002	2100010?10
Osteolepis	1001000000	010?000100	1010000110	0001101002	0101??0110
Panderichthys	0?01000000	0111100101	?000100110	0011001102	0101??11?0
Polypterus	0?00010000	000000000?	???01000?0	010?00?000	?0?001000?
Porolepis	1101000000	0102000101	?000000100	1100112002	1110010??0
Powichthys	1101000001	?1010001??	?000000?0	1000?01002	1110010???
Speonesydrion	1?0?00012?	000001????	??00001??1	2000000010??	??????????
Strepsodus	0?1?100000	0111000100	1????0011?	??0???100?	?1???0?10
Strunius	0?1?000000	1?00000100	0000000??0	0101?11002	?1?????10
Uranolophus	110100?10?	000001??1?	??00011??1	2100001003	??102120??
Ventastega	0????00000	0111100101	?0?????1??	??0????1??	?1?1??11?0
Youngolepis	1101000001	?101001100	10?0000??0	110?10100?	???0110??0
Psarolepis	11???000?1	?101000000	01?0000??0	100???????	???0010?0?
Dicksonosteus	0????00???	0000?1??1?	?????000?0	?00????01	?0?????00?
Ctenacanthus	0????00???	000001??1?	?????000?0	??0???????	?0?????0??
Acanthodes	0?0??00???	000001??1?	?????000?0	??0???????	?0?????00?

	5555555556 1234567890	6666666667 1234567890	7777777778 1234567890	8888888889 1234567890	9999999990 1234567890
Acanthostega	0101011101	11?0?1?1?0	1101?0?000	0000???110	0021010100
Allenypterus	0111001100	01?0?1?010	0011010???	111???????	101?111??0
Barameda	010??????0	?????00010	101??0????	1?????????	00?????100
Beelarongia	01010??0?0	0?????????	??????????	1?????????	0?????????
Cheirolepis	0010002?00	0000?0000?	?0??0?????	0??????0??	0000000?0
Crassigyrinus	0101011101	11?0?1?1?1	1011?0??00	0?000?????	0021010???
Diabolepis	??????????	?????????0	0010001?01	0?????????	102????10?
Diplocercides kaeseri	?111001100	01?0?1?01?	?0?1010000	1111111111	101?1111??
D. heligenstock-iensis	0111001100	01?0?1?01?	???1?10???	111???????	101?111???
Dipnorynchus	??????????	??????????	?101?01??1	00000?????	102???011?
Dipterus	0101011100	00211??01?	?101001??1	00000?????	102???011?
Elpistostege	??????????	???1001???	??????????	0?????????	0?????????
Eusthenopteron	0111011000	0021000011	1011000000	1111110111	0021010100
Glyptolepis	1101111010	001101?010	0000100110	1111110111	0121010100
Gyroptychius	0101011000	0021000011	101100??00	111111????	002?0?0100
Holoptychius	1101111010	001101?010	0000100110	11111101??	0121010100
Howqualepis	0000000000	0000?0000?	?0?000???0	0?0??0?0??	0000010000
Ichthyostega	0101011101	11?0?1?1?0	110100?000	00001?????	0021010100
Miguashaia	0111001100	01?0?1?01?	?0??0?????	1?????????	101?11??0
Mimia	0000000000	0000?0000	0000000000	000000?000	0000000000
Moythomasia	0000000000	0000?00000	0000000000	000000?000	0000000000
Onychodus	0101002?00	01?101?01?	?0?1000010	111?110???	0121010??0
Osteolepis	0101011000	002100001?	?0?1?000?0	11111101??	002?0??100
Panderichthys	0101011100	0021001011	1011?00000	0??????1??	0021010100
Polypterus	?0?00?1?00	01?0?1?010	?010?00000	000001?001	0002000100
Porolepis	1101111010	0011000010	000010?110	11111101?1	0121010100
Powichthys	???????10	00??00010	00?0001010	0111101?00	0121010101
Speonesydrion	?????????0	00?11000??	?101001???	000?0?????	102????11?
Strepsodus	???1??????	???10??01?	??????????	1?????????	0?2???????
Strunius	01?100??00	01?101?01?	??????????	1?????????	01???????0
Uranolophus	??????????	?021100010	?101?01??1	0000??????	102?????1?
Ventastega	0??101110?	?????????0	110??0??00	??????????	002201?100
Youngolepis	???10?10?0	???1000010	0010001110	0000101100	0121010101
Psarolepis	???000????	??????????	000100?010	111?1?1??0	0120010?10
Dicksonosteus	010?0?1?0?	?????????0	????000?00	0000??000	????????1?
Ctenacanthus	?1???????	?????????0	????00?00	0000???00	????????1?
Acanthodes	010?0?1?0?	?00?????0	????000?00	0000???00	??????????

	1111111111 0000000001 1234567890	1111111111 1111111112 1234567890	1111111111 2222222223 1234567890	1111111111 3333333334 1234567890	111111111 444444444 123456789
Acanthostega	????110??1	11?01?0021	11?0111???	0??1201011	100000111
Allenypterus	01?111?001	11100?11?1	????00011?	0011111100	100000??1
Barameda	0110??0??1	???1110?21	01110?????	?2?????0??	00000???
Beelarongia	???01?????	???0110?21	110???????	100???????	100000???
Cheirolepis	00?00000?0	0000000?0	???00?0???	000000???0	100100??1
Crassigyrinus	?????????1	12?01?0021	11??111???	0????01001	100000111
Diabolepis	0010???101	??????????	??????????	??????????	100000??1

Diplocercides kaeseri	0101110001	1??01001??	????0?????	?01111?100	100000111
D. heligenstock-iensis	0???11?001	1??01001??	????0?0???	0011111?100	100001?1
Dipnorynchus	0010??1111	0?????????	??????????	??????????	100000111
Dipterus	0110111111	01101101??	??11000101	01010?0001	100000111
Elpistostege	??????????	??????????	??????????	?????0100?	100000???
Eusthenopteron	0110110001	1100110021	1100000100	1001111011	100000111
Glyptolepis	1111110001	1110110101	0011000111	0001010010	100000111
Gyroptychius	0110110001	11001?0???	????0?01??	100110101?	100000111
Holoptychius	1111110001	1110110071	????0?0???	00010?????	100000111
Howqualepis	0000000000	?000000??0	???00?0???	00000?????	1001001?1
Ichthyostega	????110??1	12?0110021	11??111???	0??1211011	100000111
Miguashaia	01?011?001	11?00011??	????0?01??	00010??100	100000???
Mimia	0000000000	?0000000?0	00?00000?0	0000000000	100100111
Moythomasia	0000000000	0000000??0	00?00000?0	0000000000	100100111
Onychodus	0110110??1	0110000101	10??0?0???	?00110?010	100000111
Osteolepis	0110110001	11001100??	????0?01?0	1001001010	100000111
Panderichthys	01??110001	1100110021	1?000?1???	0001201001	100000??1
Polypterus	0110000000	1000100100	00?00000?0	0001110??1	100100111
Porolepis	1110110001	11?0110???	????0?0???	??????????	100000111
Powichthys	0110??0001	11????????	??????????	??????????	100000111
Speonesydrion	00?????111	0?????????	??????????	??????????	100000111
Strepsodus	01????????	?111110021	11010?010?	02011?????	100000??1
Strunius	011011?001	1??00001??	????0?0???	00011???0	100000??1
Uranolophus	00101??111	?110110???	????0?0???	000?01?000	10000011?
Ventastega	?????1???1	1???0?0???	?????1????	??????????	100000??1
Youngolepis	0010110101	1??0?10?11	??????????	??????????	100000111
Psarolepis	0010000001	1???0?????0	??????????	??????????	111111111
Dicksonosteus	?0???1????	??????00?0	00?000????	0??0????0	110111000
Ctenacanthus	????01?1??	??????0???	00?000????	0??0????0	0010?1?00
Acanthodes	????01?1??	??????0???	00?0000???	0??0????0	011000110

	1
	1170
	0788
	AAAA

Acanthostega	100?	
Allenypterus		00?0
Barameda		10??
Beelarongia		????
Cheirolepis		00?0
Crassigyrinus		10??
Diabolepis		?1?0
Diplocercides kaeseri	00?0	
D. heligenstock-iensis	00?0	
Dipnorynchus		???1
Dipterus		???1
Elpistostege		?0??
Eusthenopteron	1000	
Glyptolepis		010

Gyroptychius	10?0
Holoptychius	1010
Howqualepis	00?0
Ichthyostega	100?
Miguashaia	00?0
Mimia	0000
Moythomasia	0000
Onychodus	?00?
Osteolepis	1000
Panderichthys	1000
Polypterus	0000
Porolepis	1010
Powichthys	0000
Speonesydrion	???1
Strepsodus	10??
Strunius	?0?0
Uranolophus	???1
Ventastega	10??
Youngolepis	0000
Psarolepis	0000
Dicksonosteus	??0?
Ctenacanthus	??01
Acanthodes	??01

Chapter 11   A Therizinosauroid Dinosaur with Integumentary Structures from China   Xing Xu, Zhi-Lu Tang, and Xiao-Lin Wang

I. Character matrix for phylogenetic analysis of *Beipiaosaurus inexpectus*

Character states: 0, plesiomorphic state; 1, apomorphic state; a, polymorphic; ?, not preserved/unknown.

     5 10 15 20 25 30 35 40 45 50 55 60 65 70 75 80 84

Coelophysis 00000 00000 00000 00000 00000 00000 00000 00000 00000 00000 00000
            00000 00000 00000 00000 00000 0000

Ornithomimidae 01100 11000 01111 00011 01000 00011 10111 00100 00100 00000
               00000 00000 00001 01100 01111 10000 0010

Dromaeosauridae 01000 10000 01100 00010 00000 00001 0111? 10111 01101 10a01
                10111 10001 11101 10100 00101 00000 0001

Oviraptorosauria 10111 a1111 01100 11a11 0011? ????0 11111 a1?1a 01101 00a01
                 00110 00011 01011 11010 01111 a0000 10a1

Troodontidae 00?00 10110 0?111 0101? 00000 10011 10110 00111 0110? 00001 00110
             10001 11011 11100 0a101 10000 1011

Beipiaosaurus 0???? ????? ????? ????? 11?01 11010 11??? ???11 01?11 11001
                ?0??1 1101? 11101 101?1 01100 00101 1101

Alxasaurus 1???? ????? ????? ????? 11?01 11110 101?1 110?1 01011 11111 11111
             11110 0?101 00?11 01100 01111 110?

Therizinosauridae 10111 01001 11101 11111 11001 11a10 101?1 110?1 110?1 11111
                  11111 11110 0?101 00?11 11100 0111? 1101

II. Character list for phylogenetic analysis of *Beipiaosaurus inexpectus*

1. Skull-ilium length ratio large, more than 1 (0); small, less than 1 (1).
2. Premaxilla and nasal do not contact beneath nares (0) or contact (1) (Russell and Dong, 1993)
3. Premaxilla without (0) or with broad palatal shelf (1) (Sues, 1997)
4. Premaxillary teeth present (0) or absent (1) (Sues, 1997)
5. Rim around antorbital fossa reduced or absent (0) or well-developed (1) (Sues, 1997).
6. Accessory antorbital fenestra present (0) or absent (1) (Sues, 1997)
7. Maxilla with narrow palatal shelf (0) or with broad palatal shelf contributing to formation of secondary bony palate (1)
8. Dorsal surface of parietals flat, ridge borders supratemporal fenestrum (0) or parietals with sagittal crest along midline (1) (Russell and Dong, 1993)
9. Prefrontal present (0) or much reduced or absent (1) (Sues, 1997).
10. Ectopterygoid posterior to palatine (0) or lateral to it (1) (Sues, 1997).
11. Vomer limited to rostral area (0) or extends posteriorly to basicranium (1) (Russell and Dong, 1993).
12. Subsidiary fenestra between pterygoid and palatine absent (0) or present (1) (Sues, 1997)
13. Ectopterygoid unmodified (0) or expanded with deep ventral pocket (1) (Sues, 1997)
14. Parabasisphenoid unmodified (0) or expanded to form hollow 'bulla'(1) (Sues, 1997)
15. Basisphenoid unmodified (0) or greatly expanded ventrally and pneumatized (1) (Sues, 1997)
16. Basipterygoid processes well-developed (0) or short (1) (Sues, 1997)
17. Deep lateral depression in otic region of braincase absent (0) or present (1) (Sues, 1997)
18. Occipital region obliquely inclined anteroventrally (0) or deflected ventrally (1) (Sues, 1997)
19. Proximal portion of paroccipital process solid (0) or hollow (1) (Sues, 1997)
20. Distinct coronoid bone present (0) or greatly reduced or absent (1) (Sues, 1997)
21. A prominent dorsolateral shelf on the dentary absent (0); present (1).
22. Maxillary teeth present (1) or absent (1) (Sues, 1997)
23. Dentary teeth present (0) or absent (1) (Sues, 1997)
24. Mandibular symphysis straight (0) or downturned (1). (Sues, 1997)
25. Tooth homodont along the entire lengh of tooth row (0) or increase in size anteriorly (1) (Russell and Dong, 1993 and Dong, 1993; Zhao and Xu, 1998).
26. Denticles on teeth small (0) or large (1) (Clark et al, 1994)
27. Tooth crowns mediolaterally compressed (0) or basal cross section subcircular (1) (Clark et al, 1994; Zhao and Xu, 1998; adapted from Forster et al. 1998)
28. Tooth crowns recurved (0) or nearly symmetrical in labial view (1).
29. Constriction between crown and root: absent (0) or present (1) (Forster et al. 1998)
30. Unfused interdental plates present (0) or absent (1) (Forster et al. 1998)
31. Cervical neural spines tall (0) or low and anteroposteriorly inclined (1) (Sues, 1997)
32. Dorsal vertebrae without pleurocoels (0) or pleurocoels present (1) (Holtz, 1994)
33. Cervical centra with anterior articular surface higher than wide (0) or wider than high (1) (Sues, 1997).
34. Centra of sacral vertebrae nearly circular in cross section (0) or flattened (1) (Sues, 1997).
35. Pleurocoels absent on sacral vertebrae (0) or present (1) (Russell and Dong, 1993)
36. Proximal caudal centra subcircular in transverse section (0) or box-like, rectangular in transverse section (1) (Sues, 1997)
37. Centra of distal caudal vertebrae elongate (0) or short (1) (Sues, 1997)
38. Distal chevrons simple and curving in profile (0) or shafts cylindrical in profile anteriorly, and much reduced posteriorly (1) (Russell and Dong, 1993)

39. Clavicles unfused (0) or fused to form furcula (1) (Sues, 1997)
40. Coracoid rounded (0) or subrectangular (1) (Sues, 1997)
41. Anteroposterior length of ilium more than two times (0) or less than two times (1) as long as its dorsoventral depth.
42. Dorsal margin of iliac blades gently arched (0) or distinctly curved posteroventrally (1) in lateral view (Sues, 1997)
43. Dorsal margins of iliac blades well separated from (0) or almost in contact with apices of sacral neural spines (1) (Sues, 1997)
44. Iliac fossa for m. cupedicus present (0) or absent (1) (Forster et al. 1998)
45. Supraacetabular crest present on ilium (0) or reduced or absent (1) (Russell and Dong, 1993)
46. Pubic peduncle of ilium is at same level as ischiadic peduncle (0) or extends ventrally beyond it (1) (Gauthier, 1986)
47. Pubic peduncle of ilium anteroposteriorly wide (0) or narrow (1)
48. Preacetabular portion of ilium more or less equal in length to (0) or much longer than postacetabular portion (1) (Sues, 1997)
49. Dorsoventrally expanded preacetabular portion of ilium absent (0) or present (1) (Holtz, 1994)
50. Brevis fossa well developed on ilium (0), or small and very shallow (1) (Russell and Dong, 1993)
51. Pubis projects anteroventrally (propubic condition) (0) or posteroventrally (opisthopubic condition) (1) (Sues, 1997)
52. Shape of ischial shaft rod-like in part with a circular or subcircular cross section (0) mediolaterally compressed and plate-like along entire length (1) (Forster et al. 1998).
53. Obturator process in proximal position (0) or distally placed on ischium (1) (Sues, 1997).
54. Length of ischium three-quarters or more as long as pubis (0) or two-thirds or less length of pubis (1) (Sues, 1997).
55. Internal tuberosity on proximal end of humerus either not well differentiated or not angular (0) or well differentiated and angular (1) (Russell and Dong, 1993).
56. Distal condyles of humerus located mainly on the distal (0) or cranial (1) aspect (Forster et al. 1998)
57. Ventral tubercle of humerus does not project caudally, confluent with humeral head (0) or projects distincelty caudally, separated from the humeral head by a deep capital incision (1) (Forster et al. 1998).
58. Flexor groove between ulnar and radial condyles shallow and wide (0) or deep and narrow (1)
59. Ulnar facet of humerus small or absent (0) or expanded (1) (Russell and Dong, 1993)
60. Shaft of the ulna straight (0) or bowed (1) (Gauthier, 1986).
61. Manus shorter than or subequal to ulna (0) or longer than ulna by at least 20%(1)
62. Semilunate distal carpal absent (0) or present (1) (Sues, 1997)
63. Manus shorter than (0) or longer than pes (1) (Adapted from Gauthier, 1986).
64. Metacarpal I at least half length of metacarpal II (0) or only one-third length of metacarpal II (1) (Russell and Dong, 1993).
65. Proximal half of metacarpal I not appressed to (0) or closely appressed to metacarpal II (1) (Sues, 1997)
66. Metacarpal III unmodified (0) or very slender (1) (Sues, 1997)
67. Metacarpal II longer than metacarpal III (0) or subequal to metacarpal III (1) (Russell and Dong, 1993).
68. Combined lengths III-1 and III-2 relative to III-3 on manus longer (0) or subequal(1) (Forster et al. 1998).
69. Pronounced dorsal 'lip' on proximal articular surface of manual unguals absent (0) or present (1) (Sues, 1997).
70. Manual unguals proximally shallow (0) or very deep (1).

71. Ligament pits well-developed in manual phalanges (0) or weakly developed or absent (1) (Russell and Dong, 1993)

72. Greater trochanter of femur not distinct (0) or separated from proximal head of femur by cleft (1) (Sues, 1997)

73. Lesser trochanter of femur spike-like (0) or prominent wing-like (1) (Sues, 1997)

74. Fourth trochanter of femur present (0) or absent (1) (Sues, 1997)

75. Lesser trochanter large and flange-like, separated from femoral shaft, head, and greater trochanter by a deep cleft (0); nearly confluent with the greater trochanter, and separated from the proximal head of femur by only a small cleft or groove (1) (Forster et al. 1998)

76. Tibia and metatarsus not elongate relative to femur (0) or elongate (1) (Sues, 1997)

77. Tibia longer (0) or shorter (1) than femur.

78. Metatarsus more than 45% length tibia (0); less than 40%(1).

79. Theropod first metatarsal (proximal end of MT I compressed and fails to contact tarsus) present (0) or absent (1) (Gauthier, 1986)

80. Deep fossa on medial side of proximal fibular head present (0) or absent (1) (Forster et al. 1998)

81. Astragular ascending process paralleling tibia shaft (0) or curving dorsolaterally toward fibular shaft (1) (Russell and Dong, 1993).

82. Astragalus covering full distal end of tibia (0) or main body of astragulus reduced, tibia forms part of articular surface for pes (1) (Paul, 1984; Clark et al. 1994)

83. Arctometatarsal configuration of metatarsus absent (0) or present (1) (Sues, 1997; Holtz, 1995).

84. Pedal digit IV only slightly longer than II (0) or closer to III in length and distinctly longer than II (1).

Chapter 12   A Dromaeosaurid Dinosaur with a Filamentous Integument from the Yixian Formation of China   Xing Xu, Xiao-Lin Wang, and Xiao-Chun Wu

Cladistic analysis

List of 106 characters and data matrix informative for resolving the phylogenetic relationships of the taxa included in Figure 5. Characters 1–82 are derived from Forster et al. (1998); characters 83–100 are from Currie (1995); characters 101–102 are from Chiappe et al. (1998); characters 103–106 are newly added in this study. The numbers in square brackets "[ ]" are the original character numbers used in the references cited.

CHARACTERS FROM FORSTER ET AL. (1998)

1[1].   teeth in premaxilla: present (0); absent (1).

2[2].   denticles on teeth: present on anterior and posterior carinae (0), present on posterior carina, but absent or severely restricted on anterior carina (1); absent (2).

3[3].   shape of teeth: laterally compressed and recurved (0); slightly compressed, nearly conical (1).

4[4].   constriction between crown and root: absent (0); present (1).

5[5].   jugal contribution to antorbital fenestra: absent, eliminated by lacrimal-maxilla contact (0); present (1).

6[6].   maxilla contribution to naris: absent, excluded by a nasal premaxilla contact (0); present (1).

7[7].   shape of frontal: anteriorly broad (0); triangular, narrow rostrally (1).

8[8].   extreme reduction or loss of prefrontal: absent(0); present (1).

9[9].   frontal length relative to parietal smaller or sub equal (0); nearly two times as long (1).

10[10].   narrow sagittal crest on parietals: absent (0); present (1).

11[11].   jugal bar shape: robust (0); thin and straight (1).

12[13].   subsidiary foramen between palatine and pterygoid: absent (0); present (1).

13[14].   palatine-ectopterygoid contact: no contact, ectopterygoid present (0); contact present, ectopterygoid present (1); **no contact, ectopterygoid absent (2) [State 2 is not considered here because none of the included taxa display this condition]**

14[16].   distal quadrate articulation set well anterior to proximal quadrate condyle: absent (0); present (1).

15[17].   shape of paroccipital process: straight and vertically oriented (0); distal end with distinct twist as to face caudodorsally (1).

16[18].   bulbous parasphenoid capsule (cultriform process): absent (0); present (1).

17[19].   accessory lacrimal fenestration: present (0); absent (1).

18[20].   supraoccipital crest: absent or weak (0); well pronounced (1).

19[21].   fusion between parietals and laterosphenoid in adults: present (0); absent (1).

20[22].   accessory fenestra between premaxilla and maxilla: present (0); absent (1).

21[24].   pneumatic quadrate: absent (0); present (1).

22[25].   interdental palates: present (0); present, fused (1); absent (2). (**modified**).

23[26].   separate coronoid bone: absent (0); present (1).

24[28].   ventral processes (hypapophyses) on cervicodorsal vertebrae: absent or very weakly developed (0); present, well developed (1).

25[29].   ratio of height of neural canal in dorsal vertebrae to height of cranial articular face: less than 0.4 (0) more than 0.4(1).

26[30].   pneumatic foramina on dorsal vertebrae: absent (0); present (1).

27[31].   number of sacral vertebrae: 5 or less (0); 6 or more (1).

28[32].   fusion of all sacral vertebrae in adults: absent or partially fused (0); present, all completely fused into a synsacrum (1). (**slightly modified**)

29[33].   number of caudal vertebrae: 33 or more (0), 20–25 (1), **or less than 15(2) [State 2 is not considered because none of the included taxa display this condition].**

30[35].   transition point on caudal vertebrae: distal, behind caudal 10 (0); proximal, no further back than caudal 10 (1).

31[36].   middle and distal chevron shape: laterally compressed, expanded dorsoventrally (0); dorsoventrally flattened into a thin horizontal plate behind transition point (1).

32[38].   pneumatic foramen on sacral vertebrae: present (0); absent (1).

33[39].   cervical neural spines: dorsoventrally tall, axially short (0); dorsoventrally short, axially elongate (1).

34[40].   length of middle and posterior caudal vertebrae: subequal to length of proximal caudals (0); elongate, at least 130% the length of the anterior caudals (1).

35[41].   prezygopophyses on middle and posterior caudal vertebrae: elongate, extend at least half way across preceding vertebral centrum (0); reduced or absent, extend over 25% or less the length of preceding vertebral centrum (1).

36[44].   anterior projection of acromion process: absent (0); present, extends well anterior to glenoid fossa (1).

37[45].   scapulocoracoid articulation: immobile, sutural (0); mobile, cartilaginous (1).

38[46].   acrocoracoid of coracoid: absent, biceps tubercle small (0); absent, biceps tubercle large (1); **present, extends above scapulocoracoid articulation, biceps tubercle large (2) [State 2 is not considered here because none of the included taxa display this condition].**

39[47].   body of coracoid forming sharp angle with the body of scapula: absent (0); present (1).

40[49].   coracoid shape: short, rounded sternal border (0); elongate, subrectangular profile, rounded sternal border (1); **elongate, strut-like (2) [State 2 is not considered because none of the included taxa display this condition].**

41[52]. fused clavicles: absent (0); present (1); **present interclavicle angle less than 90 degrees (2) [State 2 is not considered here because none of the included taxa display this condition].**

42[53]. ossified sternum: paired (0); fused or partially fused into a single structure (1); **fused with a midline keel (2) [State 2 is not considered here because none of the included taxa display this condition].**

43[54]. distal condyles of humerus located mainly on the distal (0), or cranial (1), aspect.

44[59]. ratio of diameter of shaft of radius to ulna: more than 0.7 (0); thinner than ulna, less than 0.7 (1).

45[62]. metacarpal I greater than (0), or less than (1), one-third the length of Metacarpal II.

46[63]. metacarpal III: straight (0); bowed laterally (1).

47[64]. combined lengths of III-1 and III-2 relative to II-3 on manus: longer (0); subequal (1); absent (2).

48[65]. forelimb elongation relative to presacral length: less than one half presacral length (0); 60–90% presacral length (1); more than 100% presacral length (2).

49[66]. manus length relative to ulna length: shorter than or subequal to ulna (0); manus longer than ulna by at least 20% (1).

50[68]. ulnar distal condyle: transversely compressed and craniocaudally extended proximally in the same plane as the humero-ulnar flexion-extension movement (0); subtriangular in shape in distal view, with a dorsomedial condyle, and twisted more than 54 degree with respect to proximal end (1).

51[69]. shape of ulnar posterior margin: sigmoid (0); uniformly convex (1).

52[70]. distal radial carpal: proximodistally flattened (0); semilunate (1).

53[71]. length of preacetabular process of ilium relative to length of postacetabular process: subequal (0); twice as long (1).

54[72]. postacetabular process depth: relatively deep, more than 50% depth at acetabulum (0); shallow, less than 50% depth at acetabulum, drawn back into a low, pointed process (1).

55[73]. ratio of lengths of acetabulum to ilium: 0.15 or more (0); 0.11–0.13 (1); **0.10 or less (2) [state 2 is not considered here because none of the included taxa display this condition].**

56[75]. tubercle on dorsal margin of ilium above caudal acetabulum: absent (0); present (1).

57[76]. craniocaudal length of pubic peduncle on ilium relative to width of acetabulum: narrow, less than acetabular width (0); wide, exceeds acetabular width (1).

58[77]. pubic foot: projects cranially and caudally (0); projects caudally only (1); absent (2).

59[78]. pubic shaft relative to the long axis of sacral vertebrae: projects cranially or subvertically [but slightly anteriorly] (0); caudally or subvertically [but slightly posteriorly] (1). (modified)

60[79]. ratio of length of ischium to length of pubis: more than 0.66 (0); between 0.66–0.51 (1); less than 0.5 [probably indicating no contact between distal ischia] (2). (modified)

61[81]. obturator process on ischium: present, square (0); present, peaked and broadly triangular (1); **rudimentary or absent (2) [State 2 is not considered here because none of the included taxa display this condition].**

62[82]. placement of obturator process on ischium: proximal (0); mid-shaft (1); distal (2).

63[83]. "postacetabular" process on proximal ischium behind iliac contact: absent (0); present, nearly contacts postacetabular process of ilium (1).

64[84]. process on caudal border of midshaft of ischium: absent (0); present (1).

65[86]. shape of ischial shaft: rod-like in part with circular or subcircular cross section (0); mediolaterally compressed and plate-like along entire length (1).

66[87]. postacetabular blade on ilium: brevis shelf caudolaterally oriented, medial flange ventrally curved (0); postacetabular blade vertical, medial flange strongly reduced and perpendicular to iliac blade (1).

67[91]. configuration of lesser trochanters: large and flange like, separated from femoral shaft, head, and greater trochanter by a deep cleft (0); nearly confluent with the greater trochanter, and separated from the proximal head of femur by only a small cleft or groove (1); **joined to greater trochanter to form an undivided trochanteric crest, proximal articular surface confluent with that of femoral head (2) [state 2 is not considered here because none of the included taxa display this condition].**

68[92]. shape of fourth trochanter on femur: present (0); reduced to a low ridge or absent (1).

69[93]. "posterior trochanter" on femur: absent (0); present, cranially placed (1); present, centred on the trochanteric crest (2).

70[93]. femoral neck: present, constriction developed that separates trochanteric region from femoral head (0); absent (1).

71[98]. proximal tibia: craniocaudal length twice that of mediolateral width (0); length and width of proximal tibia subequal (1).

72[99]. midshaft diameter of fibula relative to tibia: approximately one-fifth or more than that of tibia (0); one-fifth or less than that of tibia (1).

73[100]. orientation of iliofibular tubercle on fibula: craniolateral (0); lateral (1); caudolateral or caudal (2).

74[101]. deep fossa on medial side of proximal fibular head: present (0); absent (1).

75[102]. fibular articulation with the calcaneum: present (0); absent, fibula does not reach tarsus (1).

76[103]. fusion of proximal tarsals to crus: absent (0); present, partially fused (1); **present, completely fused (2) [State 2 is not considered here because none of the included taxa display this condition].**

77[104]. fusion of astragalus to calcaneum in adults: absent (0); present, partial or complete (1).

78[105]. fusion of distal tarsals to metatarsus: absent (0); present (1).

79[108]. position of pes digit I: parallel to other digits (0); reversed to oppose other digits (1).

80[109]. relative contributions of metatarsals II, III, and IV to ankle joint: all contribute approximately equally (0); partially or completely arctometatarsalian, metatarsal III is nearly or completely eliminated from joint (1).

81[110]. relative size of pedal digit II: phalanges and ungual subequal in size and robustness to digits II and IV (0); developed into a robust, hyperextensible slashing digit with an enlarged sickle-like ungual (1).

82[111]. ratio of length of tibia to length of femur: tibia no more than 15% longer than femur (0); tibia elongate, at least 25% longer than femur (1).

CHARACTERS FROM CURRIE (1995)

83[3]. frontal anterior demarcation of supratemporal fossa: straight or slightly sinuous (0) sinusoidal with associated deep pit (1).

84 [4]. teeth—secondary premaxillary tooth: smaller than or equal in size to third and fourth premaxillary teeth (0); significantly larger than premaxillary teeth 3 and 4 (1).

85[5]. teeth—maxillary, mandibular: anterior and posterior denticles not significantly different in size (0); anterior denticles, when present, significantly smaller than posterior denticles (1).

86[7]. lacrimal: inverted L-shaped (0); T-shaped (1).

87[8]. frontal, lacrimal-prefrontal contacts: sutures on lateral, dorsal and/or ventral surfaces (0); dorsal and ventral surfaces connected by a vertical slot (1).

88[9]. supratemporal fossa: limited extension onto dorsal surfaces of postorbital and frontal (0); covers most of frontal process of postorbital and extends anteriorly on dorsal surface of frontal to at least level of the posterior orbital margin (1).

89[10]. postorbital: T-shaped (0); upturned frontal process (1).

90[11]. quadratojugal: L-shaped (0); Y- or T-shaped (1).

91[12]. quadratojugal fenestra: small foramen-like opening (0); widely open (1).

92[13]. external auditory meatus: does not extend beyond level of intertemporal bar of postorbital and squamosal (0); ventral process of squamosal and lateral extension of paroccipital process beyond head of quadrate (1).

93[15]. opisthotic-exoccipital: no periotic pneumatophore or pneumatized paroccipital process (0); periotic pneumatophore enters hollow paroccipital process (1).

94[16]. pterygoid flange: includes major contribution from pterygoid (0); is formed mostly by ectopterygoid (1).

95[19]. dentary: thick when compared to height, deep Meckelian groove, pronounced dental shelf (0); thin and high with shallow Meckelian groove and dental shelf (1).

96[20]. dentary, lateral view: tapers conspicuously anteriorly (0); upper and ventral margins subparallel (1).

97[21]. splenial: limited or no exposure of splenial on lateral surface of mandible (0); conspicuous triangular process on external surface of mandible between dentary and angular (1).

98[22]. internal mandibular fenestra: absent or small and slitlike (0); triangular and relatively large (1).

99[23]. ventral columnar process of retroarticular process: absent (0); present (1).

100[24]. ossified caudal rods extending lengths of prezygapophyses and chevrons: absent (0); present (1).

CHARACTERS FROM CHIAPPE ET AL. (1998)

101[2]. frontal process of premaxilla: short (0), relatively long to very long, at least approaching the rostral border of the antorbital fossa (1).

102[4]. caudal margin of naris farther rostral than (0), or nearly reaching or overlapping (1), the rostral border of the antorbital fossa.

CHARACTERS NEWLY ADDED IN THIS STUDY

103. scapula: longer (0), shorter (1), much shorter [less than 2/3] (2) than ulna.

104. ulna shorter (0), longer (1) than metatarsal III.

105. glenoid of pectoral girdle: facing posteroventrally or posterolaterally (0); facing laterally (1).

106. articular facet of coracoid on sternum (conditions may be determined by the articular facet on coracoid in taxa without ossified sternum): anterolateral or more lateral than anterior (0); almost anterior (1).

Data matrix

ABBREVIATIONS

Al, *Allosaurus* (data from Madson Jr., 1976; Molnar et al., 1990); Ar, *Archaeopteryx* (data from Ostrom, 1976, 1985; Wellnhofer, 1974, 1992, 1993); Ca, *Caudipteryx* (data from Ji et al., 1998; a cast [paratype] in TMP; Currie, personal communication); Co, *Compsognathus* (data from Ostrom, 1978); Dr, *Dromaeosaurus* (data from Currie, 1995); Or, Ornithomimidae (data from Barsbold and Osmólska, 1990; specimen, TMP 95.110.1 [*Ornithomimus*]); Ov, Oviraptoridae (Barsbold et al., 1990); Pr, *Protarchaeopteryx* (data from Ji and Ji, 1997; Ji et al, 1998; Currie, personal communication); Ra, *Rahona* (Forster et al., 1998); Si, *Sinornithosaurus* (data from IVPP [Institute of Vertebrate Paleontology and Paleoanthropology, Academia Sinica] V12811). Tr, Troodontidae (data from Osmólska and Barsbold, 1990; Russell and Dong, 1993 [*Sinornithoides*]; specimen, TMP 94.12.415 [*Troodon*]); Ty, Tyrannosauridae (data from Molnar et al., 1990); Un, *Unenlagia* (data from Novas and Puerta, 1997); Ve, Velociraptorinae (data from Ostrom, 1969, 1974; Witmer and Maxwell, 1996 [*Deinonychus*]; Sues, 1977; Norell and Makovicky, 1997, 1998; Norell et al., 1997; Osmólska, personal communication [*Velociraptor*]); specimens, TMP [Royal Tyrrell Museum of Palaeontology] 88.121.39, TMP 92.36.333 [*Saurornitholestes*]). The character states that are inapplicable to a certain taxon are treated as "?". We coded "?" state of character 8 for Velociraptorinae because a prefrontal is present in *Deinonychus* [Witmer and Maxwell, 1996] but absent in *Velociraptor* [Osmólska, personal communication]).

	1 1234567890	2 1234567890	3 1234567890	4 1234567890	5 1234567890
Al	0000000000	0000000000	0000000000	0000000000	0000000000
Co	0100?00000	1???0?1101	?0?0?0??0?	0??0?00?00	??000??000
Ar	0211100111	11101?1101	0010?10011	1101110111	1111111211
Ra	??????????	??????????	???11111?1	1??1111???	???1????2?1
Ov	1???110101	1110000101	0201011000	0000100100	1100000100
Ve	0100100?11	0111101101	0101010001	10010001?1	1010111100
Dr	010010001??	0111101?01	010????????	??????????	??????????
Or	0201111010	1111011101	0200000000	0010000100	0000001100
Tr	0001101111	01??011111	02?1001001	1011000100	0?11101110
Ty	0000111101	0000000010	1000010000	0000000000	0000002000
Un	??????????	??????????	????0111??	?????1????	??????????
Pr	0011?????0	0??????????	?0??01??11	?0?1??0?1	10?11?1110
Ca	0210?11?00	0??????????	00??0?0?11	10?0????10	10?11??100
Si	0100??0011	?????01???	00?1??00??	1?01010111	10?1111?1?

	1 1234567890	6 1234567890	7 1234567890	8 1234567890	9 1234567890	0 123456
Al	0000000000	0000000000	0000000000	0000000000	0000000000	000000
Co	0?000?1?00	10?00?????	????00??00	0100000?00	0???000000	000000
Ar	1111011112	1211111111	11??011110	0100000000	0110000000	112111
Ra	1?11111112	1211112121	1121101010	11????????	?????????0	??211?
Ov	0100000000	1100001100	0000000000	0100000000	0000000000	110000
Ve	1100001212	1100111110	0000000000	1011111111	1111111111	001111
Dr	??????????	??????????	??????????	??00011111	111111111?	00????
Or	0000000001	1100001100	0000000101	0000000000	0000000000	000000
Tr	11000??202	1100011120	0010011001	1100000000	0000001000	00?000
Ty	0000000000	1000000000	0000000001	0000000000	0000000000	000000
Un	??11011112	1110111101	??????????	??????????	??????????	????1?
Pr	?100??01??	????001???	00??00000	01?00?????	0????????0	0??0??
Ca	?1000?01?1	1?00001?2?	0???000000	0100?0??00	0???00???0	110000
Si	??0?0?1212	11011???1?	?1???010?1	1?11110101	1???011??1	001111

List of Synapomorphies (Ambiguous Characters in [ ]) of each Node in Figure 4

Node A: characters 5, 61 [6, 7, 47]; Node B: characters 11, 17, 18, 20 [2, 12, 13, 38, 67, 68, 82]; Node C: characters 48, 62 [2(2), 4, 22(2)]; Node D: characters 8, 24, 26, 41, 52 [27, 101, 102]; Node E: characters 11(0), 30, 31, 40, 44, 45, 58, 69(2) [22(0), 29, 43, 51, 60(2), 97]; Node F: characters 10, 66, 77, 81 [29(0), 57, 58(2)]; Node G: characters 15, 39, 46, 59, 65, 92, 93, 104, 105, 106 [2, 7(0), 36, 69, 72]; Node H: characters 53, 54, 56, 63, 70 [2(2), 3, 11, 23, 28, 29, 32, 35, 42, 48(2), 58, 71, 73(2), 74, 79, 97(0), 101, 102, 103(2)]; Node I: characters 22, 87, 89, 95 [36(0), 44(0), 49(0), 72(0), 77(0)]; Dromaeosauridae: characters 4(0), 86, 88, 90, 91, 96, 100 [14, 27(0), 82(0), 83–85, 98, 99].

A further description

The pterygoid differs little from that of other dromaeosaurids. It might have formed a mobile joint with the quadrate because the latter's pterygoid ramus possesses a columnar margin. A ring of sclerotic bones is present in the orbit. The lower jaw may not have had a hinge joint because the bifurcated posterior margin of the dentary forms an interlocked joint with the postdentary bones, as in some non-coelurosaurians, such as *Yangchuanosaurus* (Dong et al., 1978) and *Sinraptor* (Currie and Zhao, 1993). As described for *Dromaeosaurus* (Currie, 1995), the dentary, with subparallel upper and lower margins, is dorsoventrally shallow; the splenial has a conspicuous lateral exposure between the dentary and angular; and the interdental plates are present, but most of them are not fused. The second tooth of the premaxilla, as in Velociraptorinae (Currie, 1995; Osmólska, Personal Communication) is larger than the other three. The premaxillary teeth and the first dentary tooth are not serrated, a condition also seen in coelophysids (Colbert, 1989), *Compsognathus* (Ostrom, 1978), and *Caudipteryx* (Ji et al., 1998). Most of the left premaxillary teeth lie partially out of their sockets. The posterior denticles are apparently larger than the anterior ones in the maxillary teeth, except in the anteriormost teeth.

Most elements of the postcranial skeleton are known, except for the distal end of the femur and the proximal parts of the fibula and tibia. The posterior two dorsal, five fused sacral vertebrae, and the left ilium are exposed in ventral view; the left pubis and ischium in lateral view; the right ilium, pubis, and ischium in medial view; and the proximal part of the left femur in anterolateral view. The medial surface of the ilium dorsal to the public facet bears a number of pits and ridges. Margins of these pits or ridges are original and show no trace of pathology. The vertical orientation makes it impossible to determine whether the left ilium has similar ornament-like structures. The right astragalus and calcaneum are seen in posterior view. They are partially fused and disarticulated from the tibia. Metatarsal I appears to attach to the distal third of metatarsal II. It is difficult to determine whether metatarsal I ran parallel to or was positioned opposite metatarsal II because of disarticulation. Metatarsal III is splint-like (partially arctometatarsalian) in posteroventral view and has a reduced facet for articulating with the distal tarsals. Our phylogenetic analysis suggests that the arctometatarsalian metatarsal III is convergently gained by *Sinornithosaurus,* Troodontidae, Ornithomimidae, and Tyrannosauridae among theropods. Metatarsal V is almost complete, very slender and is half as long as metatarsal IV. The latter has a strong ventral flange. Phalanges are complete, except for the large, slashing, hyperextensive ungual of digit II of the foot, the distal part of which is covered by the tail.

Further references

Barsbold, R. & Osmólska, H. Ornithomimosauria, in *The Dinosauria* (eds Weishampel, D. B., Dodson, P. & Osmólska, H) 225–244 (University California Press, Berkeley, 1990).

Barsbold, R., Maryanska, T. & Osmólska, H. Oviraptorosauria. in *The Dinosauria* (eds Weishampel, D. B., Dodson, P. & Osmólska, H.) 249–258 (University California Press, Berkeley, 1990).

Dong, Z.-M., Zhou, S.-W. & Zhang, Y.-H. Note on a new carnosaur *Yangchuanosaurus shangyuensis* gen. et sp. nov. from the Jurassic of Yangchuan District, Sichuan Province. *Kexue Tongbao* 23, 298–302 (1978, In Chinese).

Chiappe, L. M., Norell, M. A. & Clark, J. M. The skull of a relative of the stem-group bird *Mononykus. Nature* 392, 275–278 (1998).

Currie, P. J. & Zhao, X.-J. A new carnosaur (Dinosauria, Theropoda) from the Jurassic of Xinjiang, People's Republic of China. *Can. J. Earth Sci.* 30, 2037–2081 (1993).

Colbert, E. H. The Triassic dinosaur *Ceolophysis. Mus. Northern Arizona. Bull.* 57, 1–60 (1989).

Madsen, J. H. Jr. *Allosaurus fragilis*: a revised osteology. *Utah Geol. Mineral Sur.* Bull. **109**, 1–163 (1976).

Molnar, R. E., Kurzanov, S. M. & Dong, Z.-M. Carnosauria. in *The Dinosauria* (eds Weishampel, D. B., Dodson, P. & Osmólska, H.) 169–209 (University California Press, Berkeley, 1990).

Osmólska, H. & Barsbold, R. Troodontidae. in *The Dinosauria* (eds Weishampel, D. B., Dodson, P. & Osmólska, H.) 259–268 (University California Press, Berkeley, 1990).

Ostrom, J. H. The osteology of *Compsognathus longipes* Wagner. *Zitteliana* **4**, 73–118 (1978).

Russell, D. A. & Dong, Z.-M. A nearly complete skeleton of a new troodontid dinosaur from the Early Cretaceous of the Ordos Basin, Inner Mongolia, People's Republic of China. *Can. J. Earth Sci.* **30**, 2163–2173 (1993).

Weishampel, D. B., Dodson, P. & Osmólska, H. The Dinosauria. 151–317 (University California Press, Berkeley, 1990).

Wellnhofer, P. Das fünfte skelettexemplar von *Archaeopteryx*. *Palaeontogr. A* **147**, 169–216 (1974).

Wellnhofer, P. Das siebte Examplar von *Archaeopteryx* aus den Solnhofener Schichten. *Archaeopteryx* **11**, 1–48 (1993).

Chapter 15   A Diapsid Skull in a New Species of the Primitive Bird
Confuciusornis   Lian-Hai Hou, Larry D. Martin, Zong-He Zhou, Alan Feduccia, and Fu-Cheng Zhang

The phylogenetic analysis was constructed by scoring the following characters from Chiappe et al., 1998 in *Caudipteryx* and *Protarchaeopteryx*.

1. Rostral portion of premaxillae unfused (0) or fused (1) in adults.
2. Frontal process of premaxilla short (0), relatively long to very long, at least approaching the rostral border of the antorbital fossa (1).
3. Cup-shaped caudal maxillary sinus: absent (0), present (1).
4. Caudal margin of naris farther rostral than (0) or, nearly reaching or overlapping (1) the, rostral border of the antorbital fossa.
5. Triradiate palatine (jugal process absent): absent (0), present (1).
6. Postorbital-jugal contact: present (0), absent (1).
7. Quadratojugal sutured to the quadrate (0), or joined by a ligament (1).
8. Quadratojugal-squamosal contact: present (0), absent (1).
9. Quadrate articulating only with the squamosal (0), or with both prootic and squamosal (1).
10. Caudal tympanic recess opens on the rostral margin of the paraoccipital process (0), or into the columellar recess (1).
11. Coronoid bone: present (0), absent (1).
12. Teeth in adults: with serrated crown (0) or unserrated crowns (1).
13. One or more pneumatic foramina piercing the centra of mid-anterior cervicals beyond the level of the parapophysis-diapophysis: present (0), absent (1).
14. Anterior cervical vertebrae heterocoelous: absent (0), present (1).
15. Carotid processes in intermediate cervicals: absent (0), present (1).
16. Prominent ventral processes on cervico-dorsal vertebrae: absent (0), present (1).
17. Cervico-dorsal vertebrae with parapophyses located at the same level as the prezygapophyses: absent (0), present (1).
18. Wide vertebral foramen in thoracic vertebrae, vertebral foramen/articular cranial facies ratio (vertical diameter) larger than 0.40: absent (0), present (1).

19. Hyposphene-hypantrum accessory intervertebral articulations in dorsal vertebrae: present (0), absent (1).
20. Synsacrum formed by 7 or fewer (0), or 8 or more vertebrae (1).
21. Cranial articular surface of synsacrum strongly concave: absent (0), present (1).
22. Caudal portion of the synsacrum forming a prominent ventral keel: absent (0), present (1).
23. Caudal articular surface of synsacrum convex: absent (0), present (1).
24. Prezygapophyses of distal caudal vertebrae: elongate (0), short or absent (1).
25. Caudal vertebrae: amphicoelous (0), or procoelous (1).
26. First caudal centrum strongly compressed ventrally: absent (0), present (1).
27. Elongated (much longer than wider) proximal haemal arches: absent (0), present (1).
28. Pygostyle: absent (0), present (1).
29. Caudal vertebral count larger than 35 (0) or fewer than 25–26 (1).
30. Coracoid and scapula articulate through a broad, sutured articulation (0), or through more localize, less extensive facets (1).
31. Scapula articulated at the shoulder (proximal) end of coracoid (0), or well below it (1).
32. Coracoid and scapula placed in the same plane (0), or forming a sharp angle (1) at the level of the glenoid cavity.
33. Coracoid shape: elongated with subrectangular profile (0), short (1), strut-like (2).
34. Bicipital tubercle (=acrocoracoidal process): present (0), absent (1).
35. Supracoracoid nerve foramen of coracoid centrally located (0), or displaced (often as an incision or even without passing through) toward the medial margin of coracoid (1).
36. Scapular caudal end blunt and usually expanded (0), or tapered to sharp point (1).
37. Prominent acromion in the scapula: absent (0), present (1).
38. Boomerang-shaped furcula, with interclavicular angle approximately 90_ (0), or U-shaped furcula, with an interclavicular angle less than 70_ (1).
39. Sternum subquadrangular to transversally rectangular (0) or longitudinally rectangular (1).
40. Ossified sternal keel: absent (0), present (1).
41. Proximal and distal humeral ends twisted (0), or expanded nearly in the same plane (1).
42. Ventral tubercle of humerus projected ventrally (0), proximally (1), or caudally, separated from the humeral head by a deep capital incision (2).
43. Humerus with distinct transverse ligamental groove: absent (0), present (1).
44. Humeral pneumatic fossa: absent (0), present (1).
45. Humeral distal condyles mainly located on distal (0), or cranial (1) aspect.
46. Humerus with two (0), or a single distal condyle (1).
47. Well-developed olecranal fossa on the caudal face of the distal end of humerus: absent (0), present (1).
48. Ulna shorter than (0), or subequal in length to, or longer than, humerus (1).
49. Diameter of ulnar shaft: radial shaft/ulnar shaft ratio larger (0), or smaller (1) than 0.70.
50. Olecranon process on ulna: relatively small (0), hypertrophied, nearly one-third (1) or one-half (2) the length of the ulna.
51. Distal end of ulna subrectangular and transversely compressed (0), or subtriangular in shape (1).
52. Semilunate ridge on ulnar dorsal condyle: absent (0), present (1).
53. Ulnare (scapholunar) of round to sub-rectangular shape (0) or U-shaped to heart-shaped (1).
54. Distal carpals and proximal portion of metacarpals unfused (0), or fused forming a carpometacarpus (1).

55. Extensor process on carpometacarpus: absent or rudimentary (0), present (1).

56. Alular metacarpal (I)/major metacarpal (II) length ratio smaller than or equivalent to 0.30: absent (0), present (1).

57. Alular metacarpal massive, depressed, and quadrangular-shaped: absent (0), present (1).

58. Alular digit (I) long, exceeding the distal end of the major metacarpal (0), or short, not surpassing this metacarpal (1).

59. Alular digit (I) large, robust, and dorsoventrally compressed: absent (0), present (1).

60. Prominent ventral projection of the latero-proximal margin of the proximal phalanx of alular digit (digit I): absent (0), present (1).

61. Alular ungual phalanx with two, ventro-proximal foramina: absent (0), present (1).

62. Pelvic elements unfused (0), or fused or partially fused (1).

63. Pubis subvertically oriented (0), or well-retroverted (1).

64. Prominent antitrochanter: absent (0), caudally directed (1), or dorso-caudally directed (2).

65. Iliac fossa for *M. cuppedicus* (=*M. iliofemoralis internus*): present (0), absent or rudimentary (1).

66. Iliac brevis fossa: present (0), absent (1).

67. Pubic pedicel ventrally or caudoventrally (0), or cranioventrally projected (1).

68. Supracetabular crest on ilium absent or rudimentary (0), extending throughout the acetabulum (1), or extending only over the rostral half of the acetabulum (2).

69. Ischiadic terminal processes in contact (0), or lacking contact (1).

70. Ischium less than two-thirds (0), or two-thirds or more of pubis length (1).

71. Obturator process of ischium: prominent (0), or reduced or absent (1).

72. Pubic apex in contact (0), or lacking contact (1).

73. Pubis shaft laterally compressed throughout its length: absent (0), present (1).

74. Pubic foot: present (0), absent (1).

75. Pubic apron more than one-third the length of the pubis (0), shorter or absent (1).

76. Laterally compressed and kidney-shaped proximal end of pubis: absent (0), present (1).

77. Femur with distinct fossa for capital ligament: absent (0), present (1).

78. Femoral anterior trochanter nearly confluent with the greater trochanter (0), or fused to it forming the trochanteric crest (1).

79. Femoral posterior trochanter: present (0), absent (1).

80. Conical and strongly distally projected lateral condyle of femur: absent (0), present (1).

81. Femoral popliteal fossa distally bounded by a complete transverse ridge: absent (0), present (1).

82. Tibiofibular crest on the lateral condyle of femur: absent (0), present (1).

83. Tibia, calcaneum, and astragalus unfused or poorly coosified (sutures still visible) (0), or complete calcaneo-astragalar-tibial fusion (1).

84. Medial border of tibiotarsus at nearly the level of lateral border (0) or strongly projected proximally (1).

85. Proximal end of fibula prominently excavated by a medial fossa (0), or nearly flat (1).

86. Fibula with tubercle for *M. iliofibularis* anterolaterally (0), laterally or caudolaterally or caudally (1) directed.

87. Fibula reaching the proximal tarsals (0), or greatly reduced distally, without reaching these elements (1).

88. Distal tarsals free (0), or completely fused to the metatarsals (1).

89. Metatarsal V: present (0), absent (1).

90. Proximal end of metatarsal III in the same plane as metatarsals II and IV (0), reduced, not reaching the tarsals (arctometatarsalian condition) (1), or plantarily displaced with respect to metatarsals II and IV (2).

Character matrix modified from Chiappe *et al.*, 1998.

Velociraptor
0000000000000000000000000000000000000000000000000000000000000000
000000000000000000000000

Archaeopteryx
01011?11?11100?00??00?010?001000000010??000000000000001000000000?00?0
100000?000?000??0000

Mononykus
?????????1?11011111?111?11???00011000?11110011000210?100101111?11?12?????
??1011100011?1??1

Shuvuia
00?011111110011111011111110100011000?11110011000210?10010111?1110121
1111111??1?0?01111001

Enantiornithines
11111111??111111011110010001111120111111102111011101111110100011201000
1000010110011101?1110

Ornithurae
1111111111111111011100010001111120111111121110111011111110100?1121100
1111111011101110111112

Protarchaeopteryx
00????0?0??0?0???0?0000?0?0010000????000??0???0100??00000000000????????
00?00?0??0?00?????0

Caudipteryx
01?1?0000??1?0???0?0????00001???0?0?00???0??00100000000000000000?00??
000?0?00??0?00??00?0

Diagnoses of the Chuniaoae and the Avialae under alternative optimizations

Characters marked with an asterisk (*) are ambiguous due to missing data.

Delayed Transformation			Accelerated Transformation	
Unnamed clade of *Caudipteryx* + Avialae				
	2	0–1	2	0–1
	4*	0–1	4*	0–1
	12	0–1	5*	0–1
			10*	0–1
			11*	0–1
			12	0–1
			15*	0–1
			19*	0–1
			24*	0–1
			37*	0–1
			69*	0–1
			85*	0–1
			86*	0–1
Avialae	5*	0–1	6*	0–1
	7	0–1	7	0–1
	8	0–1	8	0–1
	10*	0–1	9*	0–1
	11*	0–1	18*	0–1

Delayed Transformation			Accelerated Transformation	
24*	0–1		39*	0–1
71	0–1		40*	0–1
			56*	0–1
			71	0–1

References: Chiappe, L.M., M.A. Norell, and J.M. Clark. 1998. The skull of a relative of the stem-group bird *Mononykus*. Nature 392:275–278.

Chapter 16   A Chinese Triconodont Mammal and Mosaic Evolution of the Mammalian Skeleton   Qiang Ji, Zhe-Xi Luo, and Shu-An Ji

Phylogenetic Position of *Jeholodens jenkinsi* (GMV 2139) within Triconodont Mammals

I. REVIEW OF TAXONOMIC HISTORY

Simpson (1925, 1928, 1929a) classified triconodont mammals into two subfamilies: the Amphilestinae and the Triconodontinae. At that time, the fossil records of both subfamilies were only available from the Middle to Late Jurassic. Since then, the taxonomic diversity has increased manifold and their stratigraphic ranges extended, as more fossils of triconodonts have been discovered.

In the 1940's, more primitive triconodonts *Eozostrodon* and *Morganucodon* were described from the Rhaeto-Liassic of United Kingdom (Parrington, 1941; Kühne, 1949). Although these late Triassic to early Jurassic triconodont mammals are similar to the later triconodonts in some dental characters, they are far more primitive in mandibular and basicranial structures (Kermack et al., 1973, 1981; Crompton and Luo, 1993; Luo, 1994; Wible and Hopson, 1993; Rougier et al., 1996a). Diverse morganucodont-like taxa have been described from China, Africa, Europe and North America from 1960's to 1980's (Rigney, 1963; Crompton, 1974; Young, 1978, 1982; Clemens, 1980; Sigogneau-Russell, 1983; Jenkins et al., 1983)

In 1960's Patterson and Olson (1961) described a new triconodont-like mammal *Sinoconodon*. More recently, this taxon has been removed from the triconodont group after the newer and better materials show far more primitive cranial and dental characters in this taxon than in the Morganucodontidae (Crompton and Sun, 1985; Crompton and Luo, 1993; Luo, 1994; Zhang et al., 1998).

Kermack et al. (1973) proposed that *Morganucodon* and its related Rhaeto-Liassic taxa be placed in suborder Morganucodonta, in juxtaposition with the suborder Eutriconodonta (= amphilestids + triconodontids). The inclusion of the Late Triassic to Early Jurassic taxa in the Triconodonta has expanded the taxonomical range beyond what was proposed by Simpson (1925, 1928, 1929a). The inclusion of morganucodonts into Triconodonta has been broadly accepted (Jenkins and Crompton, 1979).

From 1950's to the present, more derived triconodontines and amphilestines have been discovered in North America (Patterson, 1951; 1956; Slaughter, 1969; Fox, 1969, 1976; Jenkins and Crompton, 1979; Fastovsky et al., 1987; Jenkins and Schaff, 1988; Cifelli et al., 1998; Cifelli and Madsen, 1998), Asia (Trofimov, 1968; Chow and Rich, 1984; Wang et al., 1995; Kielan-Jaworowska and Dashzeveg, 1998), Africa (Signogneau-Russell, 1995) and South America (Bonaparte, 1986, 1992).

In a recent classification of the Mammalia (McKenna and Bell, 1997), morganucodonts and *Sinoconodon* have been removed from the Triconodonta to reflect the latest systematic consensus that both morganucodonts and *Sinoconodon* are not closely related to triconodontines and amphilestines (Crompton and Sun, 1985; Rowe, 1988; Wible, 1991; Wible and

Hopson, 1993; Crompton and Luo, 1993; Luo, 1994; Rougier et al., 1996a). However, the division of "eutriconodonts" into amphilestids and tricondontids have not been re-examined by using cladistical analysis.

2. PROBLEMS WITH DIVISION OF AMPHILESTINES VS. TRICONODONTINES

In Simpson's early classification of triconodonts ("eutriconodonts" of Kermack et al., 1973 and this study), the amphilestines were characterized by more symmetrical premolars that tend to be more molarized than their counter parts in triconodontines, and a much taller mid-cusp a than cusps b and c. Triconodontines were characterized by more asymmetrical premolars with more recurved cusp a, and molars with mid-cusp a "nearly or quite" equal to cusps b, and c.

This division within the derived triconodonts ("eutriconodonts") was adopted by a long list of later publications (Kermack et al., 1973; Jenkins and Crompton, 1979; McKenna and Bell, 1997). However, it must be recognized that this scheme of classification by Simpson (1925) was based on the limited fossil records in Middle to Late Jurassic that were available in the 1920's—long before the discovery of a vast diversity of more primitive triconodont-like taxa from the Late Triassic and Early Jurassic and the more derived triconodonts from the Cretaceous.

Triconodontines have proven by later studies to be a monophyletic group with well established synapomorphies. As more fossils discovered from Cretaceous, Patterson (1951, 1956) recognized that the tongue-in-groove interlocking is a unique character for triconodontines. Patterson and Olson (1961) also suggested that absence of the anterior lingual cingular cuspule (e) could be a defining feature of triconodontines. These characters are consistently present in a wide range of relatively new triconodontines described in the last three decades (Fox, 1976; Sigogneau-Russell, 1995; Cifelli et al., 1998; Cifelli and Madsen, 1998). Triconodontines form a robust clade supported by (1) nearly equal heights of molar cusps, (2) reduction or absence of anterior and internal cuspule e, and (3) an anterior embayment or groove on main cusp b for interlocking of molars.

By contrast, the grouping of amphilestines has been poorly supported by few derived dental characters. Simpson's main diagnostic characters for amphilestines, such as the much taller cusp a than cusps b and c in molars and more symmetrical premolars have turned out to be primitive characters of more distantly related morganucodontids and *Sinoconodon*. In fact, Simpson (1928: p. 70) had already recognized that amphilestines were a heterogeneous group whose member taxa were probably not related. The main distinguishing characters for this group were "clearly primitive characters lost in the more highly specialized Triconodontinae."

Patterson and Olson (1961) were first to question whether amphilestids were closely related to triconodontids, a view that had been endorsed by Kermack (1967). Mills (1971) recognized that the pattern of wear facets were very similar in the amphilestines and in the symmetrodonts *Kuehneotherium* and *Tinodon* that have the "obtuse-angled" molar cusps. Both groups are different from triconodontines or morganucodontids. Mills (1971) suggests that amphilestines should be removed from triconodonts and be assigned to therians, a view that is not widely accepted (but see Freeman, 1979). In a recent study of *Gobiconodon* (a taxon related to amphilestines), Kielan-Jaworowska and Dashzeveg (1998) argued the *Gobiconodon*, amphilestines and the obtuse-angled symmetrodonts are closely related on the basis of their shared similarities in wear facets and in the interlocking mechanism of molars.

The controversial hypothesis that amphilestines, *Gobiconodon* and symmetrodonts are related has yet to be corroborated by additional evidence from independent character systems. However, Mills (1971) and Kielan-Jaworowska and Dashzeveg (1998) are correct to show that some dental characters of amphilestines and *Gobiconodon* are also shared by other non-triconodont mammals. This invites further examination whether these dental characters are the shared derived characters (as Mills, Kielan-Jaworowska and Dashzeveg have suggested), or primitive, or derived and convergent.

In summary, the grouping of "amphilestines" is poorly supported by few shared derived characters unique to this group. Most characters that have been previously used to diagnose this group are present outside eutriconodonts, as documented by the following parsimony analysis of the mandibular and dental characters of triconodont-like mammals.

3. RESULTS OF THE CURRENT ANALYSIS OF MAJOR TRICONODONT TAXA

We coded 33 characters for 8 taxa of amphilestines and triconodontines, plus 3 morganucodontid taxa, all of which have relatively good fossil records. For outgroups, *Sinoconodon* is selected because it is widely accepted as the sister-group to mammaliaforms. The "obtuse-angled" symmetrodonts *Kuehneotherium* and *Tinodon* are selected as they represent one the earliest known mammalian lineages that were co-eval to *Sinoconodon* and morganucodontids. They also share some dental characters of triconodonts, as well documented by Mills (1971) and Kielan-Jaworowska and Dashzeveg (1998).

A goal of this analysis of *Jeholodens jenkinsi* is to ascertain if this new taxon is closely related to the known triconodont taxa, such as amphilestids (in response to Dr. Hopson's review comments of a previous version of this manuscript). Most amphilestid taxa (except *Gobiconodon*) are only represented by mandibles and lower dentitions. Limited by the incomplete fossils of the necessary comparative taxa, we focus on the mandibular and lower dental characters in our assessment of whether *Jeholodens* is more closely related any of the eutriconodont taxa.

Monophyly of triconodontines. Our character analysis strongly supports the monophyly of the traditional group of Triconodontinae, which is diagnosed by: (1) nearly equal heights of the main cusps; (2) tongue-in-groove interlocking (without involvement of cingular cuspules); and (3) highly reduced or absence of the anterior cingular cuspule (e). However, other diagnostic dental characters established by Simpson (1925, 1928) and followed by later classifications are quite homoplastic. E.g., the more asymmetrical premolar is a primitive character shared by *Sinoconodon* (Zhang et al., 1998) and some morganucodontids.

Paraphyly of "amphilestines". Our analysis also shows that the taxa formerly assigned to amphilestids (*Phascolotherium, Amphilestes*) form an un-resolved polytomy within the eutriconodont group. *Gobiconodon* (an amphilestid-like taxon formerly assigned to the family) is excluded from amphilestids, as first proposed by Jenkins and Schaff (1988). Amphilestids are not monophyletic and should not be regarded to be a taxon under the more strict criteria of phylogenetic systematics. Most dental characters of amphilestids are primitive, including: (1) Much taller cusp a than cusps b and c (which is present outside eutriconodonts among all morganucodontids, symmetrodonts and *Sinoconodon*); (2) primary cusp occludes in the embrasure between the opposing teeth (which is present outside eutriconodonts and in symmetrodonts plus *Megazostrodon*); (3) more symmetrical premolars (present outside eutriconodonts and in some morganucodontids).

Placement of Jeholodens within triconodonts. The new triconodont *Jeholodens* is the sister taxon to the monophyletic group of triconodontines. It is more closely related to triconodontines than to any of the morganucodontid and amphilestid taxa. However, our parsimony analysis fails to place *Jeholodens* within the well defined triconodontine clade (sensu stricto). Technically, it can not be regarded to be a triconodontine either. In this paper, we refer to *Jeholodens* as a eutriconodont.

Homoplasy of dental characters. All dental characters show a considerable degree of homoplasy among the 14 taxa considered here—with the exception of two apomorphies: (1) the nearly equal height of molar cusps for triconodontines; (2) the groove/embayment interlocking without cingular cuspules for triconodontines and *Jeholodens*. For example, the interlocking mechanism of amphilestids/*Gobiconodon* and *Kuehneotherium/Tinodon* (Kielan-Jaworowska and Dashzeveg, 1998) is also present in *Dinnetherium* (Crompton and Luo, 1993, fig. 4.5). Thus this character occurs in at least three clades that are not closely related on out tree. The occlusion of primary cusp (a) in the embrasure between the opposing teeth is not only present in amphilestids and *Gobiconodon,* but also present outside eutriconodonts and in symmetrodonts and *Megazostrodon.* Almost all diagnostic characters of premolars proposed by Simpson (1925, 1928, 1929a) for separating amphilestines from triconodontines have conspicuous and numerous exceptions outside these two groups.

By contrast, systematic distribution of the mandibular features shows much less homoplasy in triconodonts and *Sinoconodon.* The mandibular apomorphies related to the development of the pterygoid shelf and absence of the postdentary trough provide strong support for monophyly of eutriconodonts (to the exclusion of morganucodontids, *Sinoconodon* and *Kuehneotherium),* although *Tinodon* is an exception.

List of characters of mandible and lower dentition of *Jeholodens jenkinsi* and other triconodonts

MANDIBLE

1. Post-dentary ridge and trough (behind the tooth row) (Rowe, 1988; Wible, 1991; Luo, 1994):
   (0) Present: *Sinoconodon, Morganucodon, Megazostrodon, Dinnetherium, Kuehneotherium;*
   (1) Absent: *Gobiconodon, Phascolotherium, Amphilestes, Tinodon, Jeholodens, Triconodon, Trioracodon, Priacodon, Astroconodon.*

2. Curvature of the meckelian groove in adults (under the tooth row) (fig. 9 of Luo, 1994):
   (0) Present and parallel to the ventral border of mandible: *Sinoconodon, Gobiconodon, Jeholodens, Triconodon, Trioracodon, Priacodon, Astroconodon;*
   (1) Present and convergent to the ventral border of mandible: *Morganucodon, Megazostrodon, Dinnetherium, Kuehneotherium, Phascolotherium* (see *P. bucklandi,* Simpson, 1928), *Amphilestes* (Simpson, 1928), *Tinodon.*

3. Degree of development of the meckelian groove:
   (0) well developed: *Sinoconodon, Morganucodon, Megazostrodon, Dinnetherium, Kuehneotherium;*
   (1) weakly developed or vestigial: *Gobiconodon, Phascolotherium, Amphilestes, Tinodon, Jeholodens, Triconodon, Trioracodon, Priacodon, Astroconodon.*

4. Angular process of dentary (Rowe, 1988; Wible, 1991; Luo, 1994; Hopson, 1994):
   (0) Present: *Sinoconodon, Morganucodon,*
   (1) Reduced or weakly developed: *Kuehneotherium* (weak); *Megazostrodon* (Gow, 1986 termed the process as the pseudo-angular process, it should be considered to be a reduced angular process); *Dinnetherium* (Jenkins et al., 1983 termed the process as the pseudo-angular process; Hopson 1994 considers this to be a true angular process of reduced size);
   (2) Absent: *Gobiconodon, Phascolotherium, Amphilestes, Tinodon, Jeholodens, Triconodon, Trioracodon, Priacodon, Astroconodon.*

5. Coronoid in adults (Rowe, 1988; Wible, 1991; Luo, 1994):
   (0) Present: *Sinoconodon, Morganucodon, Megazostrodon, Dinnetherium, Kuehneotherium, Gobiconodon, Jeholodens.*
   (1) Absent: *Phascolotherium, Amphilestes, Tinodon* (Marsh, 1887: Plate X, fig. 1 shows no coronoid, pending confirmation on the actual specimen), *Triconodon, Trioracodon, Priacodon, Astroconodon;*

6. Mandibular foramen (posterior opening of the mandibular canal) for the inferior alveolar nerve and vessels:
   (0) Located within the postdentary trough (the depression around the foramen is a part of the meckelian groove—postdentary trough): *Sinoconodon, Morganucodon, Megazostrodon, Dinnetherium, Kuehneotherium;*

(1) The foramen is not associated with either postdentary trough or the meckelian groove: *Gobiconodon, Phascolotherium, Amphilestes, Tinodon, Jeholodens, Triconodon, Trioracodon, Priacodon, Astroconodon.*

7. Alignment of the ultimate molar to coronoid process (This character is included because its character distribution shows a distinctive pattern in morganucodontids, amphilestines and triconodontines, according to Mills, 1971):

   (0) The longitudinal axis of the ultimate molar is aligned medial to the anterior edge of the coronoid process: *Sinoconodon, Morganucodon, Megazostrodon, Dinnetherium, Kuehneotherium, Tinodon, Jeholodens, Astroconodon;*

   (1) The longitudinal axis of the ultimate molar is aligned with the coronoid process: *Gobiconodon, Phascolotherium, Amphilestes, Triconodon, Trioracodon* (based on the description by Mills 1971 and Simpson, 1928 photographs), *Priacodon* (based on photograph of Rassmusen and Callison, 1981).

8. The pterygoid fossa:

   (0) Absent: *Sinoconodon, Morganucodon, Megazostrodon, Dinnetherium, Kuehneotherium;*

   (1) Present: *Gobiconodon* (shown in photographs by Kielan-Jaworowska and Dashzeveg, 1998), *Phascolotherium, Amphilestes, Tinodon, Triconodon, Trioracodon, Priacodon, Astroconodon;*

   (?) *Jeholodens.*

9. The pterygoid ridge (shelf) along the ventral border of the coronoid part of the mandible:

   (0) Absent: *Sinoconodon, Morganucodon, Megazostrodon, Dinnetherium, Kuehneotherium;*

   (1) Present: *Gobiconodon, Phascolotherium, Amphilestes, Tinodon, Jeholodens, Triconodon, Trioracodon, Priacodon, Astroconodon.*

10. The ventral ridge of masseteric fossa):

   (0) Absent: *Sinoconodon, Morganucodon, Megazostrodon* (Megazostrodon has a shallow masseteric fossa but lacks a distinctive ventral ridge, Gow, 1986), *Dinnetherium, Kuehneotherium;*

   (1) Present: *Gobiconodon, Phascolotherium* (Simpson, 1928), *Amphilestes* (no photo available, based on Simpson's brief comparison to *Phascolotherium*), *Tinodon, Triconodon, Trioracodon, Priacodon, Astroconodon;*

   (?) *Jeholodens.*

   Note: a potential character is the position of dentary condyle relative to the level of postcanine alveoli. This character is not included in this PAUP analysis of the in-group taxa of Triconodonts because it is not preserved well enough in either of the two amphilestines considered here:

   (0) Below or about the same level as the postcanine alveoli: *Sinoconodon, Jeholodens, Triconodon, Trioracodon, Astroconodon;*

   (1) Above the level of the postcanine alveoli: *Priacodon, Tinodon.*

   (?) *Phascolotherium, Amphilestes* are difficult to assess.

LOWER DENTITION

11. Number of lower incisors (this character is included in response to Dr. Hopson's review comment on the similarity in incisor number of *Jeholodens* and amphilestines):

   (0) Four: *Sinoconodon, Morganucodon, Megazostrodon, Dinnetherium, Phascolotherium, Jeholodens;*

   (1) Two or fewer: *Gobiconodon, Triconodon* (following Simpson, 1928), *Priacodon* (after Engelmann and Callison, 1998), *Astroconodon.*

   (?) *Kuehneotherium, Amphilestes, Tinodon, Trioracodon* (Simpson, 1928 reported two or more, but the precise number is unknown),

12. Replacement of canines and incisors:

   (0) More than two replacements: *Sinoconodon;*

(1) Only once: all other mammaliaforms and mammals;
(?) *Kuehneotherium.*

13. Number of premolars (larger and more mature individuals of morganucodontids and *Gobiconodon* are known to have plugged the alveoli after the loss of some premolars that were present in subadults. The number for premolar given here is the maximum number known for a taxon):

(0) Five premolars: *Morganucodon, Megazostrodon* (after Gow, 1986), *Dinnetherium, Kuehneotherium* (at least five, after Kermack et al., 1968);

(1) Four premolars: *Amphilestes* (Simpson, 1928); *Gobiconodon* (after Kielan-Jaworowska and Dashzeveg, 1998; Jenkins and Crompton [1979] identified 3 to 4 premolars; Kielan-Jaworowska and Dashzeveg 1998 show that some premolar is lost and its alveolus is plugged in the adult specimens; therefore we adopted 4 as the premolar count for *Gobiconodon*), *Triconodon, Trioracodon, Astroconodon;*

(2) Three premolars: *Tinodon, Priacodon;*

(3) Two premolars: *Sinoconodon* (after Zhang et al., 1998), *Phascolotherium, Jeholodens.*

14. The ultimate premolar—symmetry of the main cusp (a) (This character is included because the symmetry of premolar was used by Simpson, 1925, 1928 to distinguish amphilestines from triconodontines):

(0) Asymmetric (anterior edge of cusp a is longer and more convex in outline than the posterior edge): *Morganucodon, Megazostrodon, Dinnetherium, Kuehneotherium* (after Kermack et al., 1968), *Jeholodens, Triconodon, Trioracodon, Priacodon, Astroconodon.*

(1) symmetrical (anterior and posterior cutting edges are equal or subequal in length; neither edge is more convex or concave than the other): *Sinoconodon* (after Zhang et al., 1998), *Gobiconodon, Phascolotherium, Amphilestes, Tinodon.*

15. The ultimate premolar—anterior cusp (b) (Absence of anterior cusp b tends to make the premolar to appear more asymmetrical. This character is included because the asymmetry of premolar was used by Simpson, 1925, 1928 to distinguish triconodontines from amphilestines.):

(0) Present (at least subequal to cusp c of the same tooth): *Sinoconodon* (after Zhang et al., 1998), *Morganucodon, Dinnetherium, Phascolotherium, Amphilestes, Tinodon, Jeholodens;*

(1) Small (much smaller than cusp c of the same tooth), or vestigial: *Megazostrodon, Kuehneotherium* (Kermack et al., 1968), *Gobiconodon* (after Kielan-Jaworowska and Dashzeveg, 1998), *Triconodon, Trioracodon, Priacodon* (Rassmusen and Callison, 1981), *Astroconodon.*

16. The ultimate premolar—posterior cingular cuspule d in addition to main cusp c (Presence of cingular cuspule d would make the premolar look more molariform. Moreover, if cingular cuspule d is present while cuspule e is absent on the same tooth, the tooth tends to look more asymmetrical. This character is included for its relevance to the symmetry of premolar used by Simpson to separate amphilestines from triconodontines).

(0) Absent: *Sinoconodon, Morganucodon, Megazostrodon, Dinnetherium, Kuehneotherium, Tinodon, Phascolotherium, Amphilestes;*

(1) Present: *Gobiconodon, Jeholodens, Triconodon;*

(0/1) Polymorphic: *Trioracodon* (Simpson, 1929a, Plate XXII fig. 7), *Priacodon* (Simpson, 1929a, Plate XXII figs. 8, 9).

17. Number of lower molars:

(0) Five or more: *Dinnetherium, Megazostrodon* (following Gow, 1986), *Gobiconodon* (after Kielan-Jaworowska and Dashzeveg, 1998), *Phascolotherium, Amphilestes;*

(1) Four: *Jeholodens, Tinodon, Triconodon, Priacodon, Astroconodon;*

(2) Three: *Trioracodon;*

(0/1/2) Polymorphic: *Morganucodon watsoni* has a maximum of 5 lower molars, *M. oehleri* has a maximum of 4 lower molars, *M. heikoupengensis* has a maximum of 3.

(?) *Kuehneotherium* (It may range from 3 to 6, Kermack et al., 1968); (?) *Sinoconodon* (It has a variable number due to the replacement of the ultimate molars in tooth row. *Sinoconodon* is known to have five loci for molars but only 3 to 4 functional molars at a given growth stage due to the loss of anterior molars in older/larger individuals, Crompton and Luo, 1993; Zhang et al., 1998).

18. Alignment of main cusps of molars:

(0) single longitudinal row: *Sinoconodon, Morganucodon, Megazostrodon, Dinnetherium, Gobiconodon, Phascolotherium, Amphilestes, Jeholodens, Triconodon, Trioracodon, Priacodon Astroconodon;*

(1) in reversed triangle: *Kuehneotherium, Tinodon.*

19. One-to-one opposition of upper and lower molars:

(0) Absent: *Sinoconodon.*

(1) Present: *Morganucodon, Megazostrodon, Dinnetherium, Kuehneotherium, Gobiconodon, Phascolotherium, Amphilestes, Tinodon, Jeholodens, Triconodon, Trioracodon, Priacodon, Astroconodon.*

20. Occlusal relationships of the upper and lower molar cusps:

(0) Absent: *Sinoconodon.*

(1) Present, lower primary cusp a occludes the groove between upper Cusps A, B: *Morganucodon, Dinnetherium, Jeholodens, Triconodon, Trioracodon, Priacodon, Astroconodon;*

(2) Present, lower main cusp a occludes in front of the upper Cusp B and into the embrasure between the opposite upper tooth and the preceding upper tooth: *Megazostrodon, Gobiconodon, Kuehneotherium, Phascolotherium, Amphilestes* (based on Mills, 1971), *Tinodon* (Crompton and Jenkins, 1967; Mills, 1971).

21. Relative height of primary cusp a to main cusps b, c of m2 (measured as the height ratio of a and c from the bottom of the valley between the two adjacent cusps):

(0) Posterior cusp c is less than 40% of the primary cusp a: *Sinoconodon, Morganucodon, Dinnetherium, Megazostrodon, Kuehneotherium, Gobiconodon* (~25%), *Phascolotherium* (~30%), *Amphilestes* (~25%), *Tinodon, Jeholodens* (~27%).

(1) Posterior cusp c and primary cusp a are equal or subequal in height (c is 70%–100% of a): *Triconodon, Trioracodon, Priacodon, Astroconodon.*

22. Relative size of anterior cusp b to posterior cusp c (based on m2):

(0) c taller than b: *Sinoconodon, Morganucodon, Megazostrodon;*

(1) b taller than c: *Dinnetherium, Kuehneotherium;*

(2) b and c are more or less equal in height: *Gobiconodon, Phascolotherium, Amphilestes, Tinodon* (based on Crompton and Jenkins, 1967; Prothero, 1981), *Jeholodens, Triconodon, Astroconodon, Priacodon.*

23. Relative height of the cusps of anterior molar (based on m1; in some triconodontines, such as *Priacodon,* cusps are nearly equal in height in the posterior molars but not in the more anterior molars. The difference in the height of cusps a, b, and c decreases from the anterior to posterior molars. This character has different distribution from character 21, and is meant to characterize the difference in the height of cusps between the anterior molars and the posterior molars of the same taxon):

(0) Primary cusp a is taller than cusps b and c (b and c are 80% of primary cusp a, or shorter): *Sinoconodon, Morganucodon, Megazostrodon, Dinnetherium, Kuehneotherium, Gobiconodon, Phascolotherium, Amphilestes, Tinodon, Jeholodens, Priacodon;*

(1) Primary cusp a is equal to cusps b and c (b and c are 95–100% of a): *Triconodon, Trioracodon, Astroconodon.*

24. Functional development of wear facets on molars:

(0) Wear facets absent by eruption but developed later by wear: *Morganucodon, Megazostrodon, Dinnetherium, Kuehneotherium, Gobiconodon, Phascolotherium* (Mills, 1971), *Amphilestes* (Mills, 1971);

(1) Wear facets matches upon the eruption of teeth: *Tinodon, Priacodon, Triconodon, Trioracodon, Astroconodon.*

(?) *Jeholodens* (unknown), *Sinoconodon* (non-applicable due to the absence of wear facets).

25. Relationship of wear facets to the main cusps:

(0) lower cusps a, c support two different wear facets (facets 1 and 4) that contact the upper primary cusp A: *Morganucodon, Dinnetherium, Triconodon, Trioracodon, Priacodon, Astroconodon;*

(1) lower cusps a, c support a single wear facet (4) that contacts the upper primary cusp B (this facet extends onto Cusp A as wear continues, but 1 and 4 do not develop simultaneous in these taxa): *Megazostrodon* (see Gow, 1986), *Kuehneotherium, Gobiconodon* (Kielan-Jaworowska and Dashzeveg, 1998), *Phascolotherium* (Mills, 1971), *Amphilestes, Tinodon;*

(?) *Jeholodens* (unknown and not possible to be prepared out in the holotype specimen), *Sinoconodon* (non-applicable due to the absence of wear facets).

26. Buccal curvature of primary cusp a of the lower molars (at the level of cusp valley) relative to those of cusps b and c (Comment: we follow Mills 1971 and Kielan-Jaworowska and Dashzeveg, 1998 on the definition of this feature. This character has a controversial history—the more convex buccal side of amphilestids and some symmetrodonts was observed by Osborn, 1888, supported by Mills, 1971, Kielan-Jaworowska and Dashzeveg, 1998. However it was not confirmed by Goodrich, 1894. Simpson (1928) endorsed Goodrich, but provided a description that the external surface of cusp a is "gibbous"—a characterization that would be more consistent with Osborn's observation. Mills, 1971 emphasized on the dental similarities between symmetrodonts and amphilestines. This character is now found in *Gobiconodon*, although in *Gobiconodon* this pattern is less developed in the lower molars than in the uppers. Kielan-Jaworowska and Dashzeveg, 1998 argue that amphilestines are related to the obtuse-angled symmetrodonts on the basis of these characters. Therefore, this character should be coded for further test by parsimony analysis):

(0) cusp a and cusps b, c have the same degree of buccal bulging: *Sinoconodon, Morganucodon, Megazostrodon, Dinnetherium, Jeholodens, Triconodon, Trioracodon, Priacodon, Astroconodon.*

(1) cusp a is far more bulging than cusps b, c ("Gibbous" of Simpson, 1925, 1928, 1929a; "Bulging" of Mills, 1971; slight angulation of Kielan-Jaworowska and Dashzeveg, 1998): *Kuehneotherium, Gobiconodon, Tinodon, Phascolotherium, Amphilestes* (photographs by Freeman, 1979, Plate 16: figs 1 and 2).

27. Lingual curvature of main cusps a, b, c at the level to the cusp valley (see Mills, 1971, p. 53):

(0) Cusp a and cusps b, c have about the same degree of curvature: *Sinoconodon, Morganucodon, Megazostrodon, Dinnetherium, Gobiconodon, Phascolotherium, Amphilestes, Jeholodens, Triconodon, Trioracodon, Priacodon, Astroconodon.*

(1) Cusp a is slight concave (far less convex than either cusp b or cusp c): *Kuehneotherium, Tinodon.*

28. Cingulum on the lingual side of the lower molar:

(0) Absent or weak: *Sinoconodon, Jeholodens;*

(1) Distinctive and slightly crenulated: *Kuehneotherium, Gobiconodon, Phascolotherium, Amphilestes, Tinodon, Triconodon, Trioracodon, Priacodon, Astroconodon;*

(2) Strongly developed, with distinctive cuspules (such as the kuhneocone): *Morganucodon, Megazostrodon, Dinnetherium.*

29. Anterior internal (mesio-lingual) cingular cuspule (e):

(0) Present: *Morganucodon, Megazostrodon, Dinnetherium, Kuehneotherium, Gobiconodon, Phascolotherium* (from the description by Mills, 1971), *Amphilestes* (Freeman, 1979, Plate 16, figs. 1, 2) *Tinodon* (Crompton and Jenkins, 1967; Prothero, 1981);

(1) Absent: *Jeholodens, Triconodon, Priacodon, Astroconodon;*

(0/1) *Sinoconodon* (This feature is variable in *Sinoconodon* and coded polymorphic); *Trioracodon* (after the description by Simpson, 1928; Hopson comments; but published photos and illustration seem to show variability—coded polymorphic here, pending confirmation).

30. Anterior and external (mesio-buccal) cingular cuspule (f):

    (0) Absent: *Sinoconodon, Morganucodon, Megazostrodon, Jeholodens, Trioracodon, Triconodon, Priacodon, Astroconodon;*

    (1) Present: *Dinnetherium* (Crompton and Luo, 1993: fig. 4.5), *Kuehneotherium, Gobiconodon, Amphilestes* (Freeman, 1979, Plate 16, figs. 1, 2), *Tinodon* (after Crompton and Jenkins, 1967; Prothero, 1981);

    (?) Unknown: *Phascolotherium.*

31. Interlocking mechanism between two adjacent lower molars:

    (0) Absent: *Sinoconodon;*

    (1) Present, posterior cingular cuspule d fits between cingular cuspule e and cusp b of the succeeding molar: *Morganucodon, Megazostrodon;*

    (2) Present, posterior cingular cuspule d of the preceding molar fits in between cingular cuspules e and f of the succeeding molar: *Dinnetherium, Kuehneotherium, Gobiconodon, Amphilestes* (inferred from Freeman, 1979 photos), *Tinodon* (Crompton and Jenkins, 1967);

    (3) Present, posterior cingular cuspule d of the preceding molar fits into an embayment or vertical groove of the anterior aspect of cusp b the succeeding molar (without any involvement of distinctive cingular cuspules in interlocking): *Jeholodens, Priacodon, Triconodon, Trioracodon, Astroconodon.*

    (?) *Phascolotherium* (has not been studied for this character; Mills 1971 offered some self-inconsistent statement about this feature in *Phascolotherium*).

32. Length ratio of lower molars (this character reflects the proportion in size among the lower molars. It was adopted here because Simpson, 1825, 1928 used this as a diagnostic character for some triconodontine genera):

    (0) Penultimate molar is smaller than the preceding molar but larger than the ultimate molar (for a total of 5 molars: m3≥m4≥m5; for a total of 4 molars: m2≥m3≥m4, or for a total of 3 molars, m1≥m2≥m3): *Morganucodon, Megazostrodon* (Gow, 1986), *Dinnetherium, Gobiconodon* (after Kielan-Jaworowska and Dashzeveg, 1998), *Jeholodens;*

    (1) Penultimate molar is the largest of molars (m1≤m2≤m3>m4): *Sinoconodon, Phascolotherium, Amphilestes, Triconodon, Trioracodon, Priacodon, Tinodon;*

    (0/1) polymorphic: *Astroconodon;*

    (?) *Kuehneotherium.*

33. Replacement of at least some molars:

    (0) Present: *Sinoconodon, Megazostrodon* (after Gow, 1986), *Gobiconodon;*

    (1) Absent: *Morganucodon, Dinnetherium, Phascolotherium, Amphilestes, Tinodon, Jeholodens, Triconodon, Trioracodon, Priacodon, Astroconodon;*

    (?) *Kuehneotherium.*

Character list of *Jeholodens* (GMV 2139) for determining its position among major clades of mammals

SCOPE OF THIS PHYLOGENETIC STUDY

The characters used for the phylogenetic analysis of *Jeholodens jenkinsi* (holotype: GMV 2139 a,b) are selected on two criteria: (1) the characters preserved on the holotype specimen (GMV 2139 a,b); (2) the characters showing systematic variation among the following 11 taxa that are represented by relatively complete postcrania (some incomplete taxa, such as

Sinoconodon and *Adelobasileus,* are not included because too much of their postcranial anatomy remains unknown).

The anatomical study focuses on the characters that are naturally exposed, or revealed by preliminary preparation on the holotype of *Jeholodens*. A comprehensive study requires a more detailed (and very difficult) preparation of the under side of the slab specimen, and a more lengthy anatomical monograph in the future.

The phylogenetic analysis of *Jeholodens* has incorporated the published datasets from several lengthy studies on non-mammalian cynodonts and early mammals (Kemp, 1982, 1983; Sues, 1985; Hopson and Barghusen, 1986; Rowe, 1986, 1988, 1993; Wible, 1991; Lillegraven and Krusat, 1991; Luo, 1994; Hopson, 1994; Kielan-Jaworowska and Gambaryan, 1994; Kielan-Jaworowska, 1997; Sereno and McKenna, 1995; Rougier et al., 1996a; Hu et al., 1997, 1998).

It is not our intention to produce an all-encompassing matrix with all possible characters of all major clades of early mammals (see the recent discussion by Rougier et al., 1996b; Kielan-Jaworowska, 1997; Hu et al., 1998). Because papers in *Nature* have very limited space, we prefer to concentrate on the phylogenetic position of *Jeholodens* among the major mammalian lineages. Supplementary information Part I has dealt with the position of *Jeholodens jenkinsi* within eutriconodonts (among other triconodont-like mammals).

MAJOR CHARACTER LISTS AND MATRICES
Kemp (1983); Sues (1985); Hopson and Barghusen (1986); Rowe (1986, 1988); Wible (1991); Wible and Hopson (1993); Lucas and Luo (1993); Luo (1994); Kielan-Jaworowska and Gambaryan (1994); Kielan-Jaworowska (1997); Gambaryan and Kielan-Jaworowska (1997); Rougier et al. (1996a); Hu et al. (1997, 1998).

PRIMARY DESCRIPTIVE REFERENCES FOR POSTCRANIAL ANATOMY
Tritylodontids (*Oligokyphus,* Kühne, 1956; *Kayentatherium,* Sues, 1983; *Bienotheroides,* Sun and Li, 1985; also, Rowe, 1988; Szalay, 1993).
Morganucodontids (Jenkins and Parrington, 1976; Lewis, 1983; Rowe, 1988).
Gobiconodon (Jenkins and Schaff, 1988).
Jeholodens (GMV 2139 a, b is the best preserved representative for the triconodontidae. Although some isolated postcranial elements of other triconodontids were discovered, these are far from complete. See comments by Jenkins and Crompton, 1979).
Ornithorhynchus (the platypus) (Ornithorhynchidae, Monotremata) (CMNH 1788; Gregory, 1951; Klima, 1973; Lewis, 1983; Rowe, 1988).
Multituberculates (Ptilodontidae, Krause and Jenkins, 1983; *Lambdopsalis,* Kielan-Jaworowska and Qi, 1990; Meng and Miao, 1992; Taeniolabidoidea and Eucosmodontidae, Kielan-Jaworowska and Gambaryan, 1994; Gambaryan and Kielan-Jaworowska, 1997; *Bulganbaatar,* Sereno and McKenna, 1995. For general characters see also Rowe, 1988; Simmons, 1993).
Archaic therians: *Zhangheotherium* (Hu et al., 1997, 1998); *Henkelotherium* (Krebs, 1991); *Vincelestes* (Rougier, 1993).
Placentals (Kielan-Jaworowska, 1976, 1978; Rowe, 1988; Novacek et al., 1997).
Marsupials (*Didelphis,* Jenkins, 1973, 1974; Jenkins and Weijs, 1979; Lewis, 1983; Klima, 1987; CMNH c45, and several uncatalogued specimens in the CMNH VP collections; *Pulcadelphys,* Marshall and Sigogneau-Russell, 1995; *Asiatherium,* Szalay and Trofimov, 1996; *Mayulestes,* Muizon, 1998).

PRIMARY REFERENCES ON CRANIAL ANATOMY
Tritylodontids (*Oligokyphus,* Kühne, 1956; Crompton, 1964; *Kayentatherium,* Sues, 1986; *Bienotheroides,* Sun 1984; *Yunnanodon,* Luo and Cui, unpublished data; tritylodontids in general, Wible and Hopson, 1993; Luo, 1994).

Morganucodontids (Kermack et al., 1981; Crompton and Luo, 1993; Luo, 1994; Luo and
    Crompton, 1994; Luo et al., 1995).
Jeholodens (The basicranium of GMV 2139a is still under a time-consuming preparation.
    Some of its features are still not full exposed, and are coded "?" in the matrix).
Ornithorhynchus (Zeller, 1989, 1993; Archer et al., 1993; Wible and Hopson, 1995). The
    Ornithorhynchidae is chosen as the representative family for the Monotremata because it
    has a better fossil record than those of the Tachyglossidae and Zaglossidae. The earliest
    unequivocal representative of Ornithorhynchidae is *Monotrematum* of the Paleocene (Pas-
    cual et al., 1992a, b). *Steropodon,* as the earliest known monotreme from Cretaceous,
    also bears strong similarities to ornithorhynchids (Archer et al., 1985). It is considered to
    belong to the Ornithorhynchidae by some (McKenna and Bell, 1997). Cranial features of
    the Ornithorhynchidae have been also considered to be less specialized than the Tachy-
    glossidae by Zeller (1993).
Multituberculates (Kielan-Jaworowska et al., 1986; Miao, 1988; Luo, 1989; Lillegraven and
    Hahn, 1993; Meng and Wyss, 1995; Wible and Hopson, 1995; Hurum, 1997).
Archaic therians: *Zhangheotherium* (Hu et al., 1997); *Henkelotherium* (Krebs, 1991);
    Vincelestes (Rougier et al., 1992; Rougier, 1993).
Placentals (MacPhee, 1981; Novacek, 1986; Rowe, 1988; Wible, 1991).
Marsupials (*Didelphodon,* Clemens, 1963; *Didelphis,* Wible, 1990; *Pulcadelphys,* Muizon,
    1994; *Mayulestes,* Muizon, 1998).

DENTAL CHARACTERS

Tritylodontidae (Kühne, 1956; Crompton, 1972; Sues, 1985; 1986b; Luo and Sun, 1993;
    Luo, 1994).
Morganucodontids (Kermack et al., 1973, 1981; Crompton, 1974; Crompton and Luo, 1993).
Gobiconodon (Jenkins and Schaff, 1988; Kielan-Jaworowska and Dashzeveg, 1998).
Jeholodens (based on GMV 2139 a,b; additional information on the Triconodontidae as a
    whole are from: Simpson, 1925, 1928, 1929a; Patterson, 1951; Fox, 1976; Jenkins and
    Crompton, 1979; Crompton and Luo, 1993; Cifelli et al., 1998; Cifelli and Madsen, 1998).
Ornithorhynchus (dentition based on Simpson, 1929b; Hopson and Crompton, 1969;
    Archer et al., 1993).

NOTES ON DENTITION OF ORNITHORHYNCHIDS

The living platypus has teeth in the juvenile but they degenerate in the adult. As a result the
adult dentition of extant *Ornithorhynchus* is highly reduced. In response to the review
comments by a reviewer (Dr Hopson), we only code those dental characters that can be seen
on juvenile teeth of the living *Ornithorhynchus* (Simpson, 1929b; Hopson and Crompton,
1969). The molar characters are not available on these juvenile teeth and are coded "?" for
Ornithorhynchus. The purpose of restricting the coding of dental characters to juvenile teeth
of the living *Ornithorhynchus* is to preclude any potential problems of homology of molar
cusps between the fossil taxa and the highly transformed extant platypus. This insures that
any controversial homology of the molars of fossil monotremes would NOT bias our analysis.
    Ornithorhynchids have a fossil record dated back to the Paleocene (*Monotrematum
sudamericum,* Pascual et al., 1992a, b). The best-reserved extinct taxon is *Obdurodon* of the
Miocene (Archer et al., 1993), with two species (Woodburne and Tedford, 1975; Archer
et al., 1985; Archer et al., 1993; Kielan-Jaworowska et al., 1987). The earliest known
monotreme is *Steropodon* from Early Cretaceous of Australia, which is similar to *Obdurodon* in most dental features (Archer et al., 1985; Kielan-Jaworowska et al., 1987;
Pascual et al., 1992a, b). Some have classified *Steropodon* in the family Ornithorhynchidae
(e.g., McKenna and Bell, 1997). However, the earliest fossil monotremes are only
represented by teeth. *Obdurodon* is only represented by skull but not postcranium. For the
published version of this analysis we are only coding the characters from the living
Ornithorhynchus, the unit of taxonomic operation for this ms.

In several runs of our analysis, we had a permutation of our matrix in an attempt to incorporate the dental characters of the fossil ornithorhynchid *Obdurodon*. We found that inclusion of dental features of *Obdurodon* as a part of the overall character list of ornithorhynchids does not alter the position of ornithorhynchids on our phylogenetic tree (Fig. 5). Our main phylogenetic conclusions are further strengthened if dental characters of *Obdurodon* are incorporated. The fact that we did not code any molar characters from the fossil taxon *Obdurodon* does not affect the phylogenetic conclusions published in this Nature paper.

Multituberculates (Clemens, 1963; Hahn, 1969; Clemens and Kielan-Jaworowska, 1979; Krause, 1982; Wall and Krause, 1992; Greenwald, 1988; Kielan-Jaworowska and Hurum, 1997).

Therians (Simpson, 1928; 1929a; Crompton 1971; Prothero, 1981; Kielan-Jaworowska et al., 1987; Krebs, 1991; Hopson, 1994; Hu et al., 1997, 1998).

NOTES ON BRAIN ENDOCAST FEATURES

The complete skeleton of *Jeholodens* offers an unprecedented opportunity to assess the controversial phylogenetic relationship of triconodontids. A previous hypothesis (Kielan-Jaworowska, 1997) suggests that triconodontids are closely related to multituberculates, on the strength of several derived characters of the brain endocasts of *Triconodon* and multituberculates (Kielan-Jaworowska, 1997). The brain endocast is not preserved on the slab specimen in the holotype of *Jeholodens*. To assess the "triconodontid-multituberculate" hypothesis, the characters cited in support of this hypothesis were incorporated into the matrix as a test. In several permutations of phylogenetic analyses by either inclusion, or exclusion, of these features of brain endocasts, we found that the position of triconodontids (including both *Jeholodens* and *Triconodon*) on the tree of major mammalian clades (Fig. 5) is NOT affected. The triconodontid clade is consistently positioned in the basal part of the mammalian tree and separated from multituberculates, no matter whether the endocast characters of *Triconodon* are combined into the skeletal characters of *Jeholodens* for triconodontids as a whole.

Character distributions

VERTEBRAE (9 characters)

1. Proatlas neural arch as separate ossification in adults (Rowe, 1988: ch. 92):
   (0) present: outgroup, tritylodontids;
   (1) absent: *Jeholodens, Ornithorhynchus,* multituberculates, *Zhangheotherium, Vincelestes,* marsupials, placentals;
   (?) morganucodontids, *Gobiconodon, Henkelotherium.*

2. Fusion of atlas neural arch and intercentrum in adults (Rowe, 1988: ch 93):
   (0) unfused: outgroup, tritylodontids, morganucodontids, *Jeholodens,* multituberculates, *Vincelestes;*
   (1) fused: *Ornithorhynchus,* marsupials, placentals;
   (?) *Gobiconodon, Zhangheotherium, Henkelotherium.*

3. Atlas rib in adults (Jenkins and Parrington, 1976; Rowe, 1988: ch 96):
   (0) present: outgroup, tritylodontids, morganucodontids, *Jeholodens;*
   (1) absent: *Ornithorhynchus, Zhangheotherium, Vincelestes,* marsupials, placentals;
   (?) *Gobiconodon,* multituberculates, *Henkelotherium.*

4. Prezygapophysis on axis (Jenkins, 1971; Rowe, 1988: ch 97):
   (0) present: outgroup, tritylodontids;
   (1) absent: morganucodontids, *Ornithorhynchus,* multituberculates, *Zhangheotherium, Vincelestes,* marsupials, placentals;
   (?) *Jeholodens, Gobiconodon, Henkelotherium.*

5. Rib of axis in adults:

(0) present: outgroup, tritylodontids, morganucodontids, *Jeholodens, Ornithorhynchus*, multituberculates, *Zhangheotherium*,

(1) absent: placentals;

(0/1 polymorphic): marsupials (Note: extant *Didephis* has a fused axis rib and *Mayulestes*, Muizon, 1998, has a fused transverse process that is homologous to the axis rib. This condition is coded 1. However, *Pulcadelphys*, one of the earliest known marsupials with a preserved axis, appears to have a detached rib, Marshall and Sigogneau-Russell, 1995; *Asiatherium*, one of the earliest known marsupials, has not preserved the axis, Szalay and Trofimov, 1996);

(?) *Gobiconodon, Henkelotherium, Vincelestes.*

6. Postaxial cervical ribs in adults (Rowe, 1988: ch. 101):

(0) present: outgroup, tritylodontids, morganucodontids, *Jeholodens*, multituberculates, *Zhangheotherium;*

(1) absent: *Ornithorhynchus, Vincelestes,* marsupials, placentals;

(?) *Gobiconodon, Henkelotherium.*

Note: We selected ornithorhynchids as the representative family for monotremes as a whole because ornithorhynchids have a better fossil record than tachyglossids. Although tachyglossid monotremes have unfused postaxial cervical ribs, we cannot include this in the current parsimony analysis due to the lack of fossil record of this family.

7. Postaxial cervical transverse canal (Hu et al., 1997):

(0) absent: outgroup, tritylodontids, morganucodontids, *Jeholodens, Zhangheotherium;*

(1) present: *Ornithorhynchus,* multituberculates, *Vincelestes,* marsupials, placentals;

(?) *Gobiconodon, Henkelotherium.*

8. Thoracic vertebrae:

(0) 13 thoracic vertebrae: *Gobiconodon* (after the reconstruction by Jenkins and Schaff, 1988), multituberculates, *Zhangheotherium, Vincelestes,* marsupials, placentals;

(1) 15 or more: *Jeholodens* (15), *Ornithorhynchus* (16);

(?) Unknown: outgroup, tritylodontids, morganucodontids, *Henkelotherium.*

9. Lumbar ribs:

(0) unfused to the vertebra: outgroup, tritylodontids, *Jeholodens, Gobiconodon;*

(1) synostosed to the vertebra to form transverse process: morganucodontids (see the stereophotos of Jenkins and Parrington, 1976), *Ornithorhynchus,* multituberculates, *Vincelestes,* marsupials, placentals;

(?) *Zhangheotherium, Henkelotherium.*

Note: Another possible systematic character is the fusion of caudal vertebral transverse process:

(0) absent: outgroup, tritylodontids, *Jeholodens;*

(1) fused: *Ornithorhynchus,* multituberculates, *Vincelestes,* marsupials, placentals;

(?) morganucodontids, *Gobiconodon, Zhangheotherium, Henkelotherium.*

SHOULDER GIRDLE (15 characters)

10. Interclavicle in adults (Rowe, 1988: ch 110 merged with ch 113, modified by Hu et al., 1997):

(0) present: outgroup (assuming the interclavicle of *Thrinaxodon* to be a general condition for cynodonts, Jenkins, 1971), tritylodontids, morganucodontids, *Jeholodens, Ornithorhynchus,* multituberculates, *Zhangheotherium;*

(1) absent: *Henkelotherium, Vincelestes,* marsupials, placentals;

(?) *Gobiconodon.*

11. Contact relationships in adults between the interclavicle (embryonic membranous element) and the sternal manubrium (embryonic endochondral element) (assuming the homologies of these elements by Klima, 1973; 1987) (new character):

(0) two elements distinct from each other, posterior end of interclavicle abuts anterior border of manubrium: tritylodontids (Sun and Li, 1985), *Ornithorhynchus* (Klima, 1987), multituberculates (Meng and Miao, 1992; Sereno and McKenna, 1995);

(1) two elements distinct from each other, the interclavicle broadly overlaps the ventral side of the manubrium: *Zhangheotherium* (Hu et al., 1997), *Jeholodens;*

(2) complete fusion of the embryonic membranous and endochondral elements (Klima, 1987): marsupials, placentals;

(?) outgroup, morganucodontids, *Gobiconodon, Henkelotherium, Vincelestes.*

12. Cranial margin of the interclavicle (new):

(0) anterior border is emarginated or flat: tritylodontids (*Bienotheroides*, Sun and Li, 1985), *Ornithorhynchus,* multituberculates (*Bulganbaatar* "clover-shaped interclavicle", Sereno and McKenna, 1995; *Kryptobaatar,* Kielan-Jaworowska, 1989), *Zhangheotherium* (Hu et al., 1997);

(1) with a median process (assuming interclavicle is fused to the sternal manubrium in living therians, Klima, 1987): outgroup (assuming the morphology of *Thrinaxodon,* and another non-mammalian cynodont for which this is known, Jenkins, 1971), morganucodontids, *Jeholodens, Vincelestes,* marsupials, placentals;

(?) *Gobiconodon, Henkelotherium.*

13. Clavicle-sternal apparatus joint (assuming that homologous elements of the interclavicle and the manubrium are fused to each other in therians, Klima, 1973; 1987; Sereno and McKenna, 1995; Hu et al., 1997):

(0) immobile: outgroup (following Jenkins, 1971), tritylodontids (Sun and Li, 1985), *Ornithorhynchus;*

(1) mobile: *Jeholodens,* multituberculates, *Zhangheotherium, Henkelotherium, Vincelestes,* marsupials, placentals;

(?) morganucodontids (Evans, 1981; pers. comm. from Drs F. A. Jenkins and S. E. Evans); *Gobiconodon.*

14. Curvature of clavicle (new):

(0) boomerang shape: outgroup (a generalized condition for cynodonts, Jenkins, 1971), tritylodontids (Sun and Li, 1985), morganucodontids (Jenkins and Parrington, 1976), *Ornithorhynchus;*

(1) slightly curved: *Jeholodens,* multituberculates, *Zhangheotherium* (Hu et al., 1997), *Henkelotherium* (Krebs, 1991, Abh. 7), marsupials, placentals;

(?) *Gobiconodon, Vincelestes.*

15. Scapula—supraspinous fossa (Jenkins and Schaff, 1988; Rowe, 1988: ch. 114, new definition of character states by Hu et al., 1997):

(0) absent (acromion extending from the dorsal border of scapula, and is positioned anterior to the glenoid): outgroup, tritylodontids, *Ornithorhynchus;*

(1) weakly developed (present only along a part of the scapula, and acromion positioned lateral to the glenoid): morganucodontids (sensu Sereno and McKenna, 1995), *Jeholodens, Gobiconodon,* multituberculates;

(2) fully developed and present along the entire dorsal border of scapula: *Zhangheotherium, Henkelotherium, Vincelestes,* marsupials, placentals.

16. Scapula—acromion process (Rowe, 1988: ch 115, new definition of character states by Hu et al., 1997):

(0) absent or weakly developed (leveled to the glenoid): outgroup, tritylodontids, morganucodontids, *Ornithorhynchus;*

(1) strongly developed and extending below the glenoid: *Jeholodens,* multituberculates, *Zhangheotherium, Henkelotherium, Vincelestes,* marsupials, placentals;

(?) *Gobiconodon.*

17. Scapula—a distinctive fossa for the teres major muscle on the lateral aspect of the scapular plate (new):

(0) absent: outgroup (Jenkins, 1971), tritylodontids (Sun and Li, 1985), multituberculates (Sereno and McKenna, 1995), *Vincelestes*, marsupials, placentals;

(1) present: *Jeholodens, Ornithorhynchus, Zhangheotherium* (Hu et al., 1997), *Henkelotherium* (Krebs, 1991, Abh. 7);

(?) morganucodontids, *Gobiconodon*.

18. Procoracoid (as a separated element in adults) (Klima, 1973; Rowe, 1988: ch 117):

(0) present: outgroup, tritylodontids, morganucodontids, *Ornithorhynchus;*

(1) absent: *Jeholodens,* multituberculates, *Zhangheotherium, Henkelotherium, Vincelestes,* marsupials, placentals;

(?) *Gobiconodon.*

19. Coracoid (Rowe, 1988: ch. 118, new definition of the character by Hu et al., 1997):

(0) large, with posterior process: outgroup, tritylodontids, morganucodontids, *Ornithorhynchus;*

(1) small, without posterior process: *Jeholodens,* multituberculates, *Zhangheotherium, Henkelotherium, Vincelestes,* marsupials, placentals;

(?) *Gobiconodon.*

20. Fusion of medial part of the embryonic scapula-coracoid plate with the sternal manubrium (sensu Klima, 1973):

(0) scapula-coracoid plate remains as a separate element in adults: *Ornithorhynchus;*

(1) scapula-coracoid plate fused to manubrium in adults: *Jeholodens,* multituberculates, *Zhangheotherium, Vincelestes,* marsupials, placentals;

(?) outgroup, tritylodontids, morganucodontids, *Gobiconodon, Henkelotherium.*

21. Size of the anterior-most element relative to the subsequent sternebrae in the sternal apparatus in adults:

(0) large: tritylodontids, *Jeholodens, Ornithorhynchus,* multituberculates, *Zhangheotherium, Vincelestes;*

(1) small: marsupials, placentals;

(?) outgroup, morganucodontids (Evans, 1981; pers. comm. from Drs. F. A. Jenkins and S. E. Evans), *Gobiconodon, Henkelotherium.*

22. Orientation ("facing" or articular surface) of glenoid (relative to the plane or the axis of scapula) (Rougier, 1993; Hu et al., 1997):

(0) nearly parallel to the long axis of scapula and facing posterolaterally: outgroup, tritylodontids, morganucodontids, *Ornithorhynchus;*

(1) oblique to the long axis of scapula and facing more posteriorly: *Jeholodens, Gobiconodon,* multituberculates, *Zhangheotherium, Henkelotherium, Vincelestes;*

(2) articular surface of glenoid is perpendicular to the main plane of the scapular plate: marsupials, placentals.

23. Shape and curvature of the glenoid (Jenkins, 1973; Rowe, 1988; Hu et al., 1997):

(0) saddle-shaped, oval and elongate: outgroup, tritylodontids, morganucodontids, *Ornithorhynchus;*

(1) uniformly concave and more rounded in outline: *Jeholodens, Gobiconodon,* multituberculates, *Zhangheotherium, Henkelotherium, Vincelestes,* marsupials, placentals.

24. Medial surface of scapula (Sereno and McKenna 1995; Hu et al., 1997):

(0) convex: outgroup, tritylodontids, morganucodontids, *Ornithorhynchus* (if the curvature of the medial surface near the anterior scapular border is taken into account), multituberculates;

(1) flat: *Jeholodens, Zhangheotherium, Vincelestes,* marsupials, placentals;

(?) *Gobiconodon, Henkelotherium.*

FORELIMB (9 characters)

25. Humeral head (Rowe, 1988: ch. 120, new definition of character states by Hu et al., 1997):

(0) subspherical, weakly inflected: outgroup, tritylodontids, morganucodontids, *Jeholodens, Gobiconodon, Ornithorhynchus;*

(1) spherical and strongly inflected: multituberculates, *Zhangheotherium*, *Henkelotherium*, *Vincelestes*, marsupials, placentals.

26. Intertubercular groove (modified from Rowe, 1988: chs. 121, 122, 123) (Note: this character is preserved, but not visible on *Jeholodens*. It is included for being relevant to the relationships of other taxa in the analysis):

(0) pectodeltoid crest separated from lesser tubercle by shallow and broad intertubercular groove: outgroup, tritylodontids, morganucodontids, *Gobiconodon*, *Ornithorhynchus*, multituberculates, *Vincelestes;*

(1) narrow and deep intertubercular groove: *Zhangheotherium*, *Henkelotherium*, marsupials, placentals;

(?) *Jeholodens.*

27. Size of lesser tubercle of humerus (relative to the greater tubercle) (Gambaryan and Kielan-Jaworowska, 1997):

(0) wider than the greater tubercle: outgroup, tritylodontids, morganucodontids, *Ornithorhynchus*, multituberculates;

(1) narrower than the greater tubercle: *Zhangheotherium*, marsupials, placentals;

(?) *Gobiconodon*, *Jeholodens*, *Henkelotherium*, *Vincelestes.*

28. Torsion between the proximal and distal ends of humerus (Rougier et al., 1996b; Gambaryan and Kielan-Jaworowska, 1997; Kielan-Jaworowska, 1998):

(0) strong ($\geq$30°): outgroup, tritylodontids, morganucodontids, *Jeholodens*, *Gobiconodon*, *Ornithorhynchus*, *Vincelestes;*

(1) moderate (30°–15°): *Zhangheotherium*, *Henkelotherium;*

(2) weak: marsupials, placentals;

(1/2 polymorphic): multituberculates (coded 1: Kielan-Jaworowska and Qi, 1990; Kielan-Jaworowska and Gambaryan, 1994; Kielan-Jaworowska, 1998; coded 2: Sereno and McKenna, 1995).

29. Ventral extension of pectodeltoid crest (Hu et al., 1997, 1998):

(0) not extending beyond the midpoint of the humeral shaft: outgroup, tritylodontids, morganucodontids, multituberculates, *Henkelotherium*, *Vincelestes*, placentals, marsupials;

(1) extending beyond the midpoint of the shaft: *Jeholodens*, *Gobiconodon*, *Ornithorhynchus*, *Zhangheotherium.*

30. Ulnar articulation on distal humerus (Rowe, 1988: ch. 126; new definitions of character states):

(0) bulbous ulnar condyle: outgroup, tritylodontids, morganucodontids, *Ornithorhynchus*, multituberculates;

(1) cylindrical trochlea (in posterior view) with vestigial ulnar condyle in anterior view): *Jeholodens*, *Gobiconodon* (Jenkins and Schaff, 1988 described this character state—however, Prof. Kielan-Jaworowska [pers. comm. June, 1997] believes that the Mongolian *Gobiconodon* has a fully developed ulnar condyle, a primitive condition), *Zhangheotherium*, *Henkelotherium*, *Vincelestes* (Rougier, pers. comm.);

(2) cylindrical trochlea without ulnar condyle (cylindrical trochlea has extended to the anterior/ventral side): marsupials, placentals.

31. Radial articulation on the distal humerus:

(0) distinct and rounded condyle (that does not form continuous synovial surface with the ulnar articulation in the ventral/anterior view of the humerus): outgroup, tritylodontids, morganucodontids, *Jeholodens*, *Ornithorhynchus*, multituberculates, *Zhangheotherium*, *Henkelotherium*, *Vincelestes;*

(1) capitulum (radial articulating structure that forms continuous synovial surface with ulnar trochlea): marsupials, placentals;

(?) *Gobiconodon.*

32. Entepicondyle and ectepicondyle of humerus (Rowe, 1988: ch. 124):

(0) robust: outgroup, tritylodontids, morganucodontids, *Gobiconodon*, *Ornithorhynchus;*

(1) weak: *Jeholodens*, multituberculates, *Zhangheotherium, Henkelotherium, Vincelestes*, marsupials, placentals.

33. Styloid process of radius (Rowe, 1988: ch. 129):
    (0) weak: outgroup, tritylodontids, morganucodontids, *Jeholodens, Gobiconodon, Ornithorhynchus, Zhangheotherium, Vincelestes;*
    (1) strong: multituberculates, *Henkelotherium*, marsupials, placentals.

PELVIC GIRDLE (3 characters)

34. Acetabular dorsal emargination (cotyloid notch of Kühne, 1956; Rowe, 1988: ch. 134, modified by Hu et al., 1997):
    (0) open (emarginated): outgroup, tritylodontids, morganucodontids, *Jeholodens*, multituberculates, *Vincelestes;*
    (1) closed (with a complete rim): *Ornithorhynchus, Zhangheotherium*, marsupials;
    (0/1 polymorphic): placentals. The earliest known pelves of placentals have variable conditions (*Asioryctes, Kennalestes, Zalambdalestes*, Kielan-Jaworowska, 1976, 1987; *Ukhaatherium*, Novacek et al., 1997), and may be either open or closed.
    (?) *Henkelotherium, Gobiconodon.*

35. Size of the pelvic obturator foramen (Rowe, 1988: ch. 139):
    (0) smaller than that of acetabulum: outgroup, *Ornithorhynchus;*
    (1) equal to or larger than that of acetabulum: tritylodontids, morganucodontids, *Jeholodens, Gobiconodon*, multituberculates, *Zhangheotherium, Henkelotherium, Vincelestes*, marsupials, placentals.

36. Sutures of the ileum, the ischium, and the pubis within the acetabulum in adults (new):
    (0) unfused: outgroup, tritylodontids (*Oligokyphus*, Kühne, 1956), morganucodontids, *Jeholodens, Gobiconodon, Zhangheotherium, Henkelotherium;*
    (1) fused: *Ornithorhynchus*, multituberculates, *Vincelestes*, marsupials, placentals.

HINDLIMB (8 characters)

37. Inflected head of the femur set off from the shaft by a neck (Rowe, 1988: ch. 141, 142, modified by Hu et al., 1997):
    (0) neck absent (and head oriented dorsally): outgroup, tritylodontids, morganucodontids, *Jeholodens, Gobiconodon, Ornithorhynchus;*
    (1) neck present (and head inflected medially) spherical and inflected: multituberculates, *Zhangheotherium, Henkelotherium, Vincelestes*, marsupials, placentals.

38. Greater trochanter (Rowe, 1988: ch. 143)
    (0) directed dorsolaterally: outgroup, tritylodontids, morganucodontids, *Jeholodens, Gobiconodon, Ornithorhynchus;*
    (1) directed dorsally: multituberculates, *Zhangheotherium, Henkelotherium, Vincelestes*, marsupials, placentals.

39. Orientation of lesser trochanter (Rowe, 1988: ch. 144):
    (0) on medial side of shaft: outgroup, tritylodontids, morganucodontids, *Jeholodens, Gobiconodon, Ornithorhynchus, Vincelestes;*
    (1) on the ventromedial or ventral side of the shaft: multituberculates (ventral), *Zhangheotherium, Henkelotherium*, marsupials, placentals (ventromedial).

40. Size of lesser trochanter (Rowe, 1988: ch. 144):
    (0) large: outgroup, tritylodontids, morganucodontids, *Jeholodens, Gobiconodon, Ornithorhynchus*, multituberculates, *Zhangheotherium, Henkelotherium, Vincelestes;*
    (1) small: marsupials, placentals.

41. Patellar facet ("groove") of femur (Rowe, 1988: ch. 145):
    (0) absent: outgroup, tritylodontids;
    (1) shallow and weakly developed: morganucodontids (Jenkins and Parrington, 1976), *Jeholodens, Gobiconodon* (Jenkins and Schaff, 1988);

(2) well developed: *Ornithorhynchus*, multituberculates, *Zhangheotherium*, *Henkelotherium*, *Vincelestes*, marsupials, placentals.

42. Tibial malleolus:
   (0) weak: outgroup, tritylodontids, morganucodontids, *Jeholodens, Gobiconodon, Zhangheotherium;*
   (1) distinct: *Ornithorhynchus*, multituberculates, *Henkelotherium, Vincelestes*, marsupials, placentals.

43. Fibular styloid process:
   (0) weak: outgroup, tritylodontids, morganucodontids, *Jeholodens, Gobiconodon, Ornithorhynchus, Zhangheotherium;*
   (1) distinct: multituberculates, *Henkelotherium, Vincelestes*, marsupials, placentals.

44. Fibula contacting the calcaneum (= "tricontact in upper ankle joint" by Szalay, 1993):
   (0) extensive contact: outgroup, tritylodontids, morganucodontids, *Jeholodens, Gobiconodon, Ornithorhynchus*, multituberculates, *Zhangheotherium, Vincelestes;*
   (1) reduced or absent: marsupials, placentals;
   (?) *Henkelotherium.*

Note: A potential character: ratio between mediolateral width and craniocaudal depth of proximal tibial diameters (Kielan-Jaworowska and Gambaryan, 1994; Gambaryan and Kielan-Jaworowska, 1997) may be established for *Jeholodens* in future studies.

ANKLE JOINT (8 characters)

45. Superposition (overlap) of the astragalus over the calcaneum (lower ankle joint) (Jenkins, 1971; Rowe, 1988; Kielan-Jaworowska, 1997):
   (0) little or absent: outgroup, tritylodontids, morganucodontids, *Jeholodens, Ornithorhynchus;*
   (1) weakly developed: multituberculates, *Zhangheotherium;*
   (2) present: *Vincelestes*, marsupials, placentals;
   (?) *Gobiconodon, Henkelotherium.*

46. Astragalar neck (new):
   (0) absent: outgroup, tritylodontids, morganucodontids, *Jeholodens, Ornithorhynchus*, multituberculates, *Zhangheotherium* (new GMV specimen), *Vincelestes;*
   (1) present: marsupials, placentals;
   (?) *Gobiconodon, Henkelotherium.*

47. Astragalar trochlea (new):
   (0) absent: outgroup, tritylodontids, morganucodontids, *Jeholodens*, multituberculates, *Zhangheotherium* (new GMV specimen);
   (1) present: *Vincelestes*, marsupials, placentals;
   (?) Not preserved: *Gobiconodon, Henkelotherium.* Condition uncertain: *Ornithorhynchus* (The playpus has a condyle on the dorsomedial aspect of the astragalus for the trough on the distal end of the tibia. It serves to allow the tibio-astraglar joint to rotate with respect to the transverse axis of the foot for extension and flexion [Szalay, 1993]. However the astragalus lacks a true trough as in the astragalar trochlea of extant therians. For these reasons we coded *Ornithorhynchus* as "?").

48. Calcaneal tubercle (Rowe, 1988: ch. 151):
   (0) short without terminal swelling: outgroup, tritylodontids, morganucodontids, *Jeholodens, Ornithorhynchus;*
   (1) elongate with terminal swelling: multituberculates, *Zhangheotherium, Vincelestes, Henkelotherium*, marsupials, placentals;
   (?) *Gobiconodon.*

49. Peroneal process and groove of calcaneum (Modified from Kielan-Jaworowska and Gambaryan, 1994):
   (0) forming laterally directed shelf, and without a distinct process: outgroup, tritylodontids, morganucodontids, *Jeholodens, Gobiconodon;*

(1) weakly developed with shallow groove on lateral side of process: *Zhangheotherium, Vincelestes,* marsupials, placentals;

(2) with a distinct peroneal process: *Ornithorhynchus* (after the homology proposed by Szalay, 1993), multituberculates (with a distinct peroneal process: demarcated by a deep peroneal groove at the base);

(?) *Henkelotherium.*

50. Contact of the cuboid on the calcaneum (following the orientation identified by Lewis, 1983, Szalay, 1993) (new character, modified from Kielan-Jaworowska 1997; pers. comm. of 1998):

(0) on the anterior (distal) end of the calcaneum (the cuboid is aligned with long axis of the calcaneum): outgroup (the cynodont described by Jenkins, 1971), tritylodontids (*Oligokyphus* as illustrated by Szalay, 1993), *Morganucodon, Zhangheotherium, Vincelestes* (Rougier, 1993); marsupials and placentals;

(1) on the anteromedial aspect of the calcaneum: (the cuboid is skewed to the medial side of the long axis of the calcaneum): *Jeholodens, Ornithorhynchus* (Grassé, 1955; Szalay, 1993; pers. obs. of AMNH 65831) multituberculates (Krause and Jenkins, 1983; Kielan-Jaworowska and Gambaryan, 1994).

(?) *Gobiconodon, Henkelotherium.*

51. Relationships of the proximal end of metatarsal V to the cuboid:

(0) metatarsal V is off-set to the cuboid: outgroup, morganucodontids, *Jeholodens;*

(1) metatarsal V is far off-set to the cuboid that it contacts the calcaneum: *Ornithorhynchus* (metatarsal V is bound by ligament to the peroneal tubercle of calcaneum, Szalay 1993 and confirmed by our pers. obs. of AMNH 65831); multituberculates (after Kielan-Jaworowska and Gambaryan, 1994; but see the alternative by Krause and Jenkins, 1983);

(2) metatarsal V is aligned with the cuboid: *Vincelestes,* marsupials, placentals;

(?) tritylodontids, *Gobiconodon, Zhangheotherium, Henkelotherium.*

52. Angle of metatarsal III to the calcaneum (which indicates how much the sole of the foot is "bent" from the long axis of the ankle):

(0) metatarsal III is aligned with (or parallel to) the imaginary line of the long axis of the calcaneum: outgroup, morganucodontids, *Vincelestes,* marsupials, placentals;

(1) metatarsal III is arranged obliquely from the imaginary line of the long axis of the calcaneum: *Jeholodens, Ornithorhynchus;* multituberculates;

(?) tritylodontids, *Gobiconodon, Zhangheotherium, Henkelotherium.*

OTHER POSTCRANIAL CHARACTERS (2 characters)

53. Sesamoid bones in flexor tendons (Rowe, 1988: ch; 158):

(0) absent: outgroup, tritylodontids, morganucodontids, *Jeholodens;*

(1) present and unpaired: *Ornithorhynchus;*

(2) present and paired: multituberculates, *Zhangheotherium,* marsupials, placentals;

(?) *Gobiconodon, Henkelotherium, Vincelestes.*

54. External pedal (tarsal) spur (modified from Hu et al., 1997, 1998):

(0) absent: outgroup, tritylodontids, morganucodontids, *Jeholodens, Vincelestes,* marsupials, placentals;

(1) present: *Gobiconodon, Ornithorhynchus,* multituberculates (pers. comm. from Kielan-Jaworowska), *Zhangheotherium;*

(?) *Henkelotherium.*

BASICRANIUM (12 characters)

55. Cranial moiety of squamosal (Rowe, 1988; Wible, 1991; Luo 1994):

(0) narrow: outgroup, tritylodontids, morganucodontids, *Jeholodens, Gobiconodon, Ornithorhynchus,* multituberculates;

(1) broad: *Henkelotherium* (Krebs, 1993, Abh. 1), *Vincelestes,* marsupials, placentals;

(?) *Zhangheotherium.*

56. Squamosal notches for quadrate and quadratojugal (character distribution following Luo and Crompton, 1994):

(0) present: outgroup, tritylodontids;

(1) absent: morganucodontids, *Jeholodens*, *Gobiconodon*, *Ornithorhynchus*, multituberculates, *Zhangheotherium*, *Vincelestes*, marsupials, placentals;

(?) *Henkelotherium*.

57. Postglenoid depression on squamosal (= "external auditory meatus"):

(0) absent: morganucodontids, *Jeholodens*, *Gobiconodon*, *Ornithorhynchus*, multituberculates;

(1) present: *Zhangheotherium*, *Vincelestes*, marsupials, placentals;

(?) Not-applicable: outgroup, tritylodontids; not preserved: *Henkelotherium*.

58. Position of craniomandibular joint (Rowe, 1988; Wible, 1991):

(0) posterior or lateral to the level to the fenestra vestibuli: outgroup, tritylodontids, morganucodontids, *Jeholodens*, *Ornithorhynchus*, multituberculates;

(1) anterior to the level of fenestra vestibuli: *Zhangheotherium*, *Vincelestes*, marsupials, placentals;

(?) *Gobiconodon*, *Henkelotherium*.

59. Pars cochlearis (Rowe, 1988; Wible, 1991; Luo, 1994; Luo et al., 1995; Rougier et al., 1996a):

(0) without an elongate petrosal cochlear housing: outgroup, tritylodontids;

(1) with an elongate and cylindrical petrosal cochlear housing: morganucodontids, *Jeholodens*, *Ornithorhynchus*, multituberculates, *Zhangheotherium*;

(2) with a bulbous and oval shaped promontorium: *Vincelestes*, marsupials, placentals;

(?) *Henkelotherium*.

60. Cochlea (Rowe, 1988; Wible, 1991; Luo, 1994; Rougier et al., 1996a):

(0) short and uncoiled: tritylodontids (Luo and Cui, unpublished data), morganucodontids, *Jeholodens*, multituberculates, *Zhangheotherium*;

(1) elongate and partly coiled: *Ornithorhynchus*, *Vincelestes*;

(2) elongate and coiled at least 360°: marsupials, placentals;

(?) *Gobiconodon*, *Henkelotherium*;

(? Not-applicable) outgroup.

61. Crista interfenestralis (Rougier et al., 1996a):

(0) horizontal and extending to base of paroccipital process: outgroup, tritylodontids, morganucodontids, *Ornithorhynchus*;

(1) vertical, delimiting the back of the promontorium: *Zhangheotherium*, *Vincelestes*, marsupials, placentals;

(0/1 polymorphic) multituberculates;

(?) *Jeholodens*, *Gobiconodon*, *Henkelotherium*.

62. Post-tympanic recess (Rougier et al., 1996a):

(0) absent: outgroup, tritylodontids, morganucodontids, *Ornithorhynchus*, multituberculates;

(1) present: *Zhangheotherium*, *Vincelestes*, marsupials, placentals;

(?) *Jeholodens*, *Gobiconodon*, *Henkelotherium*.

63. Caudal tympanic process of petrosal (Rougier et al., 1996a):

(0) absent: outgroup, tritylodontids, morganucodontids, *Ornithorhynchus*;

(1) present: *Zhangheotherium*, *Vincelestes*, marsupials, placentals;

(0/1 polymorphic) multituberculates (Rougier et al., 1996a);

(?) *Jeholodens*, *Gobiconodon*, *Henkelotherium*.

64. Epitympanic recess:

(0) absent: outgroup, tritylodontids, morganucodontids, *Ornithorhynchus*;

(1) present (as a large depression on the crista parotica): multituberculates, *Zhangheotherium*, *Vincelestes*, marsupials, placentals;

(?) *Jeholodens*, *Gobiconodon*, *Henkelotherium*.

65. Epitympanic recess flanked laterally by squamosal (Modified from Luo, 1989; Rougier et al., 1996a: ch. 23):
    (0) absent: morganucodontids, *Ornithorhynchus*, multituberculates;
    (1) present: *Zhangheotherium, Vincelestes*, marsupials, placentals;
    (?) not applicable: outgroup, tritylodontids; not preserved: *Jeholodens, Gobiconodon, Henkelotherium*.
66. Foramen for the ramus superior of the stapedial artery:
    (0) laterally open notch (laterally open pterygo-paroccipital foramen): tritylodontids, morganucodontids;
    (1) foramen enclosed by the petrosal: *Ornithorhynchus*, multituberculates, *Vincelestes* ("ascending canal" of Rougier et al., 1992), placentals (representative condition for the group as a whole, but may not be present in all taxa, Wible, 1987);
    (2) foramen enclosed between the squamosal and the petrosal: *Zhangheotherium;*
    (3) absent: marsupials;
    (?) not preserved: *Jeholodens, Gobiconodon, Henkelotherium;* not applicable: outgroup (in non-mammalian cynodonts, the pterygo-paroccipital foramen is enclosed by different bones including the epipterygoid, the pterygoid, the lateral wing of the prootic, the squamosal, or a combination thereof. Due to the complex character distribution, we herein coded "? Not-applicable" for cynodont outgroup in this analysis. For reference see Rougier et al., 1992; Luo and Crompton, 1994).

MANDIBLE (9 characters)
67. Medial dentary ridge that overhangs the posterior segment of the postdentary trough:
    (0) broad trough with prominent ridge: outgroup, tritylodontids, morganucodontids (the groove and the medial dentary ridge are slightly more reduced in morganucodontids);
    (1) trough and ridge absent: *Jeholodens, Gobiconodon, Ornithorhynchus*, multituberculates, *Zhangheotherium, Henkelotherium*, marsupials, placentals.
68. Attachment of surangular and prearticular in adults:
    (0) Attached to the mandible: outgroup, tritylodontids, morganucodontids;
    (1) Detached from the mandible: *Jeholodens, Gobiconodon, Ornithorhynchus*, multituberculates, *Zhangheotherium, Henkelotherium, Vincelestes*, marsupials, placentals.
69. Size of anterior part of the Meckelian groove in adults (Rowe, 1988; Wible, 1991; Luo, 1994):
    (0) well developed groove: outgroup, tritylodontids, morganucodontids;
    (1) weak and faint groove: *Jeholodens, Gobiconodon, Zhangheotherium, Henkelotherium;*
    (2) absent: *Ornithorhynchus*, multituberculates, *Vincelestes*, marsupials;
    (1/2 Polymorphic): placentals (1 for *Prokennalestes* Kielan-Jaworowska and Dashzeveg, 1989; 2 in other placentals).
70. Orientation of the anterior segment of the Meckelian groove (Luo, 1994):
    (0) parallel (or nearly parallel) to and separated from the ventral edge of the mandible: outgroup, tritylodontids, *Jeholodens, Zhangheotherium, Henkelotherium;*
    (1) the groove converges toward the ventral edge of the mandible (or intersects the ventral edge of the mandible): morganucodontids, *Gobiconodon;*
    (2) groove absent: *Ornithorhynchus*, multituberculates, *Vincelestes*, marsupials, placentals (*Prokennalestes* has meckelian groove in the posterior part of the mandible, but the groove is absent anteriorly, Kielan Jaworowska and Dashzeveg, 1989; in all other placental mammals the meckelian groove is completely absent).
71. Angular process of dentary (Rowe, 1988; Wible, 1991; Luo, 1994):
    (0) present: outgroup, tritylodontids (following Sues, 1986), morganucodontids, *Vincelestes, Henkelotherium*, marsupials, placentals;
    (1) absent: *Jeholodens, Gobiconodon, Ornithorhynchus*, multituberculates, *Zhangheotherium*.

72. Distinctive insertion area for the pterygoid muscle on the medial side of the mandible (new):
    (0) absent: outgroup, tritylodontids, morganucodontids, placentals;
    (1) present: multituberculates, *Zhangheotherium, Henkelotherium, Vincelestes;*
    (2) medial pterygoid fossa has a prominent ventral shelf along the ventral border of the mandible: *Jeholodens, Gobiconodon, Ornithorhynchus, marsupials.*

73. Coronoid in adults (Rowe, 1988; Wible, 1991; Luo, 1994):
    (0) present: outgroup, tritylodontids, morganucodontids, *Jeholodens, Gobiconodon* (Jenkins and Schaff, 1988), *Zhangheotherium, Henkelotherium, Vincelestes;*
    (1) absent: marsupials, *Ornithorhynchus;*
    (0/1 polymorphic): multituberculates {(0) present in paulchoffatiids (Hahn, 1977) and (1) absent in other multituberculates}; placentals {(0) in *Prokennalestes*, Kielan-Jaworowska and Dashzeveg, 1989; (1) in other placentals}.

74. A distinct mandibular foramen for the inferior alveolar nerve and vessels:
    (0) absent (mandibular foramen contained within the postdentary trough): outgroup, tritylodontids, morganucodontids;
    (1) present: *Jeholodens, Gobiconodon,* ornithorhynchids, multituberculates, *Zhangheotherium, Henkelotherium, Vincelestes,* marsupials, placentals.

75. Mandibular symphysis:
    (0) fused: outgroup, tritylodontids;
    (1) unfused: morganucodontids, *Jeholodens, Gobiconodon,* multituberculates, *Henkelotherium, Vincelestes,* marsupials, placentals;
    (2) unfused and further reduced: *Ornithorhynchus, Zhangheotherium.*

76. Scar or depression for the splenial bone on the dentary:
    (0) present: outgroup, tritylodontids, morganucodontids, *Zhangheotherium, Henkelotherium;*
    (1) absent: *Jeholodens, Gobiconodon, Ornithorhynchus,* multituberculates, *Vincelestes,* marsupials, placentals.

DENTITION (23 characters)

77. Procumbent and enlarged first lower incisor:
    (0) present (≥125% longer than the second lower incisors): outgroup, tritylodontids, *Gobiconodon,* multituberculates, *Zhangheotherium,* placentals (predominant condition for many, but not all Cretaceous placentals);
    (1) absent (first and second lower incisors are sub-equal): morganucodontids, *Jeholodens, Vincelestes,* marsupials.
    (?) *Henkelotherium* (not preserved), *Ornithorhynchus* (not applicable).

78. Number of lower incisors:
    (0) more than four incisors: outgroup, morganucodontids, *Jeholodens, Zhangheotherium, Henkelotherium, Vincelestes,* marsupials, placentals;
    (1) reduced to two or fewer incisors, or absent: tritylodontids; multituberculates, *Gobiconodon, Ornithorhynchus.*

79. Canine size:
    (0) canine present and greatly enlarged (>150% of the last incisor in crown height): outgroup, morganucodontids, *Vincelestes,* marsupials, placentals;
    (1) present and slightly enlarged (100%–125% of the last incisor in crown height): *Jeholodens, Zhangheotherium, Henkelotherium;*
    (2) greatly reduced (<100% of the adjacent incisors) or absent: tritylodontids, *Gobiconodon, Ornithorhynchus,* multituberculates.

80. Differentiation of postcanine crowns into premolars and molars (Crompton and Sun, 1985; Luo, 1994):
    (0) absent: outgroup, tritylodontids;
    (1) present: morganucodontids, *Jeholodens, Gobiconodon,* multituberculates, *Zhangheotherium, Henkelotherium, Vincelestes,* marsupials, placentals;
    (?) Not-applicable: *Ornithorhynchus.*

81. Number of the postcanine roots (Rowe, 1988; Wible, 1991; Luo 1994):
     (0) single (undivided): outgroup;
     (1) no more than three roots: morganucodontids, *Jeholodens, Gobiconodon,* multitu-
     berculates (Clemens, 1963; Hahn, 1969; Clemens and Kielan-Jaworowska, 1979),
     Zhangheotherium, Henkelotherium, Vincelestes, marsupials, placentals;
     (2) multiple roots (more than three): tritylodontids;
     (?) Not-applicable: *Ornithorhynchus.*

82. Positional correspondence of the upper and the lower postcanines:
     (0) absent: outgroup, tritylodontids (known to have un-equal number of the upper and
     lower postcanines; also because of the posterior propalinal movement of the mandible,
     a lower postcanine may contact more than one upper postcanines: Crompton, 1972;
     Sues, 1986b), multituberculates (see Krause, 1982; Wall and Krause, 1992);
     (1) present: morganucodontids, *Jeholodens, Gobiconodon, Ornithorhynchus* (after
     Simpson, 1929b on the occlusal relationship of juvenile teeth; also see Hopson and
     Crompton 1969), *Zhangheotherium, Henkelotherium, Vincelestes,* marsupials,
     placentals.

83. Interlocking of the adjacent lower postcanines:
     (0) absent: outgroup;
     (1) linguolabially compressed molars interlock by cingulum or the cingular cuspules:
     morganucodontids (cusp d of anterior molar fits into the embayment between cusp b
     and cusp e of the posterior molar), *Jeholodens* (cusp d applied to the anteromedial
     [mesial] side of cusp b of the succeeding molar. Note that, in triconodontids, the ante-
     rior side of the cingulum has a distinctive groove that extends down the root. In
     Jeholodens, the groove is absent, but in the place of the groove there is a broadly con-
     cave area. Thus in this character *Jeholodens* resembles the known triconodontids, but
     not identical to the latter); *Gobiconodon* (cusp d fits in between cusp e and cusp f, may
     be convergent to *Kuehneotherium,* Kielan-Jaworowska and Dashzeveg, 1998; also see
     Luo 1994); *Zhangheotherium* (cusp d of the preceding molar fits into the cingular area
     anterolateral to cusp e of the succeeding molar which lacks cusp f);
     (2) linguolabially wide molars abuts each other: tritylodontids, *Ornithorhynchus* (the
     juvenile teeth are wide and abut each other after Simpson, 1929b); multituberculates,
     Henkelotherium, Vincelestes, marsupials, placentals.

84. Lingual cingula on the lower postcanines:
     (0) present, with well developed cingular cusps: outgroup, morganucodontids,
     Gobiconodon;
     (1) cingular cusps absent, cingulum vestigial or absent: *Jeholodens, Zhangheotherium,*
     Henkelotherium, Vincelestes, marsupials, placentals;
     (?) Not-applicable: tritylodontids, multituberculates, *Ornithorhynchus.*

85. Anterolingual cuspule ("cusp e") on the lingual cingulum of the lower molars:
     (0) present: morganucodontids, *Gobiconodon, Zhangheotherium;*
     (1) absent: *Jeholodens,* ornithorhynchids, *Henkelotherium, Vincelestes,* marsupials,
     placentals;
     (?) Not-applicable: tritylodontids, *Ornithorhynchus,* multituberculates.

86. Arrangement of the main cusps of the upper postcanines:
     (0) in a single longitudinal row: outgroup, morganucodontids, *Jeholodens;*
     (1) multiple cusps forming multiple longitudinal rows: tritylodontids, multituberculates;
     (2) in reversed triangle: *Gobiconodon* (Kielan-Jaworowska and Dashzeveg, 1998),
     Zhangheotherium, Henkelotherium, Vincelestes, marsupials, placentals;
     (?) Not-applicable: *Ornithorhynchus.*

87. Upper molar stylar shelf (the area between the paracone/metacone and the labial
     margin of the crown):
     (0) narrow or absent: outgroup, morganucodontids, *Jeholodens, Gobiconodon;*
     (1) broad: *Zhangheotherium, Henkelotherium, Vincelestes,* marsupials, placentals;
     (?) Not-applicable: tritylodontids, *Ornithorhynchus,* multituberculates.

88. Stylar cusps between the paracone and metacone of the upper molars:
    (0) absent or weak (=crenulations on the labial cingulum): outgroup,
    morganucodontids, *Jeholodens*, *Gobiconodon*, *Zhangheotherium*,
    placentals;
    (1) present and enlarged: *Henkelotherium*, *Vincelestes*, marsupials;
    (?) Not-applicable: tritylodontids, *Ornithorhynchus*, multituberculates.

89. Upper molar protocone:
    (0) absent: outgroup, tritylodontids, morganucodontids, *Jeholodens*, *Gobiconodon*,
    multituberculates, *Zhangheotherium*, *Henkelotherium*;
    (1) present: *Vincelestes*, marsupials, placentals;
    (?) Not-applicable: *Ornithorhynchus*.

90. Alignment of main cusps of lower postcanines:
    (0) single longitudinal row: outgroup, morganucodontids, *Jeholodens*,
    Gobiconodon;
    (1) multiple cusps in multiple rows: tritylodontids, multituberculates;
    (2) in reversed triangle: *Zhangheotherium*, *Henkelotherium*, *Vincelestes*, marsupials,
    placentals.
    (?) Not-applicable: *Ornithorhynchus*.

91. Orientation of protocristid relative to the length of the lower molar (Hu et al., 1997,
    1998; modified definitions of character states):
    (0) longitudinally orientation: outgroup, morganucodontids, *Jeholodens*,
    Gobiconodon;
    (1) oblique: *Zhangheotherium*, *Vincelestes*;
    (2) more transverse: *Henkelotherium*, marsupials, placentals;
    (?) Not-applicable: tritylodontids, *Ornithorhynchus*, multituberculates.

92. Orientation of metacristid relative to the long axis of the lower molar (new):
    (0) longitudinal: outgroup, morganucodontids, *Jeholodens*, *Gobiconodon*;
    (1) oblique: *Zhangheotherium*, *Vincelestes*;
    (2) transverse: *Henkelotherium*, marsupials, placentals;
    (?) Not-applicable: tritylodontids, *Ornithorhynchus*, multituberculates.

93. Lower molar talonid:
    (0) no talonid: outgroup, tritylodontids, morganucodontids, *Jeholodens*, *Gobicon-
    odon*, multituberculates, *Zhangheotherium*;
    (1) simple talonid with a single cusp: *Henkelotherium*, *Vincelestes*;
    (2) fully developed talonid with a basin (with more than one cusps): marsupials,
    placentals;
    (?) Not-applicable: *Ornithorhynchus*.

94. Wear facet on talonid (or on posterior cingulid of the lower molar):
    (0) absent: morganucodontids, *Jeholodens*, *Gobiconodon*, *Zhangheotherium*;
    (1) present: *Henkelotherium*, *Vincelestes*, marsupials, placentals;
    (?) Not-applicable: outgroup, tritylodontids, *Ornithorhynchus*, multituberculates.

95. Relative size of the lower molar cusps (Luo, 1994):
    (0) c larger than b: morganucodontids, *Jeholodens*;
    (1) b higher than c: *Gobiconodon*, *Zhangheotherium*, *Henkelotherium*, *Vincelestes*,
    marsupials, placentals;
    (?) Not-applicable: outgroup, tritylodontids, *Ornithorhynchus*, multituberculates.

96. Occlusion of the principal cusps of the upper and lower molars:
    (0) principal cusps of the upper and the lower molariforms lack consistent contact
    relationship: outgroup;
    (1) principal cusp a is positioned anterior to cusp *A* but posterior to cusp *B* of the same
    tooth: morganucodontids, *Jeholodens*;
    (2) principal cusp a occludes between cusp *B* of the opposite upper molar and cusp *C*
    of the preceding upper molar: *Gobiconodon*, *Zhangheotherium*, *Henkelotherium*,
    Vincelestes, marsupials, placentals;

(3) interdigital occlusion between multiple cusp rows: tritylodontids, multituberculates; (?) Not-applicable: *Ornithorhynchus*.

97. Functional development of wear facets on molars (Crompton and Luo, 1993):

(0) absent for lifetime: outgroup;

(1) absent at the eruption but developed later by wearing of the crown: morganucodontids, *Gobiconodon;*

(2) upper and lower cusps match upon eruption: tritylodontids, ornithorhynchids, multituberculates, *Zhangheotherium, Henkelotherium, Vincelestes,* marsupials, placentals;

(?) Not-applicable: *Ornithorhynchus;* not exposed on the type specimen: *Jeholodens.*

98. Relationships of wear facets to the main cusps (Luo, 1994):

(0) wear facet absent: outgroup;

(1) a single principal cusp bears two wear facets (e.g., cusp a of the lower bear two facets): morganucodontids, *Jeholodens, Gobiconodon;*

(2) two cusps bear a single wear facet (single facet supported by the lower cusps a and c occludes the facet supported by cusps A and B of the opposite tooth; single facet supported by cusps a and b occludes the facet supported by A and C of the preceding upper tooth): *Zhangheotherium, Henkelotherium, Vincelestes,* marsupials, placentals;

(3) multiple cusps, with each cusp bearing one or two transverse facets: tritylodontids, multituberculates.

(?) Not-applicable: *Ornithorhynchus.*

99. Direction of the lower jaw movement during occlusion (Crompton and Luo, 1993; Luo, 1994):

(0) orthal movement: outgroup;

(1) dorsoposterior movement: tritylodontids (Crompton, 1972), multituberculates (Krause, 1982; Wall and Krause, 1992);

(2) dorsomedial movement: morganucodontids, *Jeholodens, Gobiconodon,* ornithorhynchids, *Zhangheotherium, Henkelotherium, Vincelestes,* marsupials, placentals.

(1/2 Polymorphic): *Ornithorhynchus* (Hopson and Crompton, 1969 inferred from the wear patterns on the cusps of the juvenile teeth and the shape of the craniomandibular joint that the mandible of the platypus could have both dorsoposteriorly directed and the dorsomedially directed movement).

100. Mode of occlusion (Crompton and Luo, 1993; Luo, 1994):

(0) bilateral: outgroup, tritylodontids (following Crompton, 1974; Sues, 1985), multituberculates (following Wall and Krause, 1992);

(1) unilateral: morganucodontids, *Jeholodens, Gobiconodon, Ornithorhynchus* (after Simpson, 1929b; Hopson and Crompton, 1969; Archer et al., 1993); *Zhangheotherium, Henkelotherium, Vincelestes,* marsupials, placentals.

101. Rotation of the mandible during occlusion (Kemp, 1983; Luo, 1994, Kielan-Jaworowska, 1997), as inferred from the angles of wear facets to the vertical plane (Crompton and Luo, 1993):

(0) absent (posteriorly directed): outgroup, tritylodontids (following Crompton, 1974; Sues, 1985), multituberculates (following Krause, 1982; Wall and Krause, 1992; Kielan-Jaworowska, 1996);

(1) moderate: morganucodontids, *Jeholodens, Gobiconodon* (following Crompton and Luo, 1993), *Ornithorhynchus* (after Simpson, 1929b; Hopson and Crompton, 1969);

(2) strong: *Zhangheotherium, Vincelestes, Henkelotherium,* marsupials, placentals.

PAUP search results for *A chinese triconodont mammal and mosaic evolution of the mammalian skeleton*

PART I. IN-GROUP ANALYSIS OF TRICONODONTS

Processing of file "Jeholodens.in.mtx" begins. . .
Data matrix has 14 taxa, 33 characters
Valid character-state symbols: 0123
Missing data identified by '?'
Gaps identified by '−', treated as "missing"
Processing of file "Jeholodens.in.mtx" completed.
Input data matrix

Node/character	1111111111122222222223333
	1234567890123456789012345 67890123
Sinoconodon	000000000003100?000000??00000010
	1
Morganucodon	0100000000010000001100000002001 01
	1
	2
Megazostrodon	01010000000100100012000010020010 0
Dinnetherium	0101000000010000001101000002012 01
Kuehneotherium	0101000000??0010?11201001111012??
Gobiconodon	101201111111111100120200110101200
Phascolotherium	1112111111013100001202001101 0??11
Amphilestes	1112111111?11100001202001101 01211
Tinodon	1112110111?12100111202011111 01211
Jeholodens	1012010?1?0130011011020??00010301
Triconodon	101211111111101010111211000110311
	1
Trioracodon	1012111111?110102011121100010 0311
	1 1
Priacodon	10121111111201010111201000110311
	1
Astraconodon	101211011111101010111211000110301
	1

Branch-and-bound search settings:
Initial upper bound: unknown (compute via stepwise)
Addition sequence: furthest
Initial MAXTREES setting = 10000
Branches having maximum length zero collapsed to yield polytomies
Topological constraints not enforced
Trees are unrooted
Multi-state taxa interpreted as uncertainty

Branch-and-bound search completed:

Shortest tree found = 71
Number of trees retained = 48
Time used = 6.40 sec

Lengths and fit measures of trees in memory:
  Sum of min. possible lengths = 42
  Sum of max. possible lengths = 159

Tree #	1
Length	71
CI	0.592
RI	0.752

Strict consensus of 48 trees:

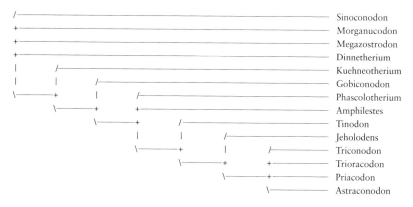

Bootstrap method with branch-and-bound search:
  Starting seed = 1
  Number of bootstrap replicates = 1000
  Initial upper bound: unknown (compute via stepwise)
  Addition sequence: furthest
  Initial MAXTREES setting = 10000
  Branches having maximum length zero collapsed to yield polytomies
  Topological constraints not enforced
  Trees are unrooted
  Multi-state taxa interpreted as uncertainty

  Time used = 01:10:59.2

Bootstrap 50% majority-rule consensus tree

PART II. PHYLOGENETIC POSITION OF JEHOLODENS AMONG MAMMALS
Processing of file "Jehol.mtx.Jan.99.1" begins. . .

Data matrix has 12 taxa, 101 characters
Valid character-state symbols: 0123
Missing data identified by '?'
Gaps identified by '−', treated as "missing"

Processing of file "Jehol.mtx.Jan.99.1" completed.

Input data matrix

```
                          1111111111222222222233333333334444444444555555555
Node/character            123456789012345678901234567890123456789012345678

outgroup                  0000000?00?10000000??0000000000000000000000000000?0
tritylodontids            0000000?00000000000?000000000000001000000000000??0000?0
morganucodontids          ?001000?10?1?010?00??00000000000001000001000000000000100
Jeholodens                100?00010011111111101110??01101001000001000000000101000100
Gobiconodon               ???????00?????1??????11?00?011?00?1000001000???0????1010?
Ornithorhynchus           11110111000000010000000000010000101000021000?02111110100
multituberculate          10?1001010001110111011010010001101111102110100121112101 00

                                                    2
Zhangheotherium           1?110000?010112111110111111111010110111 02000100110??21?111
Henkelotherium            ?????????1??1121111??11?11?101011?101110211????1?????? ?1???
Vincelestes               1011?11011?11?21011101111 0?0010100111100211020111020?01111
marsupials                11110110112111210111121111120211111111112111211110 20201111
                                                                              1
placentals                11111110112111210111121111120211101111112111211110 20201111
                                                                              1
```

Input data matrix (continued)

```
                                                      11
                          566666666666777777777788888888889999999999 00
Node                      901234567890123456789012345678901234567890 1

outgroup                  0?0000??000000000000000000?000000000??000000
tritylodontids            000000?000000000000120202??1??01??0??323100
morganucodontids          1000000000010000101001110000000000 00111211
Jeholodens                10??????111012011110111111100000000001?1211
Gobiconodon               ???????0?1111120111012111100200000000012 11211
Ornithorhynchus           11000001112212112 1?12??12???????????????111
                                                              2
multituberculate          10000101112211011101211 02??1??01??0??323100
                           1 1          1
Zhangheotherium           10111112111011012000111111 0210021100122221 2
Henkelotherium            ????????1110010110?011112112110222111222212
Vincelestes               21111111112201011100111211211121111122221 2
marsupials                2211111311220211111001112112111222211222212
placentals                22111111112000111000111211210122 2211222212
                                   2 1
```

Branch-and-bound search settings:
    Initial upper bound: unknown (compute via stepwise)
    Addition sequence: furthest
    Initial MAXTREES setting = 10000
    Branches having maximum length zero collapsed to yield polytomies
    Topological constraints not enforced
    Trees are unrooted
    Multi-state taxa interpreted as uncertainty

Branch-and-bound search completed:

    Shortest tree found = 210
    Number of trees retained = 2
    Time used = 3.03 sec

Lengths and fit measures of trees in memory:

    Sum of min. possible lengths = 134
    Sum of max. possible lengths = 409

Tree #	1	2
Length	210	210
CI	0.638	0.638
RI	0.724	0.724

Tree number 1 (rooted using default outgroup):

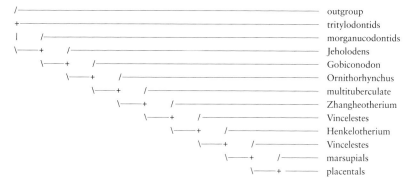

Tree number 2 (rooted using default outgroup):

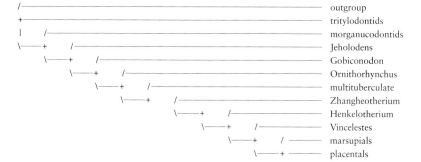

Strict consensus of 2 trees:

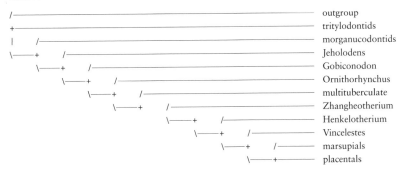

```
/─────────────────────────────────────────────  outgroup
+─────────────────────────────────────────────  tritylodontids
|    /────────────────────────────────────────  morganucodontids
\───+   /─────────────────────────────────────  Jeholodens
   \───+   /──────────────────────────────────  Gobiconodon
      \───+   /───────────────────────────────  Ornithorhynchus
         \───+   /────────────────────────────  multituberculate
            \───+   /─────────────────────────  Zhangheotherium
               \───+   /──────────────────────  Henkelotherium
                  \───+   /───────────────────  Vincelestes
                     \───+   /────────────────  marsupials
                        \───+────────────────   placentals
```

Bootstrap method with branch-and-bound search:
   Starting seed = 1
   Number of bootstrap replicates = 1000
   Initial upper bound: unknown (compute via stepwise)
   Addition sequence: furthest
   Initial MAXTREES setting = 10000
   Branches having maximum length zero collapsed to yield polytomies
   Topological constraints not enforced
   Trees are unrooted
   Multi-state taxa interpreted as uncertainty

Time used = 00:41:16.2

Bootstrap 50% majority-rule consensus tree

```
/─────────────────────────────────────────────  outgroup (1)
+─────────────────────────────────────────────  tritylodontids (2)
|    /────────────────────────────────────────  morganucodontids (3)
\-98-+   /────────────────────────────────────  Jeholodens (4)
   \-97-+   /─────────────────────────────────  Gobiconodon (5)
      \-53-+   /──────────────────────────────  Ornithorhynchus (6)
         \-56-+   /───────────────────────────  multituberculate (7)
            \-91-+   /────────────────────────  Zhangheotherium (8)
               \-79-+   /─────────────────────  Henkelotherium (9)
                  \-81-+   /──────────────────  Vincelestes (10)
                     \-59-+   /───────────────  marsupials (11)
                        \-99-+─────────────────  placentals (12)
```

References

Archer, M., Flannery, T. F., Richtie, A., and Molnar, R. E. 1985. First Mesozoic mammal from Australia—an early Cretaceous monotreme. Nature 318:363–366.

Archer, M., Jenkins, F. A., Jr., Hand, S. J., Murray, P., Godthelp, H. 1992. Description of the skull and non-vestigial dentition of a Miocene platypus (*Obdurodon dicksoni*, n. sp.) from Riversleigh, Australia and the problem of monotreme origins. Pp. 15–27. In, M. L. Augee (Ed.) Platypus and Echidnas. The Royal Zoological Society of New South Wales, Sydney.

Archer, M, Murray, P., Hand, S. J., and Godthelp, H. 1993. Reconsideration of monotreme relationships based on the skull and dentition of the Miocene *Obdurodon dicksoni* (ornithorhynchids). Pp. 75–94. In, F. S. Szalay, M. J. Novacek, and M. C. McKenna (Eds.), Mammal Phylogeny. Volume 1. Springer-Verlag, New York.

Bonaparte, J. F. 1986. Sobre *Mesungulatum houssayi* y nuevos mamíferos Cretácicios de Patagonia, Argentina. Actas IV Congreso Argentino de Paleontología y Bioestratigrafía 2:48–61.

Bonaparte, J. F. 1992. Una nueva especie de Triconodonta (Mammalia), de la Formacíon Los Alamitos, Provincia de Río Negro y comentarios sobre su fauna de mamíferos, Ameghiniana 29:99–110.

Chow, M-C. and T. H. Rich. 1984. A new triconodontan (Mammalia) from the Jurassic of China. Journal of Vertebrate Paleontology 3:226–231.

Cifelli, R. C. and Madsen, S. K. 1998. Triconodont mammals from the medial Cretaceous of Utah. Journal of Vertebrate Paleontology 18:403–411.

Cifelli, R. L., Wible, J. R. and Jenkins, F. A. 1998. Triconodont mammals from the Cloverly Formation (Lower Cretaceous), Montana and Wyoming. Journal of Vertebrate Paleontology 16:237–241.

Clemens, W. A. 1963. Fossil mammals of the type Lance Formation, Wyoming. Part I. Introduction and Multituberculata. University of California Publications in Geological Sciences 48:1–105.

Clemens, W. A. 1966. Fossil mammals of the type Lance Formation, Wyoming. Part II. Marsupialia. University of California Publications in Geological Sciences 62:1–122.

Clemens, W. A. 1980. Rhaeto-Liassic mammals from Switzerland and West Germany. Zitteliana 5:51–92.

Clemens, W. A. and Kielan-Jaworowska, Z. 1979. Multituberculata. Pp. 99–149. In, J. A. Lillegraven, Z. Kielan-Jaworowska, and W. A. Clemens (Eds.), Mesozoic Mammals: the First Two-thirds of Mammalian History. University of California Press, Berkeley.

Crompton, A. W. 1964. On the skull of *Oligokyphus*. Bulletin of British Museum of Natural History, Geology 9:70–82.

Crompton, A. W. 1971. The origin of the tribosphenic molar. Pp. 65–87. In, D. M. Kermack and K. A. Kermack (Eds.) Early Mammals. Academic Press, New York.

Crompton, A. W. 1972. Postcanine occlusion in cynodonts and tritylodonts. Bulletin of British Museum (Natural History) 21:27–71.

Crompton, A. W. 1974. The dentitions and relationships of the southern African Triassic mammals *Erythrotherium parringtoni* and *Megazostrodon rudnerae*. Bulletin of British Museum (Natural History), Geology 24:399–437.

Crompton, A. W. and F. A. Jenkins, Jr. 1967. American Jurassic symmetrodonts and Rhaetic "Pantotheres." Science 155:1006–1009.

Crompton, A. W. and Luo, Z. 1993. Relationships of the Liassic mammals, *Sinoconodon, Morganucodon oehleri,* and *Dinnetherium*. Pp. 30–44. In, F. S. Szalay, M. J. Novacek, and M. C. McKenna (Eds.), Mammal Phylogeny. Volume 1. Springer Verlag, New York.

Crompton, A. W., and Sun, A. L. 1985. Cranial Structure and relationships of the Liassic mammal *Sinoconodon*. Zoological Journal of the Linnean Society 85:99–119.

Engelmann, G. F. and G. Callison. 1998. Mammalian faunas of the Morrison Formation. Modern Geology 23:343–379.

Evans, S. E. 1981. The postcranial skeleton of the Lower Jurassic eosuchian *Gephyrosaurus bridensis*. Zoological Journal of Linnean Society 73:81–116.

Fastovsky, D. E., Clark, J. and Hopson, J. A. 1987. Preliminary report of a vertebrate fauna from an unusual paleoenvironmental setting, Huizachal Group, Early or Mid-Jurassic, Tamaulipas, Mexico. Pp. 82–88. In, P. M. Currie and E. H. Koster (Eds.), Short Papers of Fourth Symposium on Mesozoic Terrestrial Ecosystems, Tyrrell Museum, Drumheller, Alberta, Canada.

Fox, R. C. 1969. Studies of Late Cretaceous vertebrates. III. A triconodont mammal from Alberta. Canadian Journal of Zoology 47:1253–1256.

Fox, R. C. 1976. Additions to the mammalian local fauna from the upper Milk River Formation (Upper Cretaceous), Alberta. Canadian Journal of Earth Sciences 13:1105–1118.

Freeman, E. F. A middle Jurassic mammal bed from Oxfordshire. Palaeontology 22:135–166.

Gambaryan P. P. and Kielan-Jaworowska, Z. 1997. Sprawling versus parasagittal stance in multituberculate mammals. Acta Palaeontologica Polonica 42:13–44.

Goodrich, E. S. 1894. On the fossil Mammalia of the Stonefield Slate. Quarterly Journal of Microscopic Science 35:407–431.

Gow, C. E. 1986. A new skull of *Megazostrodon* (Mammalia: Triconodonta) from the Elliot Formation (Lower Jurassic) of southern Africa. Palaeontologia Africana, 26: 13–23.

Grassé, P.-P. 1955. Traité de Zoologie: Anatomie, Systématique, Biologie. Tome XVII. Mammiféres—Les Ordres: Anatomie, Éthologie, Systématique. Masson et Cie, Paris. 1170 Pp.

Greenwald, N. S. 1988. Patterns of tooth replacement in multituberculate mammals. Journal of Vertebrate Paleontology 8:265–277.

Gregory, W. K. 1951. Evolution Emerging: A Survey of Changing Patterns from Primeval Life to Man. The MacMillan Company, New York. 730 Pp.

Hahn, G. 1969. Beiträge zur Fauna der Grube Guimarota Nr. 3. Die Multituberculata. Palaeontographica 133:1–100.

Hahn, G. 1977. Das Coronoid der Paulchoffatiidae (Multituberculata, Ober-Jura). Paläontologische Zeitschrift 51:246–253.

Hopson, J. A. 1994. Synapsid evolution and the radiation of non-eutherian mammals. Pp. 190–219. In, Prothero, D. R. and R. M. Schoch (Eds.), Major features of Vertebrate Evolution. Short Courses in Paleontology 7. Paleontological Society, Knoxville, Tennessee.

Hopson, J. A. and Barghusen, H. R. 1986. An analysis of therapsid relationships. Pp. 83–106 In, N. Hotton, III, P. D. MacLean, J. J. Roth, and E. C. Roth (Eds.), The Ecology and Biology of Mammal-like Reptiles. Smithsonian Institution Press, Washington, D.C.

Hopson, J. A. and A. W. Crompton. 1969. Origin of mammals. Evolutionary Biology 3:15–72.

Hu, Y., Wang, Y. C. Li, and Z. Luo. 1998. Morphology of dentition and forelimb of *Zhangheotherium*. Vertebrata PalAsiatica 38:102–125.

Hu, Y., Wang, Y., Luo, Z. and C. Li. 1997. A new symmetrodont mammal from China and its implications for mammalian evolution. Nature 390:137–142.

Hurum, J. H. 1997. Cranial structure and relationships of Mongolian Late Cretaceous multituberculate mammals. Thesis for Doctor Scientiarum. Paleontological Museum, University of Oslo.

Jenkins, F. A., Jr. 1971. The postcranial skeleton of African cynodonts. Bulletin of Peabody Museum of Natural History, Yale University 36:1–216.

Jenkins, F. A., Jr. 1973. The functional anatomy and evolution of the mammalian humero-ulnar joint. The American Journal of Anatomy 137:281–298.

Jenkins, F. A., Jr. 1974. The movement of the shoulder in claviculate and aclaviculate mammals. Journal of Morphology 144:71–84.

Jenkins, F. A. Jr. and Crompton, A. W. 1979. Triconodonta. Pp. 74–90. In, Lillegraven, J. A., Kielan-Jaworowska, Z. and Clemens, W. A. (eds.), Mesozoic Mammals: The First Two-thirds of Mammalian History. Berkeley, University of California Press.

Jenkins, F. A. Jr., A. W. Crompton, and W. R. Downs, 1983. Mesozoic mammals from Arizona: new evidence on mammalian evolution. Science, 222: 1233–1235.

Jenkins, F. A., Jr. and Parrington, F. R. 1976. Postcranial skeleton of the Triassic mammals *Eozostrodon, Megazostrodon,* and *Erythrotherium.* Philosophical Transactions of the Royal Society of London 273B:387–431.

Jenkins, F. A., Jr. and Schaff, C. R. 1988. The Early Cretaceous mammal *Gobiconodon* (Mammalia, Triconodonta) from the Cloverly Formation in Montana. Journal of Vertebrate Paleontology 6:1–24.

Jenkins, F. A., Jr. and Weijs, W. A. 1979. The functional anatomy of the shoulder in the Virginia opossum (*Didelphis virginiana*). Journal of Zoology (London) 188:379–410.

Kemp, T. S. 1982. Mammal-like reptiles and origin of mammals. Academic Press, London. 363 Pp.

Kemp, T. S. 1983. The interrelationships of mammals. Zoological Journal of the Linnean Society (London) 77:353–384.

Kermack, D. M., K. A. Kermack and F. Mussett. 1968. The Welsh pantothere *Kuehneotherium praecursoris*. Zoological Journal of Linnean Society, London 47:407–423.

Kermack, K. A. 1967. Interrelations of early mammals. Zoological Journal of Linnean Society 47:241–249.

Kermack, K. A., F. Mussett and H. W. Rigney. 1973. The lower jaw of *Morganucodon*. Zoological Journal of Linnean Society 53:87–175.

Kermack, K. A., Mussett, F., and Rigney, H. W. 1981. The skull of *Morganucodon*. Zoological Journal of Linnean Society (London) 71:1–158.

Kielan-Jaworowska, Z. 1976. Evolution of the therian mammals in the Late Cretaceous of Asia Part II. Postcranial skeleton in *Kennalestes* and *Asioryctes*. Palaeontologia Polonica 38:3–41.

Kielan-Jaworowska, Z. 1978. Evolution of the therian mammals in the Late Cretaceous of Asia Part III. Postcranial skeleton in Zalambdalestidae. Palaeontologia Polonica 38:3–41.

Kielan-Jaworowska, Z. 1989. Postcranial skeleton of a Cretaceous multituberculate mammal. Acta Palaeontologica Polonica 34:75–85.

Kielan-Jaworowska, Z. 1997. Characters of multituberculates neglected in phylogenetic analyses of early mammals. Lethaia 29:249–266.

Kielan-Jaworowska, Z. 1998. Humeral torsion in multituberculate mammals. Acta Palaeontologica Polonica 43:131–134.

Kielan-Jaworowska, Z., Crompton, A. W., and Jenkins, F. A., Jr. 1987. The origin of egg-lying mammals. Nature 326:871–873.

Kielan-Jaworowska, Z. and Dashzeveg, D. 1998. New Early Cretaceous amphilestid ('triconodont') mammals from Mongolia. Acta Palaeontologica Polonica 43:413–438.

Kielan-Jaworowska, Z. and Gambaryan, P. P. 1994. Postcranial anatomy and habits of Asian multituberculate mammals. Fossils and Strata 36:1–92.

Kielan-Jaworowska, Z. and Hurum, J. H. 1997. Djadochtatheria—a new suborder of multituberculate mammals. Acta Palaeontologica Polonica 42:201–242.

Kielan-Jaworowska, Z., Presley, R., and Poplin, C. 1986. The cranial vascular system in taeniolabidoid multituberculate mammals. Philosophical Transactions of the Royal Society of London 313B:525–602.

Kielan-Jaworowska, Z. and Qi, T. 1990. Fossorial adaptations of a taeniolabidoid multituberculate from Eocene of China. Vertebrata PalAsiatica 28:81–94.

Klima, M. 1973. Die Frühentwicklung des Schültergürtels und des Brustbeins bei den Monotremen (Mammalia: Prototheria). Advances in Anatomy, Embryology and Cell Biology 47:1–80.

Klima, M. 1987. Early development of the shoulder girdle and sternum in marsupials (Mammalia: Metatheria). Advances in Anatomy, Embryology and Cell Biology 109:1–91.

Krause, D. W. 1982. Jaw movement, dental function, and diet in the Paleocene multituberculate *Ptilodus*. Paleobiology 8:265–281.

Krause, D. W. and Jenkins, F. A. Jr. 1983. The postcranial skeleton of North American multituberculates. Bulletin of the Museum of Comparative Zoology 150:199–246.

Krebs, B. 1991. Das Skelett von *Henkelotherium guimarotae* gen. et sp. nov. (Eupantotheria, Mammalia) aus dem Oberen Jura von Portugal. Berliner Geowissenschaftliche Abhandlungen A 133:1–110.

Kühne, W, G. 1949. On a triconodont tooth of a new pattern from a fissure-filling in South Glamorgan. Proceedings of the Zoological Society of London 119:345–350.

Kühne, W, G. 1956. The Liassic therapsid *Oligokyphus*. British Museum (Natural History), London. 149 Pp.

Lewis, O. J. 1983. The evolutionary emergence and refinement of the mammalian pattern of foot architecture. Journal of Anatomy 137:21–45.

Lillegraven, J. A. and G. Hahn. 1993. Evolutionary analysis of the middle and inner ear of Late Jurassic multituberculates. Journal of Mammalian Evolution 1:47–74.

Lillegraven, J. A. and G. Krusat. 1991. Cranio-mandibular anatomy of *Haldanodon exspectatus* (Docodonta; Mammalia) from the Late Jurassic of Portugal and its implications to the evolution of mammalian characters. Contributions to Geology, University of Wyoming 28:39–138.

Lucas, S. G. and Z. Luo. 1993. *Adelobasileus* from the Upper Triassic of West Texas: the oldest mammal. Journal of Vertebrate Paleontology 13:309–334.

Luo, Z. 1989. The petrosal structures of Multituberculata (Mammalia) and the molar morphology of the early arctocyonids (Condylarthra: Mammalia). Ph.D. Dissertation, Department of Paleontology, University of California at Berkeley. 422 Pp.

Luo, Z. 1994. Sister taxon relationships of mammals and the transformations of the diagnostic mammalian characters. Pp. 98–128. In, Fraser, N. C. and H.-D. Sues (Eds.), In the Shadow of Dinosaurs—Early Mesozoic Tetrapods. Cambridge University Press, Cambridge, Cambridge and New York.

Luo, Z. and Crompton, A. W. 1994. Transformations of the quadrate (incus) through the transition from non-mammalian cynodonts to mammals. Journal of Vertebrate Paleontology 14:341–374.

Luo, Z., Crompton, A. W., and Lucas, S. G. 1995. Evolutionary origins of the mammalian promontorium and cochlea. Journal of Vertebrate Paleontology 15:113–121.

Marsh, O. C. 1887. American Jurassic mammals. American Journal of Science (series 3) 33:326–348.

Marshall, L. G. and Sigogneau-Russell, D. 1995. Part. III. Postcranial skeleton. In, Muizon, C. De (Ed.), *Pucadelphys andinus* (Marsupialia, Mammalia) from the early Paleocene of Bolivia. Mémoire du Muséum National d'Histoire Naturelle 165:91–164.

McKenna, M. C. and S. K. Bell. 1997. Classification of mammals above the species level. Columbia University Press, New York. 631 Pp.

Meng, J. and Miao, D. 1992. The breast-shoulder apparatus of *Lambdopsalis bulla* (Multituberculata) and its systematic and functional implications. Journal of Vertebrate Paleontology 12(3):43A.

Meng, J. and Wyss, A. 1995. Monotreme affinities and low-frequency hearing suggested by multituberculate ear. Nature 377:141–144.

Mills, J. R. E., 1971. The dentition of *Morganucodon*. Pp. 26–63. In, D. M. Kermack and K. A. Kermack (Eds.), *Early Mammals*. Zoological Journal of Linnean Society Vol. 50, Supplement No.1.

Muizon, C. de. 1998. *Mayulestes ferox*, a borhyaenoid (Metatheria, Mammalia) from the early Palaeocene of Bolivia. Phylogenetic and palaeobiological implications. Geodiversitas 20:19–142.

Muizon, C. de. 1994. A new carnivorous marsupial from the Paleocene of Bolivia and the problem of marsupial monophyly. Nature 370:208–211.

Novacek, M. J. 1986. The skull of leptictid insectivorans and the higher-level classification of eutherian mammals. Bulletin, American Museum of Natural History 183:1–112.

Novacek, M. J., Rougier, G. W., Wible, J. R., McKenna, M. C., Dashzeveg, D. and Horovitz, I. 1997. Epipubic bones in eutherian mammals from the Late Cretaceous of Mongolia. Nature 389:483–486.

Osborn, H. F. 1888. Additional observations upon the structure and classification of the Mesozoic Mammalia. Proceedings of Academy of Natural Sciences, Philadelphia 1888:292–301.

Pascual, R., Archer, M., Jaureguizar, E. O., Prado, J. L., Godthelp, H. and Hand, S. J.
  1992a. First discovery of monotreme in South America. Nature 356:704–706.

Pascual, R. Archer, M., Jaureguizar, E. O., Prado, J. L., Godthelp, H. and Hand, S. J. 1992b.
  The first non-Australian monotreme: an early Paleocene South American platypus
  (Monotremata, Ornithorhynchidae). Pp. 2–15. In, M. L. Augee (Ed.), Platypus and Echid-
  nas. The Royal Zoological Society of New South Wales, Sydney.

Parrington, F. R. 1941. On two mammalian teeth from the Lower Rhaetic of Somerset.
  Annals and Magazine of Natural History 11:140–144.

Patterson, B. 1951. Early Cretaceous mammals from Northern Texas. American Journal of
  Science 249:31–46.

Patterson, B. 1956. Early Cretaceous mammals and the evolution of mammalian molar
  teeth. Fieldiana (Geology) 13:1–105.

Patterson, B. and E. O. Olson, 1961. A triconodontid mammal from the Triassic of Yunnan.
  Internal Colloquium in the evolution of lower and non-specialized mammals. 129–191.
  Brussels: Koninklijke Vlaamse Academiie voor Wetenschapen, Letteren en Schone
  Kunsten van Belgie.

Prothero, D. R. 1981. New Jurassic mammals from Como Bluff, Wyoming, and the
  interrelationships of non-tribosphenic Theria. Bulletin of American Museum of Natural
  History 167:277–326.

Rassmusen, T. E. and G. Callison. 1981. A new species of triconodont mammal from the
  upper Jurassic of Colorado. Journal of Paleontology 55:628–634.

Rigney, H. W. 1963. A specimen of *Morganucodon* from Yunnan. Nature 197:1122–1123.

Rougier, G. W. 1993. *Vincelestes neuquenianus* Bonaparte (Mammalia, Theria), un primitivo
  mammifero del Cretacico Inferior de la Cuenca Neuqina. Ph.D. Thesis, Universidad Nacional
  de Buenos Aires. Facultad de Ciencias Exactas y Naturales, Buenos Aires. 720 Pp.

Rougier, G. W., Wible, J. R., and Hopson, J. A. 1992. Reconstruction of the cranial vessels in
  the Early Cretaceous mammal *Vincelestes neuquenianus*: implications for the evolution of
  the mammalian cranial vascular system, Journal of Vertebrate Paleontology 12:188–216.

Rougier, G. W., Wible, J. R., and Hopson, J. A. 1996a. Basicranial anatomy of *Priacodon
  fruitaensis* (Triconodontidae, Mammalia) from the Late Jurassic of Colorado, and a reappraisal of mammaliaform interrelationships. American Museum Novitates 3183:1–28.

Rougier, G. W., Wible, J. R., and Novacek, M, J. 1996b. Scientific correspondence "multituberculate phylogeny." Nature 379:406.

Rowe, T. 1986. Osteological diagnosis of Mammalia. L. 1758, and its relationship to extinct
  Synapsida. Ph.D. Dissertation, University of California at Berkeley. 446 Pp.

Rowe, T. 1988. Definition, diagnosis, and origin of Mammalia. Journal of Vertebrate
  Paleontology 8:241–264.

Rowe, T. 1993. Phylogenetic systematics and the early history of mammals. Pp. 129–145.
  In, F. S. Szalay, M. J. Novacek, and M. C. McKenna (Eds.), Mammal Phylogeny.
  Volume 1. Springer-Verlag, New York.

Sereno, P. and McKenna, M. C. 1995. Cretaceous multituberculate skeleton and the early
  evolution of the mammalian shoulder girdle. Nature 377:144–147.

Signogneau-Russell, D. 1983. Nouveaux taxons de Mammiferes rhétiens. Acta
  Paleontologica Polonica 28:233–249.

Signogneau-Russell, D. 1995. Two possibly aquatic triconodont mammals from the Early
  Cretaceous of Morocco. Acta Palaeontologica Polonica 40:149–162.

Simmons, N. B. 1993. Phylogeny of Multituberculata. Pp. 147–164. In, F. S. Szalay, M. J.
  Novacek, and M. C. McKenna (Eds.), Mammal Phylogeny. Volume 1. Springer Verlag,
  New York.

Simpson, G. G. 1925. American triconodonts. American Journal of Science (series 5) 10:
  145–165.

Simpson, G. G. 1928. A catalogue of the Mesozoic mammals in Geological Department of the British Museum, London. 215 Pp.

Simpson, G. G. 1929a. American Mesozoic Mammals. Memoirs of Peabody Museum of Yale University 3:1–171.

Simpson, G. G. 1929b. The dentition of Ornithorhynchus as evidence of its affinities. American Museum Novitates 390:1–15.

Slaughter, B. 1969. Astroconodon, the Cretaceous triconodont. Journal of Mammalogy 50:102–107.

Sues, H.-D. 1983. Advanced mammal-like reptiles from the Early Jurassic of Arizona. Ph. D. Dissertation, Harvard University.

Sues, H.-D. 1985. The relationships of the Tritylodontidae (Synapsida). Zoological Journal of the Linnean Society (London) 85:205–217.

Sues, H.-D. 1986a. The skull and dentition of two tritylodontid synapsids from the Lower Jurassic of Western North America. Bulletin of the Museum of Comparative Zoology, Harvard University 151:217–268.

Sues, H.-D. 1986b. Relationships and biostratigraphic significance of the Tritylodontidae (Synapsida) from the Kayenta Formation of northeastern Arizona. Pp. 279–285. In, Kevin Padian (Ed.) The beginning of the age of dinosaurs—faunal changes across the Triassic-Jurassic boundary. Cambridge University Press, Cambridge and New York.

Sun, A-L. and Li, Y-H. 1985. The postcranial skeleton of the late tritylodont *Bienotheroides*. Vertebrata PalAsiatica 23:135–151.

Szalay, F. S. 1993. Pedal evolution of mammals in the Mesozoic: tests for taxi relationships. Pp. 108–128. In, F. S. Szalay, M. J. Novacek, and M. C. McKenna (Eds.), Mammal Phylogeny, Volume 1. Springer-Verlag, New York.

Szalay, F. S. and Trofimov, B. A. 1996. The Mongolian Late Cretaceous *Asiatherium*, and the early phylogeny and paleobiogeography of Metatheria. Journal of Vertebrate Paleontology 16:474–509.

Trofimov, B. A. 1968. The first triconodonts (Mammalia, Triconodonta) from Mongolia. Reports, Academy of USSR 243:213–216. (In Russian).

Wall, C. E. and Krause, D. W. 1992. A biomechanical analysis of the masticatory apparatus of *Ptilodus* (Multituberculata). Journal of Vertebrate Paleontology 12:172–187.

Wang, Y.-Q., Y.-M. Hu, M.-C. Chow and C.-K. Li. 1995. Mesozoic mammal localities in western Liaoning, Northeast China. Pp. 221–228. In, A-L. Sun and Y-Q. Wang (Eds.), Sixth Symposium on Mesozoic Terrestrial Ecosystems and Biota, Short Papers. Beijing, China Ocean Press.

Wible, J. R. 1987. The eutherian stapedial artery: character analysis and implications for superordinal relationships. Zoological Journal of Linnean Society (London) 91:107–135.

Wible, J. R. 1991. Origin of Mammalia: the craniodental evidence reexamined. Journal of Vertebrate Paleontology 11:1–28.

Wible, J. R. and Hopson, J. A. 1993. Basicranial evidence for early mammal phylogeny. Pp. 45–62. In, F. S. Szalay, M. J. Novacek, and M. C. McKenna (Eds.), Mammal Phylogeny. Volume 1. Springer Verlag, New York.

Wible, J. R. and Hopson, J. A. 1995. Homologies of the prootic canal in mammals and non-mammalian cynodonts. Journal of Vertebrate Paleontology 15:331–356.

Woodburne, M. O. and Tedford, R. H. 1975. The first Tertiary monotreme from Australia. American Museum Novitates 2588:1–11.

Young, C. C. 1978. New material of *Eozostrodon*. Vertebrate PalAsiatica 16:1–3.

Young, C. C. 1982. Selected Works of Yang Zhongjian (Young Chung-Chien). (edited by M.-C. Chow et al.). Science Press, Beijing, China.

Zeller, U. 1989. Die Entwicklung und Morphologie des Schadels von *Ornithorhynchus anatinus* (Mammalia: Prototheria: Monotremata), Abhandlungen der Senchenbergischen Naturforschenden Gesellschaft, Frankfurt am Main 545:1–188.

Zeller, U. 1993. Ontogenetic evidence for cranial homologies in monotremes and therians, with special reference to *Ornithorhynchus*. Pp. 95–107. In, F. S. Szalay, M. J. Novacek, and M. C. McKenna (Eds.), Mammal Phylogeny, Volume 1. Springer-Verlag, New York.

Zhang, F-K., A. W. Crompton, Z. Luo and C. R. Schaff. 1998. Pattern of dental replacement of *Sinoconodon* and its implications for evolution of mammals. Vertebrata PalAsiatica 36:197–217.

Chapter 18   A New Symmetrodont Mammal from China and Its Implications for Mammalian Evolution   Yao-Ming Hu, Yuan-Qing Wang, Zhe-Xi Luo, and Chuan-Kui Li

Character list of *Zhangheotherium* (IVPP 7466)

SCOPE OF THIS PHYLOGENETIC STUDY

Characters on this list are selected on three criteria: (1) the characters preserved on *Zhangheotherium* (holotype: IVPP7466); (2) with systematic variation among the following 11 taxa that are represented by more or less complete postcrania (some incomplete taxa, such as *Sinoconodon* and triconodontids, are not included because too much of their postcranial anatomy remains unknown); and (3) the characters that are relevant to the forelimb posture.

The list represents only a part of a much larger published dataset from several lengthy studies on non-mammalian cynodonts and early mammals (Kemp, 1982, 1983; Hopson and Barghusen, 1986; Rowe, 1986, 1988, 1993; Wible, 1991; Luo, 1994; Kielan-Jaworowska and Gambaryan, 1994; Kielan-Jaworowska, 1996; Rougier, 1993; Rougier et al., 1996a; Sereno and McKenna, 1995).

Other parts of the holotype specimen of *Zhangheotherium*, such as the wear facets on the dentition and the hindlimbs, also have very interesting information. This will have to be treated in a more lengthy anatomical monograph in the future. For the current study, we decide to concentrate on the main characters.

Because of the limited space for the short papers in *Nature*, we the authors decided to concentrate on how to establish the phylogenetic position of *Zhangheotherium* among the major lineages of mammals. It is not the intention of the authors to produce an all-encompassing matrix with all possible character of all major clades of cynodonts and early mammals (see the recent discussion by Rougier et al., 1996b; Kielan-Jaworowska, 1996).

MAJOR CHARACTER LISTS AND MATRICES

Kemp (1983); Hopson and Barghusen (1986); Rowe (1988); Wible (1991, Wible & Hopson, 1993); Rougier (1993); Luo (1994); Kielan-Jaworowska and Gambaryan (1994; Kielan-Jaworowska, 1996; Gambaryan and Kielan-Jaworowska, 1997); Rougier et al. (1996a).

PRIMARY DESCRIPTIVE REFERENCES FOR POSTCRANIAL ANATOMY

Tritylodontids (Oligokyphus, Kühne, 1956; Kayentatherium, Sues, 1983; tritylodontids in general, Rowe, 1988; Szalay, 1993).

Morganucodontids (Jenkins and Parrington, 1976; Lewis, 1983; Rowe, 1988)

Gobiconodon (Jenkins and Schaff, 1988).

Ornithorhynchidae (Monotremata) (CMNH 1788; Gregory, 1951; Klima, 1973; Lewis, 1983; Rowe, 1988).

Multituberculates (Ptilodontidae, Krause and Jenkins, 1983; Lambdopsalis, Kielan-Jaworowska and Qi, 1990; Meng and Miao, 1992; Taeniolabidoidea and Eucosmodontidae, Kielan-Jaworowska and Gambaryan, 1994; Gambaryan and Kielan-Jaworowska,

1997; Bulganbaatar, Sereno and McKenna, 1995; multituberculates in general, Rowe, 1988; Simmons, 1993).

Henkelotherium (Krebs, 1991).

Vincelestes (Rougier, 1993).

Placentals (Kielan-Jaworowska, 1976, 1978; Rowe, 1988).

Marsupials (Didelphis, Jenkins, 1973, 1974; Jenkins and Weijs, 1979; Lewis, 1983; Klima, 1987; CMNH c45, and several uncatalogued specimens in VP collections; Pucadelphys, Marshall and Sigogneau-Russell, 1995; Asiatherium, Szalay and Trofimov, 1996).

PRIMARY REFERENCES ON BASICRANIAL AND DENTAL CHARACTERS

Basicrania (Kermack et al., 1981; Kielan-Jaworowska et al., 1986; Djadochtatheria, Kielan-Jaworowska and Hurum, 1997; Rougier et al., 1992; Rougier et al., 1996a; Wible and Hopson, 1993, 1995; Muizon, 1994; Luo, 1994; Luo et al., 1995; Meng and Wyss, 1995).

Dentition: Ornithorhynchidae (Archer et al., 1985, 1993; Kielan-Jaworowska et al., 1987; Pascual et al., 1992); multituberculates (Clemens, 1963; Hahn, 1969; Clemens and Kielan-Jaworowska, 1979; Kause, 1982; Wall and Krause, 1992); therians (Crompton 1971; Prothero, 1981; Hopson, 1994).

CERVICAL VERTEBRAE

1. Proatlas neural arch as separate ossification in adults (Rowe, 1988: ch. 92):
    (0) present: tritylodontids.
    (1) absent: multituberculates, Ornithorhynchidae, *Zhangheotherium*, marsupials, placentals, *Vincelestes*.
    (?) Gobiconodon, morganucodontids, *Henkelotherium*.

2. Fusion of atlas neural arch and intercentrum in adults (Rowe, 1988: ch. 93):
    (0) unfused: tritylodontids, morganucodontids, multituberculates, *Vincelestes*.
    (1) fused: Ornithorhynchidae, marsupials, placental.
    (?) *Henkelotherium*, Gobiconodon, *Zhangheotherium*.

3. Atlas rib in adults (Jenkins and Parrington, 1976; Rowe, 1988: ch. 96):
    (0) present: tritylodontids, morganucodontids.
    (1) absent: Ornithorhynchidae, placentals, marsupials, *Zhangheotherium*, *Vincelestes*.
    (?) *Henkelotherium*, Gobiconodon, multituberculates.

4. Prezygapophysis on axis (Rowe, 1988: ch. 97):
    (0) present: tritylodontids.
    (1) absent: morganucodontids, multituberculates, Ornithorhynchidae, *Zhangheotherium*, placentals, marsupials, *Vincelestes*.
    (?) *Henkelotherium*, Gobiconodon.

5. Rib of axis in adults:
    (0) present: tritylodontids, morganucodontids, Ornithorhynchidae, *Zhangheotherium*, multituberculates, marsupials (Pucadelphys? and/or Asiatherium).
    (1) absent: placentals.
    (?) *Henkelotherium*, Gobiconodon, *Vincelestes*.

6. Postaxial cervical rib in adults (Rowe, 1988: ch. 101):
    (0) present: tritylodontids, morganucodontids, *Zhangheotherium*, multituberculates.
    (1) absent: Ornithorhynchidae, *Vincelestes*, placentals, marsupials.
    (?) *Henkelotherium*, Gobiconodon.

Note: Tachyglossid monotremes also have un-fused postaxial cervical ribs. Because tachyglossids do not have as good a fossil record as ornithorhynchids do, we selected ornithorhynchids as the representative family for monotremes as a whole.

7. Postaxial cervical transverse canal (new):
    (0) absent: tritylodontids, morganucodontids, *Zhangheotherium*.
    (1) present: Ornithorhynchidae, placentals, marsupials, multituberculates, *Vincelestes*.
    (?) *Henkelotherium*, Gobiconodon.

SHOULDER GIRDLE

8. Interclavicle in adults (Rowe, 1988: ch. 110 merged with ch. 113):
   (0) present: tritylodontids, morganucodontids, Ornithorhynchidae, multituberculates, *Zhangheotherium*.
   (1) absent: placentals, marsupials, *Henkelotherium, Vincelestes*.
   (?) *Gobiconodon*.

9. Clavicle-sternal apparatus joint (assuming that homologous elements of interclavicle is fused to manubrium in therians, Klima, 1987; Sereno and McKenna, 1995):
   (0) immobile: tritylodontids, morganucodontids, Ornithorhynchidae.
   (1) mobile: multituberculates, *Zhangheotherium*, placentals, marsupials, *Henkelotherium, Vincelestes*.
   (?) *Gobiconodon*.

10. Scapula—supraspinous fossa (Jenkins and Schaff, 1988; Rowe, 1988: ch. 114, new definition of character states; Kielan-Jaworowska, pers. comm):
    (0) absent (acromion extending from the dorsal border of scapula, and is positioned anterior to the glenoid): tritylodontids, Ornithorhynchidae, morganucodontids.
    (1) weakly developed (present only along a part of the scapula, and acromion positioned lateral to the glenoid): multituberculates, *Gobiconodon* (the assignment of this character state to *Gobiconodon* as its scapula is incomplete [Jenkins and Schaff, 1988].
    Prof. Kielan-Jaworowska questioned the association of the scapula with the skeleton in her pers. comm. to the authors, June, 1997).
    (2) fully developed and present along the entire dorsal border of scapula: *Zhangheotherium, Henkelotherium*, placentals, marsupials, *Vincelestes*.

11. Scapula—acromion process (Rowe, 1988: ch. 115, new definition of character states):
    (0) weakly developed, and levelled to the glenoid: tritylodontids, morganucodontids, Ornithorhynchidae
    (1) strongly developed and extending below the glenoid: multituberculates, *Zhangheotherium, Henkelotherium*, placentals, marsupials, *Vincelestes*.
    (?) *Gobiconodon*.

12. Procoracoid (as a separated element in adults) (Klima, 1973; Rowe, 1988: ch. 117):
    (0) present: tritylodontids, morganucodontids, Ornithorhynchidae.
    (1) absent: multituberculates, *Henkelotherium, Zhangheotherium*, placentals, marsupials, *Vincelestes*.
    (?) *Gobiconodon*.

13. Coracoid (Rowe, 1988: ch. 118, new definition of the character):
    (0) large, with posterior process: tritylodontids, morganucodontids, Ornithorhynchidae.
    (1) small, without posterior process: multituberculates, *Henkelotherium, Zhangheotherium*, placentals, marsupials, *Vincelestes*.
    (?) *Gobiconodon*.

14. Fusion of medial part of the embryonic scapula-coracoid plate with the sternal manubrium (sensu Klima, 1973):
    (0) scapula-coracoid plate remains as a separate element in adults: Ornithorhynchidae.
    (1) scapula-coracoid plate fused to manubrium in adults: *Zhangheotherium*, Multituberculates. *Vincelestes*, marsupials, placentals.
    (?) tritylodontids, morganucodontids *Gobiconodon, Henkelotherium*.

15. Size of the anterior-most element relative to the sternebrae in the sternal apparatus in adults:
    (0) large: tritylodontids, Ornithorhynchidae, multituberculates, *Zhangheotherium, Vincelestes*.
    (1) small: marsupials, placentals.
    (?) morganucodontids, *Gobiconodon, Henkelotherium*.

16. Orientation ("facing" or the long axis) of glenoid (relative to the plane or the axis of scapula):
    (0) nearly parallel to the long axis of scapula and facing posterolaterally: tritylodontids, morganucodontids, Ornithorhynchidae.

(1) more perpendicular to the long axis of scapula and facing more posteriorly: multituberculates, *Zhangheotherium, Henkelotherium,* placentals, marsupials, *Vincelestes, Gobiconodon.*

17. Shape and curvature of the glenoid:

(0) oval, elongate, and more or less saddle-shaped: tritylodontids, morganucodontids, Ornithorhynchidae.

(1) more or less round in outline and uniformly concave: multituberculates, *Zhangheotherium, Henkelotherium,* placentals, marsupials, *Vincelestes, Gobiconodon.*

18. Convex medial surface of scapula:

(0) present: tritylodontids, morganucodontids, Ornithorhynchidae, multituberculates.

(1) Absent: : *Zhangheotherium, Vincelestes,* marsupials, placentals.

(?) *Gobiconodon, Henkelotherium.*

HUMERUS

19. Humeral head (Rowe, 1988: ch. 120, new definition of character states):

(0) subspherical, weakly inflected: tritylodontids, morganucodontids, Ornithorhynchidae, *Gobiconodon.*

(1) spherical and strongly inflected: multituberculates, *Zhangheotherium, Henkelotherium,* placentals, marsupials, *Vincelestes.*

20. Intertubercular groove (modified from Rowe, 1988: ch. 121, ch.122, ch. 123):

(0) Pectodeltoid crest separated from lesser tubercle by shallow and broad intertubercular groove: tritylodontids, morganucodontids, Ornithorhynchidae, *Gobiconodon,* multituberculates.

(1) Narrow and deep intertubercular groove: *Zhangheotherium, Henkelotherium,* placentals, marsupials.

(?) *Vincelestes.*

21. Size of lesser tubercle of humerus (relative to the greater tubercle) (Gambrayan and Kielan-Jaworowska, 1997):

(0) Wider than the greater tubercle: tritylodontids, morganucodontids, Ornithorhynchidae, multituberculates.

(1) Narrower than the greater tubercle: *Zhangheotherium,* marsupials, placentals.

(?) *Gobiconodon, Henkelotherium, Vincelestes.*

22. Torsion between the proximal and distal ends of humerus (Rougier et al., 1996b; Gambaryan and Kielan-Jaworowska, 1997):

(0) strong ($\geq 30°$): tritylodontids, morganucodontids, Ornithorhynchidae, *Gobiconodon, Vincelestes.*

(1) moderate ($30°–15°$): *Zhangheotherium, Henkelotherium.*

(2) weak: placentals, marsupials.

(Polymorphic 0/1): multituberculates (coded 0: Kielan-Jawrowska and Qi, 1990; Kielan-Jawrowska and Gambaryan, 1994; coded 1: Sereno and McKenna, 1995).

23. Pectodeltoid crest:

(0) not extending beyond the midpoint of the humeral shaft: tritylodontids, morganucodontids, multituberculates, placentals, marsupials, *Henkelotherium, Vincelestes.*

(1) extending beyond the midpoint of the shaft: Ornithorhynchidae, *Gobiconodon, Zhangheotherium.*

24. Ulnar articulation on distal humerus (Rowe, 1988: ch. 126):

(0) bulbous ulnar condyle: tritylodontids, morganucodontids, multituberculates, Ornithorhynchidae.

(1) cylindrical trochlea with vestigial ulnar condyle: *Zhangheotherium, Henkelotherium, Vincelestes* (Rougier, pers. comm), *Gobiconodon* (we follow Jenkins and Schaff, 1988 in assigning this character state. Prof. Kielan-Jaworowska [pers. comm. June, 1997] believes that the Mongolian *Gobiconodon* has a fully develped ulnar condyle, a primitive condition. But this information has not been published).

(2) cylindrical trochlea without ulnar condyle: placentals, marsupials.

25. Entepicondyle and ectepicondyle of humerus (Rowe, 1988: 124):

    (0) robust: tritylodontids, morganucodontids, Ornithorhynchidae, *Gobiconodon*.

    (1) weak: multituberculates, *Zhangheotherium, Henkelotherium, Vincelestes*, placentals, marsupials.

LOWER FORELIMB

26. Styloid process of radius (Rowe, 1988: ch. 129):

    (0) weak: tritylodontids, morganucodontids, Ornithorhynchidae, *Gobiconodon*.

    (1) strong: multituberculates, *Henkelotherium*, placentals, marsupials.

    (?) *Vincelestes, Zhangheotherium*.

PELVIC GIRDLE

27. Acetabular dorsal emargination (modified from Rowe, 1988: ch. 134):

    (0) open: tritylodontids, morganucodontids, multituberculates, *Vincelestes*.

    (1) closed: Ornithorhynchidae, *Zhangheotherium*, placentals, marsupials.

    (?) *Henkelotherium, Gobiconodon*.

28. Size of pelvic obturator foramen (Rowe, 1988: ch. 139):

    (0) smaller than that of acetabulum: Ornithorhynchidae, outgroup.

    (1) equal to or larger than that of acetabulum: tritylodontids, morganucodontids, multituberculates, *Vincelestes, Zhangheotherium*, placentals, marsupials, *Henkelotherium, Gobiconodon*.

FEMUR

29. Inflected head of the femur set off from the shaft by a neck (modified from Rowe, 1988: ch. 141, 142):

    (0) neck absent (and head oriented dorsally): tritylodontids, morganucodontids, Ornithorhynchidae, *Gobiconodon*.

    (1) neck present (and head inflected medially) spherical and inflected: multituberculates, *Henkelotherium, Zhangheotherium*, placentals, marsupials, *Vincelestes*.

30. Greater trochanter (Rowe, 1988: ch. 143)

    (0) directed dorsolaterally: tritylodontids, morganucodontids, *Gobiconodon*, Ornithorhynchidae.

    (1) directed dorsally: multituberculates, *Zhangheotherium, Henkelotherium*, placentals, marsupials, *Vincelestes*.

31. Orientation of lesser trochanter (Rowe, 1988: ch. 144, not visible on *Zhangheotherium*, but useful for a better resolution among the comparative taxa):

    (0) on medial side of shaft: tritylodontids, morganucodontids, Ornithorhynchidae, *Gobiconodon, Vincelestes*.

    (1) on the ventromedial or ventral side of the shaft: *Henkelotherium*, placentals, marsupials (ventromedial) multituberculates (ventral).

    (?) *Zhangheotherium*.

32. Size of lesser trochanter (Rowe, 1988: ch. 144, not visible on *Zhangheotherium*, but useful for a better resolution among the comparative taxa):

    (0) large: tritylodontids, morganucodontids, *Gobiconodon*, Ornithorhynchidae, multituberculates, *Henkelotherium, Vincelestes*.

    (1) small:, placentals, marsupials.

    (?) *Zhangheotherium*.

33. Patellar facet ("groove") of femur (Rowe, 1988: ch. 145):

    (0) absent: tritylodontids.

    (1) shallow and weakly developed: morganucodontids (Jenkins and Parrington, 1976), *Gobiconodon* (Jenkins and Schaff, 1988).

    (2) well developed: Ornithorhynchidae, multituberculates, *Henkelotherium, Zhangheotherium*, placentals, marsupials, *Vincelestes*.

LOWER HINDLIMB

34. Tibial malleolus and fibular styloid process:

(0) weak: tritylodontids, morganucodontids, *Gobiconodon,* Ornithorhynchidae.

(1) distinct: multituberculates, *Henkelotherium, Vincelestes,* placentals, marsupials.

(?) *Zhangheotherium.*

35. Fibula contacting the calcaneum (tricontact in upper ankle joint: Szalay, 1993):

(0) present: tritylodontids, morganucodontids, multituberculates, *Zhangheotherium,* Ornithorhynchidae, *Vincelestes.*

(1) absent: marsupials, placentals.

(?) *Gobiconodon, Henkelotherium.*

ANKLE JOINT

36. Superposition (overlap) of the astragalus over the calcaneum (lower ankle joint):

(0) little or absent: tritylodontids, morganucodontids, Ornithorhynchidae.

(1) weakly developed: multituberculates, *Zhangheotherium.*

(2) present: marsupials, placentals, *Vincelestes.*

(?) *Gobiconodon, Henkelotherium.*

37. Calcaneal tubercle (Rowe, 1988: ch. 151):

(0) short without terminal swelling: tritylodontids, morganucodontids, Ornithorhynchidae.

(1) elongate with terminal swelling: *Vincelestes, Henkelotherium,* multituberculates, *Zhangheotherium,* marsupials, placentals.

(?) *Gobiconodon.*

38. Peroneal process and groove of calcaneum (Modified from Kielan-Jaworowska and Gambaryam, 1994):

(0) forming laterally directed shelf, and without a distinct process: tritylodontids, morganucodontids, Ornithorhynchidae, *Gobiconodon.*

(1) weakly developed with shallow groove on lateral side of process: *Zhangheotherium, Vincelestes,* marsupials, placentals.

(2) with a distinct peroneal process demarcated by a deep peroneal groove at the base: multituberculates.

(?) *Henkelotherium.*

*A potential character: ratio between mediolateral width and craniocaudal depth of proximal tibial diameters (Kielan-Jaworowska and Gambaryan, 1994; Gambaryan and Kielan-Jaworowska, 1997) may be established for *Zhangheotherium* after reconstruction of the tibia from its silicone rubber molding.

OTHER POSTCRANIAL CHARACTERS

39. Sesamoid bones in flexor tendons (Rowe, 1988: ch. 158):

(0) absent: tritylodontids, morganucodontids.

(1) present and unpaired: Ornithorhynchidae.

(2) present and paired: multituberculates, *Zhangheotherium,* marsupials, placentals.

(?) *Henkelotherium, Gobiconodon, Vincelestes.*

40. External pedal (tarsal) spur:

(0) absent: tritylodontids, morganucodontids, multituberculates, placentals, marsupials, *Vincelestes, Henkelotherium.*

(1) present: Ornithorhynchidae, *Gobiconodon, Zhangheotherium.*

BASICRANIUM

41. Cranial moiety of squamosal (Rowe, 1988; Wible, 1991; Luo 1994):

(0) narrow: tritylodontids, multituberculates, morganucodontids, Ornithorhynchidae, *Gobiconodon.*

(1) broad: *Vincelestes,* marsupials, placentals.

(?) *Zhangheotherium, Henkelotherium.*

42. Squamosal notches for quadrate and quadratejugal (character distribution following Luo and Crompton, 1994):

(0) present: tritylodontids.

(1) absent: morganucodontids, *Gobiconodon,* Ornithorhynchidae, Multituberculates, *Zhangheotherium, Vincelestes,* marsupials, placentals, *Henkelotherium.*

43. Postglenoid depression on squamosal (= "external auditory meatus"):

(0) absent: morganucodontids, *Gobiconodon,* Ornithorhynchidae, Multituberculates.

(1) present: *Zhangheotherium, Vincelestes,* marsupials, placentals

(?) *Henkelotherium.*

(?non-applicable): Tritylodontids.

44. Position of craniomandibular joint (Rowe, 1988; Wible, 1991):

(0) lateral to fenestra vestibuli: tritylodontids, morganucodontids, Ornithorhynchidae, multituberculates.

(1) anterior to the level of fenestra vestibuli: *Zhangheotherium, Vincelestes,* marsupials, placentals.

(?) *Gobiconodon, Henkelotherium.*

45. Promontorium (Rowe, 1988; Wible, 1991; Luo, 1994; Rougier et al. 1996a):

(0) absent: tritylodontids.

(1) present, elongate and cylindrical: morganucodontids, multituberculates, Ornithorhynchidae, *Zhangheotherium.*

(2) present, bulbous and oval shaped: *Vincelestes,* marsupials, placentals.

(?) *Henkelotherium.*

46. Cochlea (Rowe, 1988; Wible, 1991; Luo, 1994; Rougier et al. 1996a):

(0) short and uncoiled: tritylodontids, morganucodontids, multituberculates.

(1) elongate and partly coiled: Ornithorhynchidae, *Vincelestes.*

(2) elongate and coiled at least 360°: marsupials, placentals.

(?) *Gobiconodon, Zhangheotherium, Henkelotherium.*

47. Crista interfenestralis (Rougier et al., 1996a):

(0) horizontal and extending to base of paroccipital process: tritylodontids, Ornithorhynchidae, multituberculates, morganucodontids.

(1) vertical, delimiting the back of the promontorium: *Zhangheotherium, Vincelestes,* marsupials, placentals.

(?) *Gobiconodon, Henkelotherium.*

48. Post-tympanic recess (Rowe, 1988, Rougier et al., 1996a):

(0) absent: tritylodontids, Ornithorhynchidae, multituberculates, morganucodontids.

(1) present: *Zhangheotherium, Vincelestes,* marsupials, placentals.

(?) *Gobiconodon, Henkelotherium.*

49. Caudal tympanic process of petrosal (Rougier et al., 1996a):

(0) absent: tritylodontids, Ornithorhynchidae, morganucodontids.

(1) present: *Zhangheotherium, Vincelestes,* marsupials, placentals. (Polymorphic, 0/1) multituberculates (Rougier et al., 1996a).

(?) *Gobiconodon, Henkelotherium.*

50. Epitympanic recess:

(0) absent: tritylodontids, Ornithorhynchidae, morganucodontids.

(1) present (as a large degression on the crista parotica): multituberculates, *Zhangheotherium, Vincelestes,* marsupials, placentals.

(?) *Gobiconodon, Henkelotherium.*

51. Epitympanic recess flanked laterally by squamosal (Modified from Luo, 1989; Rougier et al. 1996a: ch. 23):

(0) absent: tritylodontids, Ornithorhynchidae, morganucodontids, multituberculates.

(1) present: *Zhangheotherium, Vincelestes,* marsupials, placentals

(?) *Gobiconodon, Henkelotherium.*

52. Foramen for the ramus superior of the stapedial artery:

(0) laterally open notch (laterally open pterygo-paroccipital foramen): tritylodontids, morganucodontids.

(1) foramen enclosed by the petrosal: outgroup, Ornithorhynchidae, multituberculates, *Vincelestes* ("ascending canal" of Rougier et al. 1992), placentals (representative condition for the group as a whole, but may not be present in all taxa, Wible, 1987).

(2) foramen enclosed between the squamosal and the petrosal: *Zhangheotherium*.

(3) absent: marsupials.

(?) *Gobiconodon, Henkelotherium.*

MANDIBLE

53. Meckelian groove in adults (Rowe, 1988; Wible, 1991; Luo, 1994):

(0) forming medial trough or groove: tritylodontids, morganucodontids.

(1) as weak and faint groove: *Gobiconodon, Zhangheotherium, Henkelotherium.*

(2) absent: Ornithorhynchidae, multituberculates, marsupials, placentals.

(?) *Vincelestes.*

54. Angular process of dentary (Rowe, 1988; Wible, 1991; Luo, 1994; Rougier et al. 1996a):

(0) present: tritylodontids, morganucodontids, *Vincelestes, Henkelotherium,* marsupials, placentals.

(1) absent: *Zhangheotherium, Gobiconodon,* multituberculates. (polymorphic 0/1): Obdurodon (0) (Archer et al., 1993); Ornithorhynchus (1).

55. Coronoid in adults (Rowe, 1988; Wible, 1991; Luo, 1994):

(0) present: tritylodontids, morganucodontids, *Gobiconodon, Zhangheotherium, Henkelotherium.*

(1) absent: marsupials, placentals, Ornithorhynchidae.

Polymorphic (0/1): (0) present in Paulchoffatiids (Hahn, 1977), and (1) absent in other multituberculates.

(?) *Vincelestes.*

56. A distinct mandibular foramen for the inferior alveolar nerve and vessels:

(0) absent: tritylodontids, morganucodontids.

(1) present: Ornithorhynchidae, *Gobiconodon,* multituberculates. *Zhangheotherium, Henkelotherium, Vincelestes,* marsupials, placentals.

*A potential character: presence vs. absence of the splenial as a separate element in adult.

DENTITION

57. Mode of occlusion (Crompton and Luo, 1993; Luo, 1994)

(0) bilateral: tritylodontids (following Crompton, 1974; Sues, 1985), multituberculates (following Wall and Krause, 1992).

(1) unilateral: morganucodontids, *Gobiconodon,* Ornithorhynchidae, *Zhangheotherium, Henkelotherium, Vincelestes,* marsupials, placentals.

58. Rotation of the mandible during occlusion (Kemp, 1983; Luo, 1994, Kielan-Jaworowska, 1996):

(0) absent (posteriorly directed): tritylodontids (following Crompton, 1974; Sues, 1985), multituberculates (following Krause, 1982; Wall and Krause, 1992; Kielan-Jaworowska, 1996).

(1) moderate: morganucodontids, *Gobiconodon* (following Crompton and Luo, 1993).

(2) Strong: Ornithorhynchidae, *Zhangheotheriun, Vincelestes, Henkelotherium,* marsupials, placentals.

59. Differentiation of postcanine crowns into premolars and molars (Crompton and Sun, 1985; Luo, 1994):

(0) absent: tritylodontids.

(1) present: morganucodontids, *Gobiconodon,* multituberculates, *Zhangheotherium, Henkelotherium, Vincelestes,* marsupials, placentals, Ornithorhynchidae (*Obdurodon,* Archer et al. 1993).

60. Number of the postcanine roots (Rowe, 1988; Wible, 1991; Luo 1994):

(0) single (undivided): outgroup.

(1) no more than three roots: morganucodontids, *Gobiconodon,* multituberculates (Clemens, 1963; Hahn, 1969; Clemens and Kielan-Jaworwoska, 1979), *Zhangheotherium, Henkelotherium, Vincelestes,* marsupials, placentals.

(2) multiple roots (more than three): tritylodontids, Ornithorhynchidae (Obdurodon).

61. Alignment of main cusps of upper postcanines:

(0) single longitudinal row: morganucodontids, *Gobiconodon.*

(1) multiple cusps in multiple rows: tritylodontids, multituberculates.

(2) in reversed triangle: Ornithorhynchidae *Zhangheotherium, Henkelotherium, Vincelestes,* marsupials, placentals.

62. Upper molar stylar shelf (the area between the paracone/metacone and the labial row of stylar cusps):

(0) absent: tritylodontids, morganucodontids, *Gobiconodon,* multituberculates, Ornithorhynchidae (Pascual et al., 1992).

(1) present: *Zhangheotherium, Henkelotherium, Vincelestes,* marsupials, placentals.

63. Orientation of protocristid relative to the length of the molar:

(0) longitunally orientation: outgroup, morganucodontids, *Gobiconodon*

(1) more transverse: *Zhangheotherium, Vincelestes* Ornithorhynchidae, *Henkelotherium,* marsupials, placentals.

(?Non-applicable): tritylodontids, multituberculates.

64. Upper molar protocone:

(0) absent: tritylodontids, morganucodontids, *Gobiconodon,* multituberculates, Ornithorhynchidae, *Zhangheotherium, Henkelotherium.*

(1) present: *Vincelestes,* marsupials, placentals.

65. Lower molar talonid:

(0) No talonid: tritylodontids, morganucodontids, *Gobiconodon,* multituberculates; *Zhangheotherium.*

(1) simple talonid with a single cusp: *Henkelotherium, Vincelestes.*

(2) fully developed talonid with a basin (with more than one cusps): marsupials, placentals.

(3) antero-posteriorly compressed talonid basin: Ornithorhynchidae.

66. Wear facet on talonid (or on posterior cingulid of the lower molar):

(0) absent: *Morganucodon, Gobiconodon, Zhangheotherium.*

(1) present: Ornithorhynchidae, *Henkelotherium, Vincelestes,* Marsupials, Placentals.

(?Non-applicable): outgroup, tritylodontids, multituberculates.

References

Archer, M., Flannery, T. F., Richtie, A., and Molnar, R. E. 1985. First Mesozoic mammal from Australia—an early Cretaceous monotreme. Nature 318:363–366.

Archer, M, Murray, P., Hand, S. J., and Godthelp, H. 1993. Reconsideration of monotreme relationships based on the skull and dentition of the Miocene Obdurodon dicksoni (Ornithorhynchidae). Pp. 75–94. In, F. S. Szalay, M. J. Novacek, and M. C. McKenna (Eds.), Mammal Phylogeny. Volume 1. Springer-Verlag, New York.

Clemens, W. A. 1963. Fossil mammals of the Type Lance Formation, Wyoming. Part I. Introduction and Multituberculata. University of California Publications in Geological Sciences 48:1–105.

Clemens, W. A. and Kielan-Jaworowska, Z. 1979. Multituberculata. Pp. 99–149. In, J. A. Lillegraven, Z. Kielan-Jaworowska, and W. A. Clemens (Eds.), Mesozoic

Mammals: the First Two-thirds of Mammalian History. University of California Press, Berkeley.

Crompton, W. A. 1971. The origin of the tribosphenic molar. Pp. 65–87. In. D. M. Kermack and K. A. Kermack (Eds.) Early Mammals. Academic Press, New York, New York.

Crompton, W. A. 1972. Postcanine occlusion in cynodonts and tritylodonts. Bulletin of British Museum (Natural History) 21:27–71.

Crompton, A. W., and Sun, A. L. 1985. Cranial Structure and relationships of the Liassic mammal Sinoconodon. Zoological Journal of the Linnean Society 85: 99–119.

Crompton, A. W. and Luo, Z. 1993. Relationships of the Liassic mammals, Sinoconodon, Morganucodon oehleri, and Dinnetherium. Pp. 30–44. In, F. S. Szalay, M. J. Novacek, and M. C. McKenna (Eds.), Mammal Phylogeny. Volume 1. Springer Verlag, New York.

Gambaryan P. P. and Kielan-Jaworowska, Z. 1997. Sprawling versus parasagittal stance in multituberculate mammals. Acta Palaeontologica Polonica 42:13–44.

Gregory, W. K. 1951. Evolution Emerging: A Survey of Changing Patterns from Primeval Life to Man. The MacMillan Compan, New York. 730 Pp.

Hahn, G. 1969. Beiträge zur Fauna der Grube Guimarota Nr. 3. Die Multituberculata. Palaeontographica 133:1–100.

Hahn, G. 1977. Das Coronoid der Paulchoffatiidae (Multituberculata, Ober-Jura). Paläontologische Zeitschrift 51:246–253.

Hopson, J. A. 1994. Synapsid evolution and the radiation of non-eutherian mammals. Pp. 190–219. In Prothero, D. R. and R. M. Schoch (Eds.), Major features of Vertebrate Evolution. Short Courses in Paleontology 7. Paleontological Society.

Hopson, J. A. and Barghusen, H. R. 1986. An analysis of therapsid relationships. Pp. 83–106 In, N. Hotton, III, P. D. MacLean, J. J. Roth, and E. C. Roth (Eds.), The Ecology and Biology of Mammal-like Reptiles. Smithsonian Institution Press, Washington, D.C.

Jenkins, F. A., Jr. 1973. The functional anatomy and evolution of the mammalian humero-ulnar joint. The American Journal of Anatomy 137:281–298.

Jenkins, F. A., Jr. 1974. The movement of the shoulder in claviculate and aclaviculate mammals. Journal of Morphology 144:71–84.

Jenkins, F. A., Jr. and Parrington, F. R. 1976. Postcranial skeleton of the Triassic mammals Eozostrodon, Megazostrodon, and Erythrotherium. Philosophical Transactions of the Royal Society of London 273B:387–431.

Jenkins, F. A., Jr. and Schaff, C. R. 1988. The Early Cretaceous mammal Gobiconodon (Mammalia, Triconodonta) from the Cloverly Formation in Montana. Journal of Vertebrate Paleontology 6:1–24.

Jenkins, F. A., Jr. and Weijs, W. A. 1979. The functional anatomy of the shoulder in the Virginia opossum (Didelphis virginiana). Journal of Zoology (London) 188:379–410.

Kemp, T. S. 1982. Mammal-like reptiles and origin of mammals. Academic Press, London. 363pp.

Kemp, T. S. 1983. The interrelationships of mammals. Zoological Journal of the Linnean Society (London) 77:353–384.

Kermack, K. A., Mussett, F., and Rigney, H. W. 1981. The skull of Morganucodon. Zoological Journal of Linnean Society (London) 71:1–158.

Kielan-Jaworowska, Z. 1976. Evolution of the therian mammals in the Late Cretaceous of Asia Part II. Postcranial skeleton in Kennalestes and Asioryctes. Palaeontologia Polonica 38:3–41.

Kielan-Jaworowska, Z. 1978. Results of the Polish-Mongolian Palaeontological expeditions—Part III. Evolution of the therian mammals in the Late Cretaceous of Asia Part III. Postcranial skeleton in Zalambdalestidae. Palaeontologia Polonica 38:3–41.

Kielan-Jaworowska, Z. 1996. Characters of multituberculates neglected in phylogenetic analyses of early mammals. Lethaia 29:249–266.

Kielan-Jaworowska, Z., Crompton, A. W., and Jenkins, F. A., Jr. 1987. The origin of egg-lying mammals. Nature 326:871–873.

Kielan-Jaworowska, Z. and Gambaryan, P. P. 1994. Postcranial anatomy and habits of Asian multituberculate mammals. Fossils and Strata 36:1–92.

Kielan-Jaworowska, Z. and Hurum, J. H. 1997. Djadochtatheria—a new suborder of multituberculate mammals. Acta Palaeontollogica Polonica 42:201–242.

Kielan-Jaworowska, Z., Presley, R., and Poplin, C. 1986. The cranial vascular system in taeniolabidoid multituberculate mammals. Philosophical Transactions of the Royal Society of London 313B:525–602.

Kielan-Jaworowska, Z. and Qi, T. 1990. Fossorial adaptations of a taeniolabidoid multituberculate from Eocene of China. Vertebrata PalAsiatica 28:81–94.

Klima, M. 1973. Die Frühentwicklung des Schültergürtels und des Brustbeins bei den Monotremen (Mammalia: Prototheria). Advances in Anatomy, Embryology and Cell Biology 47:1–80.

Klima, M. 1987. Early development of the shoulder girdle and sternum in marsupials (Mammalia: Metatheria). Advances in Anatomy, Embryology and Cell Biology 109:1–91.

Krause, D. W. 1982. Jaw movement, dental function, and diet in the Paleocene multituberculate Ptilodus. Paleobiology 8:265–281.

Krause, D. W. and Jenkins, F. A. Jr. 1983. The postcranial skeleton of North American multituberculates. Bulletin of the Museum of Comparative Zoology 150:199–246.

Krebs, B. 1991. Das Skelett von *Henkelotherium* guimarotae gen. et sp. nov. (Eupantotheria, Mammalia) aus dem Oberen Jura von Portugal. Berliner Geowissenschaftliche Abhandlungen A 133:1–110.

Kühne, W, G. 1956. The Liassic therapsid Oligokyphus. British Museum (Natural History), London, 149 pp.

Lewis, O. J. 1983. The evolutionary emergence and refinement of the mammalian pattern of foot architecture. Journal of Anatomy 137:21–45.

Luo, Z. 1989. The petrosal structures of Multituberculata (Mammalia) and the molar morphology of the early arctocyonids (Condylarthra: Mammalia). Ph.D. Dissertation, Department of Paleontology, University of California at Berkeley, 422 Pp.

Luo, Z. 1994. Sister taxon relationships of mammals and the transformations of the diagnostic mammalian characters. Pp.98–128. In, Fraser, N. C. and H.-D. Sues (Eds.), In the Shadow of Dinosaurs—Early Mesozoic Tetrapods. Cambridge University Press, Cambridge.

Luo, Z. andCrompton, A. W. 1994. Transformations of the quadrate (incus) through the transition from non-mammalian cynodonts to mammals. Journal of Vertebrate Paleontology 14:341–374.

Luo, Z., Crompton, A. W., and Lucas, S. G. 1995. Evolutionary origins of the mammalian promontorium and cochlea. Journal of Vertebrate Paleontology 15:113–121.

Marshall, L. G. and Sigogneau-Russell, D. 1995. Part. III. Postcranial skeleton. In, Muizon, C. De (Ed.), Pucadelphys andinus (Marsupialia, Mammalia) from the early Paleocene of Bolivia. Mémoire du Muséum National d'Histoire Naturelle 165:91–164.

Meng, J. and Miao, D. 1992. The breast-shoulder apparatus of Lambdopsalis bulla (Multituberculata) and its systematic and functional implications. Journal of Vertebrate Paleontology 12(3):43A.

Meng, J. and Wyss, A. 1995. Monotreme affinities and low-frequency hearing suggested by multituberculate ear. Nature 377:141–144.

Muizon, C. de. 1994. A new carnivorous marsupial from the Paleocene of Bolivia and the problem of marsupial monophyly. Nature 370:208–211.

Pascual, R., Archer, M., Jaureguizar, E. O., Prado, J. L., Godthelp, H. and Hand, S. J. 1992. First discovery of monotreme in South America. Nature 356:704–706.

Prothero, D. R. 1981. New Jurassic mammals from Como Bluff, Wyoming, and the interrelationships of non-tribosphenic Theria. Bulletin of American Museum of Natural History 167:277–326.

Rougier, G. W. 1993. *Vincelestes* neuquenianus Bonaparte (Mammalia, Theria), un primitivo mammifero del Cretaccico Inferior de la Cuenca Neuqina. Ph.D. Thesis, Universidad Nacional de Buenos Aires. Facultad de Ciencias Exactas y Naturales. Buenos Aires, 720 pp.

Rougier, G. W., Wible, J. R., and Hopson, J. A. 1992. Reconstruction of the cranial vessels in the Early Cretaceous mammal *Vincelestes* neuquenianus: implications for the evolution of the mammalian cranial vascular system, Journal of Vertebrate Paleontology 12:188–216.

Rougier, G. W., Wible, J. R., and Hopson, J. A. 1996a. Basicranial anatomy of Priacodon fruitaensis (Triconodontidae, Mammalia) from the Late Jurassic of Colorado, and a reappraisal of mammaliaform interrelationships. American Museum Novitates 3183:1–28.

Rougier, G. W., Wible, J. R., and Novacek, M, J. 1996b. Scientific correspondence "multituberculate phylogeny." Nature 379:406.

Rowe, T. 1986. Osteological diagnosis of Mammalia. L. 1758, and its relationship to extinct Synapsida. Ph.D. Dissertation, University of California at Berkeley. 446 Pp.

Rowe, T. 1988. Definition, diagnosis, and origin of Mammalia. Journal of Vertebrate Paleontology 8:241–264.

Rowe, T. 1993. Phylogenetic systematics and the early history of mammals. Pp. 129–145. In, F. S. Szalay, M. J. Novacek, and M. C. McKenna (Eds.), Mammal Phylogeny. Volume 1. Springer-Verlag, New York.

Sereno, P. and McKenna, M. C. 1995. Cretaceous multituberculate skeleton and the early evolution of the mammalian shoulder girdle. Nature 377:144–147.

Simmons, N. B. 1993. Phylogeny of Multituberculata. Pp.147–164. In, F. S. Szalay, M. J. Novacek, and M. C. McKenna (Eds.), Mammal Phylogeny. Volume 1. Springer Verlag, New York.

Sues, H.-D. 1983. Advanced mammal-like reptiles from the Early Jurassic of Arizona. Ph. D. Dissertation, Harvard University.

Sues, H.-D. 1985. The relationships of the Tritylodontidae (Synapsida). Zoological Journal of the Linnean Society (London) 85:205–217.

Sues, H.-D. 1986. The skull and dentition of two tritylodontid synapsids from the Lower Jurassic of Western North America. Bulletin of the Museum of Comparative Zoology, Harvard University 151:217–268.

Szalay, F. S. 1993. Pedal evolution of mammals in the Mesozoic: tests for taxi relationships; pp.108-128 in F. S. Szalay, M. J. Novacek, & M. C. McKenna (eds.), Mammal Phylogeny, Volume 1. Springer-Verlag, New York.

Szalay, F. S. and Trofimov, B. A. 1996. The Mongolian Late Cretaceous Asiatherium, and the early phylogeny and paleobiologeography of Metatheria. Journal of Vertebrate Paleontology 16:474–509.

Wall, C. E. and Krause, D. W. 1992. A biomechanical analysis of the masticatory apparatus of Ptilodus (Multituberculata). Journal of Vertebrate Paleontology 12:172–187.

Wible, J. R. 1987. The eutherian stapedial artery: character analysis and implications for superordinal relationships. Zoological Journal of Linnean Society (London) 91:107–135.

Wible, J. R. 1991. Origin of Mammalia: the craniodental evidence reexamined. Journal of Vertebrate Paleontology 11:1–28.

Wible, J. R. and Hopson, J. A. 1993. Basicranial evidence for early mammal phylogeny. Pp. 45–62. In, F. S. Szalay, M. J. Novacek, and M. C. McKenna (Eds.), Mammal Phylogeny. Volume 1. Springer Verlag, New York.

Wible, J. R. and Hopson, J. A. 1995. Homologies of the prootic canal in mammals and non-mammalian cynodonts. Journal of Vertebrate Paleontology 15:331–356.

Chapter 22   Cretaceous Age for the Feathered Dinosaurs
of Liaoning, China   Carl C. Swisher III, Yuan-Qing Wang, Xiao-Lin Wang, Xing Xu, and Yuan Wang

Supplementary Data Table 1. Individual ^{40}Ar/^{39}Ar single crystal dates

			Tuff P4T-1				
#	Ca/K	^{36}Ar/^{39}Ar	^{40}Ar*/^{39}Ar	%^{40}Ar*	Age (Ma)	±	1SD
01	0.0176	0.00004	10.723	99.9	124.17	±	0.20
02	0.0259	0.00032	10.796	99.1	124.99	±	0.19
03	0.0234	0.00012	10.783	99.7	124.84	±	0.20
04	0.0128	0.00004	10.773	99.9	124.74	±	0.20
05	0.0162	0.00011	10.733	99.7	124.29	±	0.20
06	0.0233	0.00007	10.752	99.8	124.50	±	0.20
07	0.0112	0.00005	10.757	99.9	124.56	±	0.20
08	0.0149	0.00009	10.766	99.7	124.66	±	0.20
09	0.0272	0.00008	10.756	99.8	124.55	±	0.19
10	0.0153	0.00006	10.764	99.8	124.63	±	0.20
11	0.0124	0.00009	10.762	99.8	124.61	±	0.19
12	0.0195	0.00009	10.755	99.8	124.53	±	0.20
13	0.0234	0.00005	10.761	99.9	124.60	±	0.20
14	0.0168	0.00007	10.754	99.8	124.52	±	0.20
15	0.0169	0.00007	10.747	99.8	124.45	±	0.20
16	0.0203	0.00016	10.770	99.6	124.70	±	0.20
17	0.0220	0.00005	10.777	99.9	124.78	±	0.20
18	0.0103	0.00014	10.731	99.6	124.26	±	0.20
19	0.0323	0.00005	10.772	99.9	124.72	±	0.20
20	0.0177	0.00003	10.775	99.9	124.76	±	0.20
21	0.0244	0.00010	10.768	99.7	124.68	±	0.21
22	0.0149	0.00007	10.803	99.8	125.07	±	0.20
23	0.0419	0.00007	10.777	99.8	124.78	±	0.20
24	0.0196	0.00007	10.763	99.8	124.63	±	0.20
25	0.0185	0.00011	10.770	99.7	124.70	±	0.20
26	0.0172	0.00007	10.754	99.8	124.52	±	0.20
27	0.0192	0.00007	10.779	99.8	124.80	±	0.20
28	0.0309	0.00009	10.774	99.8	124.75	±	0.21
29	0.0267	0.00005	10.754	99.9	124.53	±	0.20
30	0.0116	0.00010	10.725	99.7	124.20	±	0.20
31	0.0129	0.00004	10.768	99.9	124.68	±	0.20
32	0.0228	0.00005	10.747	99.9	124.45	±	0.20
33	0.0102	0.00004	10.753	99.9	124.51	±	0.20
34	0.0195	0.00005	10.757	99.9	124.56	6	0.20

Supplementary Data Table 1. Individual ^{40}Ar/^{39}Ar single crystal dates

			Tuff P1T-2				
#	Ca/K	^{36}Ar/^{39}Ar	^{40}Ar*/^{39}Ar	%^{40}Ar*	Age (Ma)	±	1SD
01	0.0227	0.00005	10.713	99.9	124.68	±	0.20
02	0.0213	0.00005	10.715	99.9	124.70	±	0.20
03	0.0237	0.00007	10.685	99.8	124.36	±	0.21
04	0.0170	0.00005	10.779	99.9	125.41	±	0.20

Supplementary Data Table 1 *continued*

#	Ca/K	^{36}Ar/^{39}Ar	^{40}Ar*/^{39}Ar	%^{40}Ar*	Age (Ma)	±	1 SD
05	0.0206	0.00007	10.717	99.8	124.72	±	0.20
06	0.0207	0.00007	10.706	99.8	124.60	±	0.20
07	0.0207	0.00007	10.695	99.8	124.47	±	0.20
08	0.0200	0.00008	10.683	99.8	124.34	±	0.20
09	0.0219	0.00009	10.708	99.8	124.62	±	0.21
10	0.0142	0.00004	10.706	99.9	124.60	±	0.20
11	0.0254	0.00013	10.706	99.7	124.60	±	0.22
12	0.0251	0.00011	10.698	99.7	124.50	±	0.20
12	0.0337	0.00007	10.737	99.8	124.94	±	0.19
14	0.0260	0.00009	10.704	99.8	124.58	±	0.20
15	0.0260	0.00010	10.719	99.7	124.74	±	0.20
16	0.0171	0.00007	10.691	99.8	124.43	±	0.20
17	0.0199	0.00008	10.710	99.8	124.65	±	0.20
18	0.0350	0.00004	10.717	99.9	124.72	±	0.21
19	0.0187	0.00007	10.714	99.8	124.69	±	0.20
20	0.0200	0.00011	10.693	99.7	124.45	±	0.20
21	0.0177	0.00009	10.746	99.8	125.04	±	0.20
22	0.0169	0.00005	10.695	99.9	124.47	±	0.20
23	0.0235	0.00011	10.677	99.7	124.28	±	0.22
24	0.0183	0.00005	10.700	99.9	124.53	±	0.20
25	0.0251	0.00007	10.710	99.8	124.64	±	0.21
26	0.0199	0.00008	10.664	99.8	124.12	±	0.20
27	0.0231	0.00007	10.694	99.8	124.47	±	0.20
28	0.0120	0.00009	10.711	99.8	124.66	±	0.19
29	0.0191	0.00006	10.723	99.8	124.79	±	0.20
30	0.0318	0.00008	10.670	99.8	124.20	±	0.20
31	0.0231	0.00010	10.680	99.7	124.31	±	0.21
32	0.0400	0.00102	10.731	97.3	124.88	±	0.21
33	0.0195	0.00008	10.681	99.8	124.32	±	0.20
34	0.0240	0.00003	10.720	99.9	124.76	±	0.20
35	0.0232	0.00022	10.693	99.4	124.46	±	0.20

Supplementary Data Table 2. ^{40}Ar/^{39}Ar incremental heating data

Tuff P4T-1

Step	Laser Watts	%39	Cum.%^{39}Ar	%^{40}Ar*	Ca/K	Apparent Age (Ma 6 1SD)
A	0.5	0.03	0.03	47.4	0.0490	109.98 ± 10.6
B	1.0	0.1	0.1	56.9	0.0142	122.82 ± 4.02
C	1.5	0.2	0.3	99.1	0.0165	128.12 ± 1.52
D	2.0	0.7	1.0	99.8	0.0124	125.25 ± 0.55
E	2.5	1.9	2.9	99.9	0.0111	125.15 ± 0.39
F	2.8	1.8	4.7	99.9	0.0097	124.51 ± 0.45
G	3.0	2.1	6.8	99.5	0.0107	123.70 ± 0.37
H	3.3	3.5	10.3	99.7	0.0098	125.02 ± 0.36
I	3.4	4.3	14.6	99.9	0.0094	124.26 ± 0.33
J	3.7	4.8	19.4	99.8	0.0092	124.07 ± 0.32
K	3.9	5.5	24.9	99.9	0.0086	124.25 ± 0.31
L	4.1	4.5	29.4	99.7	0.0086	124.33 ± 0.36
M	4.3	5.5	34.9	99.8	0.0079	124.50 ± 0.31

Supplementary Data Table 2 *continued*

Step	Laser Watts	%39	Cum.%^{39}Ar	%^{40}Ar*	Ca/K	Apparent Age (Ma 6 1SD)
N	4.5	5.3	40.3	99.8	0.0082	124.66 ± 0.31
O	4.7	6.2	46.5	99.9	0.0074	124.47 ± 0.31
P	4.9	5.3	51.8	99.9	0.0085	124.50 ± 0.31
Q	5.1	4.4	56.2	99.8	0.0100	124.76 ± 0.33
R	5.3	5.0	61.1	100.0	0.0104	124.63 ± 0.31
S	5.5	4.4	65.5	99.8	0.0121	124.80 ± 0.32
T	5.7	5.5	71.1	100.0	0.0137	124.78 ± 0.32
U	5.9	4.2	75.2	99.9	0.0169	125.19 ± 0.33
V	6.1	3.1	78.3	99.9	0.0199	124.72 ± 0.35
W	6.3	3.3	81.7	99.8	0.0177	124.04 ± 0.37
X	6.5	2.4	84.1	99.8	0.0317	124.20 ± 0.39
Y	6.7	2.0	86.0	99.7	0.0362	124.34 ± 0.41
Z	6.9	2.3	88.4	99.8	0.0454	124.05 ± 0.36
ZA	7.1	2.2	90.6	99.8	0.0316	124.96 ± 0.40
ZB	7.3	1.6	92.2	99.6	0.0547	124.31 ± 0.44
ZC	7.5	2.0	94.2	99.8	0.0067	124.84 ± 0.40
ZD	8.0	4.3	98.6	99.9	0.0158	124.90 ± 0.33
ZE	10.0	1.4	100.0	99.6	0.0587	125.14 ± 0.48

Contributors

PER ERIK AHLBERG
Department of Palaeontology
The Natural History Museum
Cromwell Road
London, SW7 5BD
United Kingdom

PAUL M. BARRETT
Department of Zoology
University of Oxford
Oxford OX1 3PS
United Kingdom

STEFAN BENGTSON
Department of Paleozoology
Swedish Museum of Natural History
Box 50007
SE-104 05 Stockholm
Sweden

JUN-YUAN CHEN
Nanjing Institute of Palaeontology
and Geology
Nanjing 210008
China

LIANG CHEN
Department of Geology
Northwest University
Xi'an 710069
China

LIANG-ZHONG CHEN
Yunnan Institute of Geological Sciences
131 Baita Road
Kunming
China

PEI-JI CHEN
Nanjing Institute of Geology and Palaeontology
Nanjing
China

PHILIP J. CURRIE
Royal Tyrrell Museum of Palaeontology
Box 7500, Drumheller
Alberta T0J 0Y0
Canada

ZHI-MING DONG
Institute of Vertebrate Paleontology
   and Paleoanthropology
Chinese Academy of Science
Beijing 100044
China

ALAN FEDUCCIA
Department of Biology
The University of North Carolina
Chapel Hill, North Carolina 27599-3280

JIAN HAN
Department of Geology
Northwest University
Xi'an 710069
China

LIAN-HAI HOU
Institute of Vertebrate Paleontology and Paleoanthropology
Chinese Academy of Sciences
Beijing 100044
China

DI-YING HUANG
Nanjing Institute of Paleontology and Geology
Nanjing 210008
China

SHI-XUE HU
Yunnan Institute of Geological Sciences
131 Baita Road
Kunming
China

YAO-MING HU
Institute of Vertebrate Paleontology
  and Paleoanthropology
Chinese Academy of Sciences
Beijing 100044
China

SHINJI ISAJI
Chiba Prefectural Museum of Natural History
Chiba 260-0682
China

PHILIPPE JANVIER
Laboratoire de Paleontologie
Museum National d'Histoire Naturelle
1, rue René Panhard
75013 Paris
France

QIANG JI
National Geological Museum of China
Yangrouhutong No. 15, Xisi
Beijing 100034
China
and
China University of Geosciences
Beijing
China

SHU-AN JI
National Geological Museum of China
Yangrouhutong No. 15, Xisi,
Beijing 100034
China

ANDREW H. KNOLL
Botanical Museum
Harvard University

26 Oxford Street
Cambridge, MA 02138

CHIA-WEI LI
Department of Life Science
National Tsing Hua University
101, Section 2 Kuang Fu Road
Hsinchu, Taiwan 300
China

CHUAN-KUI LI
Institute of Vertebrate Paleontology
    and Paleoanthropology
Chinese Academy of Sciences
Beijing 100044
China

YONG LI
Department of Geology
Northwest University
Xi'an 710069
China

HUI-LIN LUO
Yunnan Institute of Geological Sciences
131 Baita Road
Kunming
China

ZHE-XI LUO
Section of Vertebrate Paleontology
Carnegie Museum of Natural History
4400 Forbes Avenue
Pittsburgh, PA 15213-4048

MAKOTO MANABE
Department of Geology
National Science Museum
7-20 Ueno Koen, Taito-ku
Tokyo 110-8718
Japan

LARRY D. MARTIN
Natural History Museum and the

Department of Ecology and Evolutionary Biology
University of Kansas
Lawrence, KS 66045-2454

S. CONWAY MORRIS
Department of Earth Sciences
University of Cambridge
Downing Street
Cambridge CB2 3EQ
United Kingdom

MARK A. NORELL
American Museum of Natural History
Central Park West at 79th Street
New York, NY 10024-5192

KEVIN PADIAN
Department of Integrative Biology
   and the Museum of Paleontology
University of California, Berkeley
3060 Valley
Berkeley, CA 94720-3140

TIMOTHY ROWE
Department of Geological Sciences and
the Vertebrate Paleontology Laboratory
University of Texas at Austin
Austin, TX 78712

DE-GAN SHU
Department of Geology
Northwest University
Xi'an 710069
China

CARL C. SWISHER III
Berkeley Geochronology Center
2455 Ridge Road
Berkeley, CA 94709

ZHI-LU TANG
Institute of Vertebrate Paleontology
   and Paleoanthropology
Chinese Academy of Science

Beijing 100044
China

DAVID M. UNWIN
Department of Geology
University of Bristol
Will's Memorial Building
Queen's Road
Bristol BS8 1RJ
United Kingdom

XIAO-LIN WANG
Institute of Vertebrate Paleontology
  and Paleoanthropology
Chinese Academy of Science
Beijing 100044
China
and
Natural History Museum
Changchun University of Science and Technology
Changchun
China

YUAN WANG
Institute of Vertebrate Paleontology
  and Paleoanthropology
Chinese Academy of Sciences
Beijing 100044
China

YUAN-QING WANG
Institute of Vertebrate Paleontology
  and Paleoanthropology
Chinese Academy of Sciences
Beijing 100044
China

XIAO-CHUN WU
Vertebrate Morphology Research Group
Department of Biological Sciences
The University of Calgary
Calgary, AB T2N 1N4
Canada

SHU-HAI XIAO
Botanical Museum
Harvard University
26 Oxford Street
Cambridge, MA 02138

XING XU
Institute of Vertebrate Paleontology
   and Paleoanthropology
Chinese Academy of Science
Beijing, 100044
China

XIAO-BO YU
Department of Biological Sciences
Kean University
Union, NJ 07083

FU-CHENG ZHANG
Institute of Vertebrate Paleontology and Paleoanthropology
Chinese Academy of Science
Beijing 100044
China

XING-LIANG ZHANG
Department of Geology
Northwest University
Xi'an 710069
China

YUN ZHANG
College of Life Sciences
Beijing University
Beijing, China

SHUO-NAN ZHEN
Beijing Natural History Museum
no.126, South Qiao Nan Road 1000005
Xuan Wu District
Beijing
China

ZHONG-HE ZHOU
Institute of Vertebrate Paleontology
   and Paleoanthropology

Chinese Academy of Sciences
Beijing 100044
China
and
Natural History Museum and The
Department of Ecology and Evolutionary Biology
University of Kansas
Lawrence, KS 66045

MIN ZHU
Institute of Vertebrate Paleontology and Paleoanthropology
Chinese Academy of Science
Beijing 10004
China

Index